Advances in Environmental Microbiology

Volume 4

Series Editor
Christon J. Hurst
Cincinnati, Ohio
USA

and

Universidad del Valle
Santiago de Cali, Valle
Colombia

More information about this series at http://www.springer.com/series/11961

Christon J. Hurst
Editor

Modeling the Transmission and Prevention of Infectious Disease

Springer

Editor
Christon J. Hurst
Cincinnati, OH
USA

Universidad del Valle
Santiago de Cali, Valle
Colombia

ISSN 2366-3324 ISSN 2366-3332 (electronic)
Advances in Environmental Microbiology
ISBN 978-3-319-60614-9 ISBN 978-3-319-60616-3 (eBook)
DOI 10.1007/978-3-319-60616-3

Library of Congress Control Number: 2017948674

Cover illustration: "Volvox reimagined" is a montage image created in 2015 and is being used with permission of the artist, Christon J. Hurst. Those images incorporated into this montage were: center image "Aspergillus flavus" (author: Hossein Mirhendi, image supplied by and used with author's permission); clockwise from upper right "Micrasterias" (author: Rogelio Moreno Gill, image supplied by and used with author's permission), "BrownGiantKelp3600ppx" (author: FASTILY, Creative Commons Attribution-Share Alike 3.0 Unported license), "Stumpfungus" (author: Ecornerdropshop, public domain image), "Letharia vulpina JHollinger crop" (author: Jason Hollinger, Creative Commons Attribution-Share Alike 3.0 Unported license), "Phytoplankton - the foundation of the oceanic food chain", (author: NOAA MESA Project, public domain image), "Spirogyra cell" (author: Jasper Nance, Creative Commons Attribution-Share Alike 3.0 Unported license), and "Morelasci" (author: Peter G. Werner, Creative Commons Attribution 3.0 Unported license).

Printed on acid-free paper

This Springer imprint is published by Springer Nature
The registered company is Springer International Publishing AG
The registered company address is: Gewerbestrasse 11, 6330 Cham, Switzerland

Dedication

I dedicate this book to my doctoral advisor and dear friend Charles Peter Gerba. When I started working in Chucks laboratory at Baylor College of Medicine in Houston, Texas, I was his first graduate student and he said that he kept me around because my mother periodically mailed homemade cookies to the laboratory. Usually those were chocolate chip cookies that contained chow mein noodles which my mother added as reinforcement against the cookies breaking during handling by the postal service. Both Chuck and I did of course manage to survive those years together at Baylor. I learned a lot about science and performing research. The cookies from my mother were good, and Chuck defended me against the departmental politics. Chuck still keeps going at the task of training new generations of scientists and has advised more than 129 additional graduate students. His wife Peggy keeps him grounded as he navigates the frustrating path of research and advising. I owe Chuck an appreciative "Thank you".

Charles P. Gerba

Series Preface

The light of natural philosophy illuminates many subject areas including an understanding that microorganisms represent the foundation stone of our biosphere by having been the origin of life on Earth. Microbes therefore comprise the basis of our biological legacy. Comprehending the role of microbes in this world which together all species must share, studying not only the survival of microorganisms but as well their involvement in environmental processes, and defining their role in the ecology of other species, does represent for many of us the Mount Everest of science. Research in this area of biology dates to the original discovery of microorganisms by Antonie van Leeuwenhoek, when in 1675 and 1676 he used a microscope of his own creation to view what he termed "animalcula," or the "little animals" which lived and replicated in environmental samples of rainwater, well water, seawater, and water from snow melt. Van Leeuwenhoek maintained those environmental samples in his house and observed that the types and relative concentrations of organisms present in his samples changed and fluctuated with respect to time. During the intervening centuries we have expanded our collective knowledge of these subjects which we now term to be environmental microbiology, but easily still recognize that many of the individual topics we have come to better understand and characterize initially were described by van Leeuwenhoek. van Leeuwenhoek was a draper by profession and fortunately for us his academic interests as a hobbyist went far beyond his professional challenges.

It is the goal of this series to present a broadly encompassing perspective regarding the principles of environmental microbiology and general microbial ecology. I am not sure whether Antonie van Leeuwenhoek could have foreseen where his discoveries have led, to the diversity of environmental microbiology subjects that we now study and the wealth of knowledge that we have accumulated. However, just as I always have enjoyed reading his account of environmental microorganisms, I feel that he would enjoy our efforts through this series to summarize what we have learned. I wonder, too, what the microbiologists of still future centuries would think of our efforts in comparison with those now unimaginable discoveries which they will have achieved. While we study the many wonders of microbiology, we also further our recognition that the microbes are our biological critics, and in the end they undoubtedly will have the final word regarding life on this planet.

Christon J. Hurst in Heidelberg

Indebted with gratitude, I wish to thank the numerous scientists whose collaborative efforts will be creating this series and those giants in microbiology upon whose shoulders we have stood, for we could not accomplish this goal without the advantage that those giants have afforded us. The confidence and very positive encouragement of the editorial staff at Springer DE has been appreciated tremendously and it is through their help that my colleagues and I are able to present this book series to you, our audience.

Cincinnati, OH

Christon J. Hurst

Volume Preface

This volume addresses two of the principle subject areas that must be considered as we research the goal of eliminating infectious diseases, those are blocking environmental transmission and understanding the ecological perspective of pathogens and their pathogenic processes. The first section of the book addresses environmental transmission and contains chapters that discuss the procedures used to assure microbiological safety of space flight habitats, a review of biocides and biocide resistance mechanisms, plus health and safety requirements for preventing aerosol related transmission of infections within health care treatment facilities. The second section of the book contains chapters which offer insight regarding ecological aspects of infectious disease. These ecological insights introduce us to the role of indigenous gut microbiota in maintaining human health; present discussion on environmentally ecountered bacterial and fungal pathogens associated with soil and water including those species which variously cause the necrotizing skin disease Buruli ulcer and coccidioidomycosis; and consider influenza A virus as an example for understanding how zoonoses, those infectious illnesses typically affecting other animals, spillover into human populations.

Cincinnati, OH — Christon J. Hurst

Contents

Part I
Preventing the Environmental Transmission of Infections

Chapter 1
Preventing Infectious Diseases in Spacecraft and Space Habitats

Wing C. Wong, Cherie Oubre, Satish K. Mehta, C. Mark Ott, and Duane L. Pierson

Abstract Spacecraft crewmembers live and work in a closed environment that is monitored to ensure health and safety. Lessons learned from previous spaceflight missions have been incorporated into the design and development of the International Space Station (ISS). The microbial control actions on the ISS include engineering designs, such as high efficiency particulate air filtering of the air, microbial monitoring of the air, surfaces, and water, as well as remediation procedures when needed. This chapter will describe an overview of microbial risks of spaceflight focusing on measures to prevent infectious disease. The information discussed in this chapter is focused on the microbial monitoring activities in United States Operating Segment (USOS) of the ISS and experimental data obtained on USOS crewmembers.

1.1 Introduction

Spaceflight creates a unique environment where humans live and work for extended periods in a crowded spacecraft breathing reconditioned air and using reconditioned water. This closed environment creates difficult challenges for sustained human habitation. Early spaceflight missions, including Apollo, Skylab, NASA-Mir, and Space Shuttle missions, provided valuable knowledge about human spaceflight that is used to develop future spacecraft and habitats. In particular, the cumulative

---

W.C. Wong
National Institute of Standards and Technology, Gaithersburg, MD, USA

C. Oubre
Wyle, Science, Technology and Engineering Group, Houston, TX 77058, USA

S.K. Mehta
JES Tech, Houston, TX, USA

C.M. Ott • D.L. Pierson (✉)
National Aeronautics and Space Administration, Houston, TX, USA
e-mail: Duane.l.pierson@nasa.gov

C.J. Hurst (ed.), *Modeling the Transmission and Prevention of Infectious Disease*, Advances in Environmental Microbiology 4, DOI 10.1007/978-3-319-60616-3_1

lessons learned from these early programs were applied to the ISS resulting in the most sophisticated and safe space habitat yet.

There are many physical impacts to humans living and working in space habitats. Spaceflight research has documented microgravity-related physiologic changes (Nicogossian et al. 1994) including bone mineral density and muscle mass loss (Trappe et al. 2009; Nagaraja and Risin 2013; Whitson et al. 1997; Aubert et al. 2005; Schneider and LeBlanc 1995), cardiovascular deconditioning (Aubert et al. 2005), neurovestibular imbalances (Bacal et al. 2003; Clément and Reschke 2008), and dysfunctional immunity (Crucian et al. 2013; Mehta et al. 2013).

In addition to documented physiologic effects, spaceflight introduces a variety of stressors including isolation, containment, psychosocial, physical exertion, anxiety, variable acceleration forces, elevated radiation, sleep deprivation, environmental contaminants, and others (De la Torre 2014). Further confounding the effects of these stressors is the subjective nature of individual crewmember's responses.

As future spaceflight missions expand outside of low-Earth orbit and increase in duration, these physiologic challenges will potentially increase in severity. During the extended duration multi-year missions, communication delays of up to 40 min roundtrip from Earth to Mars are anticipated, and there is a limited potential to return a crewmember to Earth for emergency treatment. Current research is identifying ways to prevent or minimize the impacts of the microgravity environment on the human.

1.2 Microbiological Risks and Their Adverse Effects

Infectious diseases are perhaps the best recognized adverse effects of microbial contamination affecting crew health and performance. Microorganisms accompany all spacecraft and habitats occupied by humans. The closed environments of these spacecraft increase the importance of infection control measures to keep crewmembers healthy, safe, and productive. Major potential microbiological routes of infection to crewmembers include food, surfaces, water, payloads, air, other crewmembers, animals, and biohazardous materials (Fig. 1.1).

Evidence gained from previous human spaceflight programs suggests that infectious diseases and allergic responses may increase on long-duration missions due to sustained immune dysfunction (Kaur et al. 2004, 2005; Mehta et al. 2001; Ott et al. 2004; Stowe et al. 2001; Crucian et al. 2013) and increased virulence of some microorganism (Wilson et al. 2007, 2008; Crabbe et al. 2011). In combination, a dysfunctional immune response and a potential for increased virulence could lead to an increased risk for infection of crewmembers during longer-duration missions. Other environmental contaminants, such as mold spores and animal dander, may

Fig. 1.1 Crewmembers gathering and eating in confined spaces in ISS

lead to allergic responses that can also jeopardize mission objectives. The release of volatile compounds from microorganisms can produce objectionable odors or, in the worst case, a toxic environment. Production of nonvolatile toxins, such as aflatoxins from fungi, can result in acute or delayed toxin effects. In addition, microbes can cause food spoilage that could affect the sustainability of a closed environment. Some viruses, such as cytomegalovirus, can diminish cell-mediated immune defenses. Cytomegalovirus (CMV) is a double-stranded DNA virus and is a member of the *Herpesviridae* family that has been associated with immunosuppression. Reactivation of CMV from latency results in serious morbidity and mortality in immunocompromised transplant recipients (Cook 2007; Hummel and Abecassis 2002).

Adverse effects of microorganisms vary in severity and can range from simple skin irritations to illnesses that can jeopardize either crewmember health and safety or mission objectives. Not all adverse effects of microorganisms result in crew illness. Microorganisms can cause biodegradation of critical materials, including system components and space suits, that can result in system failures, thus endangering crewmembers (Fig. 1.2).

Medical incidents in crewmembers have occurred since the onset of the US human space program. Reported illnesses and symptoms related to immune function during the Space Shuttle and ISS programs are listed in Table 1.1. Common infectious diseases and immune-related symptoms including allergic reactions, ear and fungal infections, rashes and skin disorders, upper respiratory infections, hypersensitivity reactions, and delayed wound healing have been identified. The physical impacts of spaceflight require engineering controls, close microbiological monitoring, and medical support to ensure crewmember health and safety so that mission objectives are accomplished.

A

B

Fig. 1.2 Examples of adverse effects of microorganisms during and after spaceflight. (**a**) Fungal contamination on a greenhouse experiment crop on the ISS, ISS020E014565. (**b**) Fungal contamination of fabric panel on ISS NASA image ISS008E05950. (**c**) Bacterial biofilm on a water processor system filter assembly, photo credit: Boeing Huntsville Laboratory at NASA/Marshall Space Flight Center. (**d**) ISS NASA image S129E010312

C

D

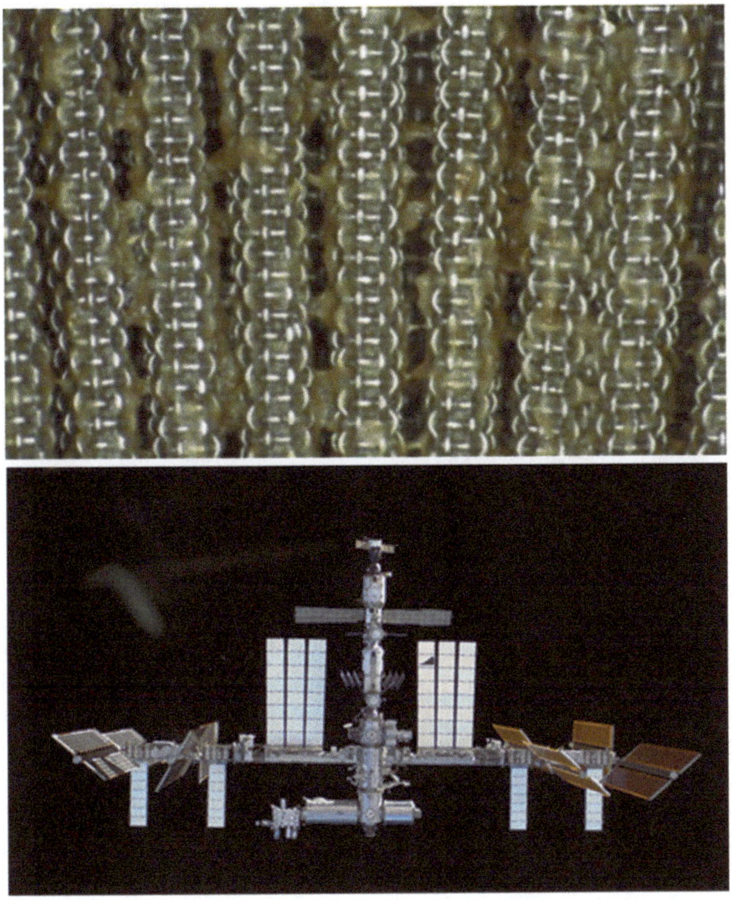

Fig. 1.2 (continued)

1.3 Strategy to Mitigate Microbiological Risks

The overall strategy to mitigate the adverse medical events identified in Table 1.1 is *prevention*. The cornerstone of this mitigation strategy is infection control practices. To be successful in preventing disease, appropriate steps should begin early in the design phase of new spacecraft and habitats. Microbiologists and engineering teams work together to achieve a microbiologically safe habitat. Spacecraft design must include requirements to control accumulation of water including humidity, leaks, and condensation on surfaces. Materials used in habitable volumes should not be conducive (i.e., serve as a food source) to microbial growth and should be easily cleanable. High efficiency particulate air (HEPA) filters can ensure low levels of bioaerosols and particulates in the breathing air. The use of physical and chemical steps to disinfect drinking water coupled with filtration can provide safe drinking water.

Table 1.1 Examples of medical events reported during spaceflight operations for Space Shuttle missions and ISS expeditions

Examples of medical events reported during spaceflight operations
• Allergic reactions
• Ear infections
• Eye irritations/issues (Sty)
• Fungal infections
• Herpes zoster
• Herpes simplex (cold sores)
• Gastroenteritis
• Rash/skin disorders (including acne, cold sore/fever blister, dermatitis, eczema, psoriasis, rosacea, scalp pruritus, and dry skin with redness)
• Upper respiratory infections (such as the common cold and sore throat)
• Other infections (including lymph node swelling)
• Immune symptoms (such as allergic rhinitis, hypersensitivity, coughing/sneezing, infection of cuts, delayed wound healing, and sinus pressure/congestion)

Source: Lifetime Surveillance of Astronaut Health, NASA, August 2013

Infection control approaches should be integral to design and spacecraft operations. In addition to engineering practices, operational procedures can minimize the negative impacts of microorganisms. For example, thorough preflight examination of flight crews to screen for infectious or communicable diseases and restrictions on the interactions of crewmembers with family members and the public before flight can prevent exposure to a debilitating illness that could affect mission success (JSC 50667, MED Volume B Preflight and Postflight Medical Evaluation Requirements for Long-Duration ISS Crewmembers, NASA, 2014 and JSC 22538, Flight Crew Health Stabilization Program, NASA, 2010). Preflight evaluation of the microbial load on cargo and hardware items and evaluation of the environments (air and surfaces) of the spacecraft can detect the presence of pathogens and microorganisms that can biodegrade materials, allowing the ability to disinfect on the ground before introduction into the spacecraft environment occurs (SSP 50260, International Space Station Medical Operation Requirement Document, NASA, 2013). Food items and water are tested for the presence of microorganisms that can cause disease before shipment to the spacecraft and are eliminated from the cargo load in the event that pathogens are detected (SD-T-0251, Microbiological Specification and Testing Procedure for Foods which are not Commercially Sterile, NASA, 2014). Use of cleanable construction materials and the implementation of robust housekeeping requirements that implement periodic cleaning (vacuuming) and disinfection will prevent high levels of microbial growth on surfaces. Personal hygiene activities are also essential to mitigate both the psychological and physical effects of living in a closed environment. The capability to remediate contaminated systems and the environment must be available, such as residual biocide or "shock" treatment availability for water delivery systems. Ultimately, control of the temperature, ventilation rate, and humidity level in the spacecraft is essential in controlling unwanted microbial growth, proliferation, and accumulation.

Fig. 1.3 Debris collected on air filters in Spacelab module in the Space Shuttle. Note large debris including small bristle brush, wooden stick, and various paper wrappers

Spaceflight microbial monitoring was initiated during the NASA-Mir Program to develop and optimize collection methods and procedures for long-duration mission spacecraft and has been performed routinely over the lifetime of the ISS. An example of air monitoring during a Space Shuttle mission measured airborne particulates in the crew compartment and identified an unusual size distribution over time in the crew compartment (Lui et al. 1991). The study confirmed that as expected larger particulates that settle quickly on Earth remained suspended much longer in the microgravity environment of space. Debris of assorted material was collected on the 70–300 μm stainless-mesh filters (Fig. 1.3). As a result of this type of environmental information, a cabin air cleaner was installed on extended duration missions to remove particulates and microorganisms. This reduced airborne particulates and microorganisms and decreased crew complaints about air quality.

When a rare microbial contamination occurs, the microbiologists and engineers work together to identify and mitigate the root cause of the contamination. These events have included fungal growth on transportable water containers, payload water reservoirs, and fabric-covered wall panels as well as bacterial contamination of the potable water dispenser system. Some of the events have been resolved using chemical treatment, while others have been resolved procedurally. Typically following any chemical treatment remediation activity, follow-up monitoring is performed to verify the efficacy of the treatment.

1.4 Crew Health

The Astronaut Health Stabilization Program (JSC 22538, Flight Crew Health Stabilization Program, 2015) was implemented midway through the Apollo program with the goal of minimizing infectious disease during spaceflight by isolating crewmembers from ill individuals and other high-risk groups before spaceflight. The program was based on the idea that if crewmembers went to closed and monitored living quarters in good health without infectious disease, then the risk of developing an illness just before flight would be minimized. So to maintain health, crewmembers are encouraged to take communicable disease vaccinations before spaceflight. For example, during influenza season, the flight crews are urged to take the appropriate influenza vaccine. In addition to immunizations, crewmembers are tested for tuberculosis annually to ensure that they are free of this disease. As described earlier in the chapter, a safe and healthy spaceflight environment is maintained by routine housekeeping and in-flight verification monitoring and any necessary remediation of the environment and drinking water. This combined prevention, monitoring, and remediating approach has proven to be effective, with only one Apollo and one Space Shuttle launch delayed due to illness or exposure to an ill individual.

During early phases of the Space Shuttle program, the microbial evaluation of flight crews included pre- and postflight collection and analyses of feces, urine, nasal swabs, throat swabs, and swabs of some selected skin sites (Pierson et al. 1999). Antibiotic susceptibility testing of medically important microorganisms was conducted to provide data to flight surgeons in cases of infectious illness among crewmembers. The crewmembers were exceptionally healthy and well conditioned, and the microorganisms recovered by pre- and postflight examinations showed microbial profiles consistent with healthy individuals (Pierson et al. 1993a, 1999). Detailed investigations showed that while bacteria and fungi migrated among the crew, permanent colonization was rare (Pierson et al. 1993b, 1995, 1996).

The current microbial evaluations of crewmembers performed for ISS operations are not as comprehensive as those implemented in the early days of the Apollo and Space Shuttle programs. Currently, all crewmembers participate in a general physical evaluation before flight and are screened for methicillin-resistant *Staphylococcus aureus* (MRSA) approximately 50 days before flight. If MRSA is detected, mupirocin can be administered along with disinfectant body washes. No other microbiologic steps are taken unless clinically indicated. Mermel (2013) provided an excellent review of current infection-control measures performed to ensure crew health and safety and provided additional recommendations.

Presently, scientists from the Johnson Space Center's Microbiology Laboratory are working with NASA epidemiologists to identify potential correlations between environmental monitoring data and crew illness.

Table 1.2 Acceptability limits for microbial counts from air, surface, and water for the ISS

		Acceptability limit	
Monitoring category	Organism	Before flight	During flight
Air	Total bacteria	300 CFU/m^3	1000 CFU/m^3
	Total fungi	50 CFU/m^3	100 CFU/m^3
Surface	Total bacteria	500 CFU/cm^2	10,000 CFU/100 cm^2
	Total fungi	10 CFU/cm^2	100 CFU/100 cm^2
Water (potable)	Total bacteria	50 CFU/mL	50 CFU/mL
	Coliform	None detected in 100 mL	None detected in 100 mL

CFU = colony-forming units

1.5 Acceptability Limits and Sampling Schedules

To ensure a safe spacecraft environment and safe drinking water and food, requirements for acceptable levels of bacteria and fungi have been established, and verification testing is routinely conducted before and during a mission. Requirements for the ISS air, surfaces, and drinking water are shown in Table 1.2. These requirements were established through scientific and medical expert panel meetings, using professional experience and federal standards where available to determine conservative safe levels of microorganisms. Through time as new data emerged, sampling schedules and limits have been altered to frequencies and levels still considered acceptable to minimize risk to the crew and spacecraft systems. A current list of acceptability limits for bacteria and fungi associated with air, surfaces, and drinking water can be found in NASA's ISS Medical Operations Requirements Document (SSP 50260, 2013).

1.6 Environmental Microbial Monitoring

Microbial monitoring is routinely performed before and during spaceflight and is an important step in minimizing the risk to crew and to ensure that a safe spacecraft environment is obtained before flight following all ground processing operations. Preflight air and surface sampling is conducted for all cargo and crew spacecraft traveling to the ISS, including the cargo spacecraft provided by private commercial suppliers. Sampling of air and surfaces continues during spaceflight operations to ensure a healthy living and working environment are maintained.

Preflight surface samples are collected from the interior cabin of all spacecraft approximately 2 months before launch. Samples are also collected from randomly selected hardware items and cargo bags approximately 2 months before launch.

Samples are collected using standard environmental sampling methods and identified by Vitek® (bioMérieux, Inc., Durham, NC) analysis, molecular methods, and microscopy (fungal). Preflight acceptability limits for surface samples are designated in Table 1.2.

Surface samples are collected during spaceflight operations on a quarterly basis. Collection of surface samples is performed using contact plate slides containing appropriate growth media for bacterial and fungal growth. After incubation, crewmembers evaluate media and determine approximate microbial counts. The acceptability limits for in-flight surface samples are indicated in Table 1.2. Contact slides are returned to Earth for completion of analysis that consists of performing microbial identification of the isolates recovered.

Air samples are collected before spaceflight from the habitable volume of spacecraft approximately 2 months before launch. A commercial air sampler (SAS Super 180 air sampler; Bioscience International, Rockville, MD) is used to impact air onto bacterial and fungal media. Following standard incubation times, laboratory scientists perform enumeration and microbial identification of samples. Isolates are identified by Vitek® analysis, molecular methods, and microscopy (fungal). The acceptability limits for preflight air samples are specified in Table 1.2.

Air sample collection on board the ISS is performed in a similar method to ground-based air sample collection. A modified commercial air sampler (Burkard Manufacturing Co. Ltd., Hertfordshire, UK) uses an air pump to draw a defined volume of air through the sampler, and the airborne microbes are impacted onto standard agar plates for bacterial and fungal growth. The agar plates are incubated and then approximate microbial counts are determined by crewmembers in a similar method as surface samples. The acceptability limits for spaceflight air samples are indicated in Table 1.2. Air sample plates are returned to the ground for completion of analysis, which consists of performing microbial identification of the isolates recovered.

If preflight or in-flight surface microbial counts exceed the established limits, then subject matter experts review the data and determine the appropriate actions that include remediation of the sample item, location, or area. Historically, the viable bacterial genera recovered from the air and surfaces of the ISS were typically associated with humans (e.g., *Staphylococcus* spp., *Micrococcus* spp., and *Corynebacterium* spp.) and the environment (*Bacillus* spp.) as shown in Fig. 1.4. Similarly, the most commonly recovered fungal genera recovered from the air and surface samples included *Aspergillus* spp., *Penicillium* spp., and *Rhodotorula* spp. (Pierson et al. 2012). Similar data were observed during the Space Shuttle program (Pierson et al. 2011).

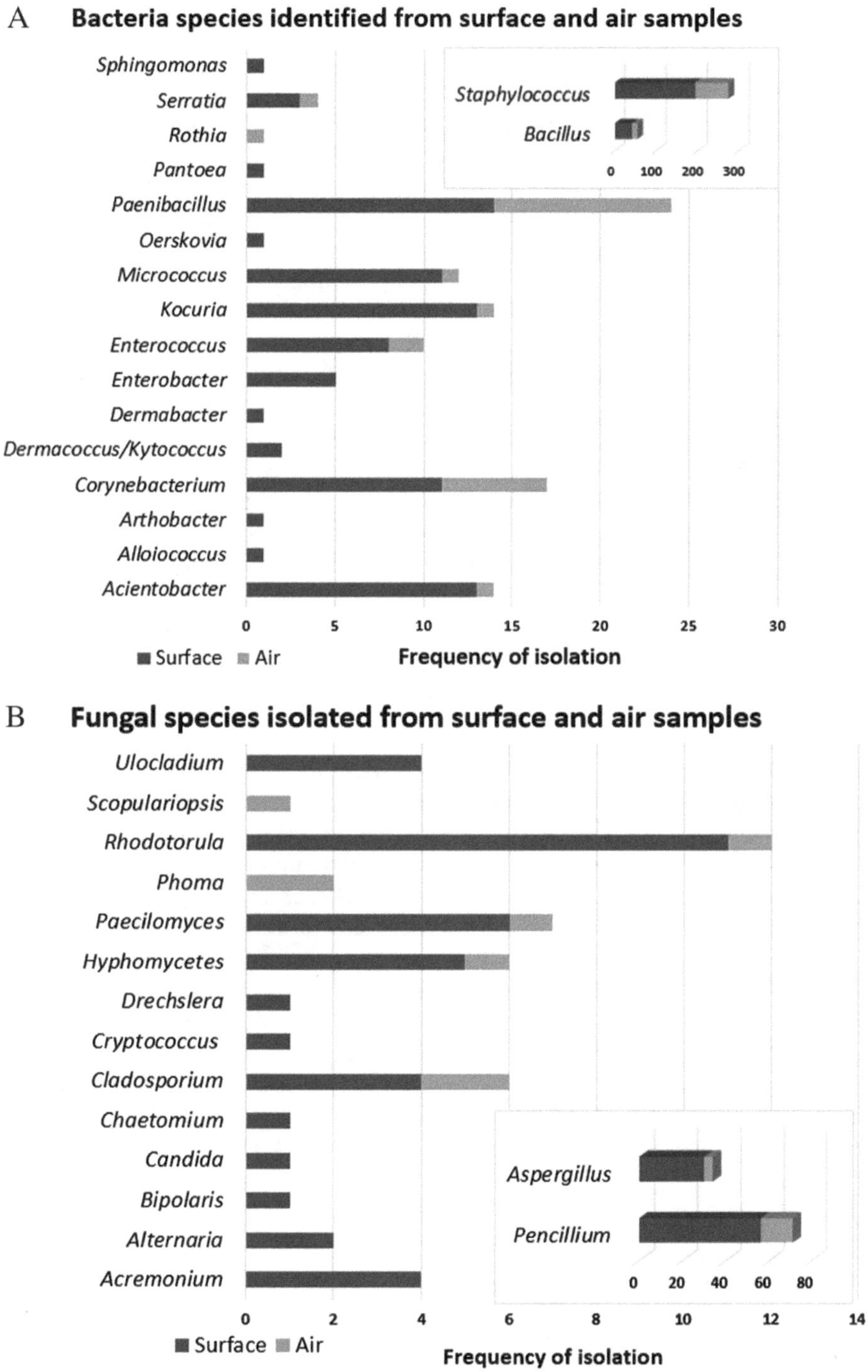

Fig. 1.4 Bacterial (**a**) and fungal (**b**) species identified from environmental monitoring of the ISS. Air samples collected during ISS Expeditions 1 through 38 (December 2000 through April 2014). Surface samples collected during ISS Expeditions 2 through 38 (July 2001–April 2014)

1.7 Water Microbial Monitoring

Safe drinking water is critical for successful spaceflight missions. The water used in the USOS of the ISS is provided by reclaiming water from the humidity condensate and urine distillate combined into a waste water tank, with the capability to provide additional water from ground-supplied water sources. The reclaimed water goes through a series of steps to clean and purify the water, including treatment in a catalytic oxidizer that reaches 135 °C, addition of 2–3 ppm iodine as a residual disinfectant, and filtration through a 0.22-μm microbial filter before consumption. The potable product water can be supplied at ambient or hot temperatures to rehydrate some foods and drinks. Microbial monitoring of the water system is performed monthly during spaceflight operations for total bacteria and coliforms. To be in compliance, the total bacteria cannot exceed 50 CFU/mL, and coliforms cannot be detected as indicated in Table 1.2.

Historically, the bacterial genera recovered from the potable water supply of ISS are typical waterborne, Gram-negative bacteria, with the most common genera recovered being *Ralstonia* and *Burkholderia*. Coliform bacteria have never been recovered from the potable water in over 12 years of monitoring.

1.8 Food Microbial Monitoring

Foods and beverages for consumption by crewmembers are analyzed before flight and delivery to the launch vehicle for total aerobic bacteria, yeast and molds, and selected pathogens (e.g., *Salmonella*). Air and surface monitoring of the food preparation area reduces risks of microbial contamination during the preparation and packaging process. Representative samples of each food lot prepared for crew consumption are rigorously analyzed to ensure safe products for the crewmembers. Raw food materials are analyzed before packaging to ensure the safety of the bulk food products used in food preparation. Food items are also analyzed after packaging to verify safety of final food products. If the total aerobic bacterial count, *Enterobacteriaceae*, *Salmonella*, or yeast and mold acceptability limits are exceeded (Table 1.3), then food lots are disqualified for flight. Acceptability limits for microbial counts are defined in NASA's "microbiology specification and testing procedure for foods which are not commercially sterile" (SD-T-0251, 2014).

Table 1.3 Acceptable limits for microbial counts

Food microbiology[a]	Acceptability limit
Total aerobic count	<20,000/g
Enterobacteriaceae	<100 CFU/g
Salmonella	0 CFU/25 g
Yeast and molds	<1000 CFU/g

[a]Food samples are tested before flight only

1.9 Payload Reviews

The NASA Biosafety Review Board reviews all payloads as part of the overall safety evaluation conducted by the Payload Safety Review Panel. Biohazardous materials are identified and assessed, and a biosafety level is assigned (JSC 63828, Biosafety Review Board Operations and Requirements Document, 2010). The biosafety level determines the levels of containment required to prevent leakage or crewmember exposure to safeguard the crew and the integrity of the spacecraft.

Animals are used for biomedical research as surrogates to determine effects of space environment on biological systems. To reduce the risks of zoonotic disease, specific pathogen-free microbiology requirements were established for animals used for experimentation on spacecraft. These efforts were coordinated through the NASA Biosafety Review Board, the NASA Institutional Review Board, and NASA veterinarian.

1.10 Conclusions

During spaceflight, crewmembers experience a uniquely stressful environment that adversely affects various physiologic systems. In particular, the general dysfunction of the immune system (Crucian et al. 2013; Risin et al. 2013) and an increase in virulence observed in some microbial pathogens, such as *Salmonella* (Wilson et al. 2007), are concerns for infection control during spaceflight. A robust infection-control program is essential in maintaining the health, safety, and productivity of crewmembers during spaceflight. The major sources of infectious risks have been identified, and engineering and operational controls are in place to manage the risks to acceptable levels. Engineering controls include HEPA filtration of the breathing air, processing drinking water, and containment of biohazardous materials associated with operational systems, nominal monitoring equipment, and experimental payloads. Operational controls include comprehensive preflight microbial evaluation of foods and beverages and pre- and in-flight monitoring of air, surfaces, and water. When excessive microbial contamination of internal environments is identified, remediation is performed to restore healthy conditions. Clinical examinations including thorough microbial evaluations of crewmembers for pathogens or signs of infection and vaccinations against likely infectious agents are encouraged. Antimicrobials against bacteria, fungi, and viruses are available on spacecraft for use when infections occur during spaceflight.

Microbiological risks are expected to increase for future long-duration exploration missions. However, most infections and other adverse effects of microorganisms can be prevented by microbiologists and engineering teams working together early in the design phase of spacecraft. The design of spacecraft and space habitats must control available water (humidity, condensate, leaks, and spills) and temperature to minimize microbial growth. Implementation of strict preflight requirements that may include a quarantine to prevent exposures of crewmembers to infectious individuals is an effective tool that should be implemented for exploration missions that may require 2 years or more away from Earth. In addition, expansion of vaccinations against additional pathogens and a more rigorous preflight microbial evaluation of flight crews should be considered for long-duration space missions.

Compliance with Ethical Standards
Conflict of Interest: Wing C. Wong declares that he/she has no conflict of interest. Cherie Oubre declares that he/she has no conflict of interest. Satish K Mehta declares that he/she has no conflict of interest. C. Mark Ott declares that he/she has no conflict of interest. Duane L. Pierson declares that he/she has no conflict of interest.

Ethical approval: This chapter does not contain any studies with human participants or animals performed by any of the authors.

References

Aubert AE, Beckers B, Verheyden F (2005) Cardiovascular function and basics of physiology in microgravity. Acta Cardiol 60:129–151

Bacal K, Billica R, Bishop S (2003) Neurovestibular symptoms following space flight. J Vestib Res 13:93–102

Clément G, Reschke MF (2008) Neuroscience in space. Springer, New York

Cook CH (2007) Cytomegalovirus reactivation in "immunocompetent" patients: a call for scientific prophylaxis. J Infect Dis 196:1273–1275

Crabbe A, Schurr MJ, Monsieurs P et al (2011) Transcriptional and proteomic responses of *Pseudomonas aeruginosa* PAO1 to spaceflight conditions involve Hfq regulation and reveal a role for oxygen. Appl Environ Microbiol 77:1221–1230

Crucian B, Stowe R, Mehta S et al (2013) Immune system dysregulation occurs during short duration spaceflight on board the Space Shuttle. J Clin Immunol 33:456–465

De la Torre G (2014) Cognitive neuroscience in space. Life 4:281–294

Hummel M, Abecassis MM (2002) A model for reactivation of CMV from latency. J Clin Virol 25 (Suppl 2):S123–S136

Kaur I, Simons ER, Castro VA, Ott CM, Pierson DL (2004) Changes in neutrophil functions in astronauts. Brain Behav Immun 18:443–450

Kaur I, Simons ER, Castro VA, Ott CM, Pierson DL (2005) Changes in monocyte functions in astronauts. Brain Behav Immun 19:547–554

Lui B, Rubow K, McMurry P, Kotz T, Russo D (1991) Airborne particulate and spacecraft internal environments, SAE Technical Paper Series No. 911476. Society of Automotive Engineers, Warrendale, PA

Mehta SK, Kaur I, Grimm EA, Smid C, Feeback D, Pierson DL (2001) Decreased non-MHC-restricted (CD56+) killer cell cytotoxicity after spaceflight. J Appl Physiol 91:1814–1818

Mehta SK, Crucian B, Stowe R et al (2013) Reactivation of latent viruses is associated with increased plasma cytokines in astronauts. Cytokine 61:205–209

Mermel LA (2013) Infection prevention and control during prolonged human space travel. Clin Infect Dis 56:123–130

Nagaraja MP, Risin D (2013) The current state of bone loss research: data from spaceflight and microgravity simulators. J Cell Biochem 114:1001–1008

Nicogossian AE, Huntoon A, Pool SL (1994) Space physiology and medicine. Lea and Febiger, Philadelphia

Ott CM, Bruce RJ, Pierson DL (2004) Microbial diversity aboard spacecraft: evaluation of the Mir Space Station. Microb Ecol 47:133–136

Pierson DL, Bassinger VJ, Molina TC et al (1993a) Preflight and post flight microbiological results from Space Shuttle crews, SAE Technical Paper Series No. 932139. Society of Automotive Engineers, Warrendale, PA

Pierson DL, Nicogossian AE, Huntoon CL, Pool SL (1993b) Microbiology. In: Space physiology and medicine, 3rd edn. Lea and Febiger, Philadelphia, pp 157–166

Pierson DL, Mehta SK, Magee BB, Mishra SK (1995) Person-to-person transfer of *Candida albicans* in the spacecraft environment. J Med Vet Mycol 33:145–150

Pierson DL, Chidambaram M, Heath JD, Mallary L, Mishra SK, Sharma B (1996) Epidemiology of *Staphylococcus aureus* during space flight. FEMS Immunol Med Microbiol 16:273–281

Pierson DL, James J, Russo D, Limero T, Beck S, Groves T (1999) Environmental health. In: Sawin CF, Taylor GR, Smith WL, Brown JT (eds) Extended duration orbiter medical project 1989–1995, Final report. National Aeronautic and Space Administration, Washington, DC

Pierson DL, Ott CM, Bruce RJ, Castro VA, Mehta SK (2011) Microbiological lessons learned from the space shuttle. In: 41st International meeting on environmental systems, Portland, OR. AIAA 2011-5266

Pierson DL, Botkin DJ, Bruce RJ et al (2012) Microbial monitoring of the International Space Station. In: Moldenhauer J (ed) Environmental monitoring: a comprehensive handbook, 1st edn. DHI Publishing, Bethesda, MD

Risin D, Crucian BE, Mehta SK, Pellis NR, Pierson DL, Sams C (2013) Immune system. In: Risin D, Stepaniak PC (eds) Biomedical results of the space shuttle program. U.S. Government Printing Office, Washington, DC, pp 245–257

Schneider VO, LeBlanc A (1995) Bone and body mass changes during space flight. Acta Astronaut 36:463–466

Stowe RP, Mehta SK, Ferrando AA, Feeback D, Pierson D (2001) Immune responses and latent herpesvirus reactivation in spaceflight. Aviat Space Environ Med 72:884–891

Trappe SC, Costill D, Gallagher P et al (2009) Exercise in space: human skeletal muscle after 6 months aboard the International Space Station. J Appl Physiol 106(4):1159–1168

Whitson PA, Pietrzyk RA, Pak CY (1997) Renal stone risk assessment during space shuttle flights. J Urol 158:2305–2310

Wilson JW, Ott CM, zu Bentrup KH et al (2007) Space flight alters bacterial gene expression and virulence and reveals a role for global regulator Hfq. Proc Natl Acad Sci USA 104:16299–16304

Wilson JW, Ott CM, Quick L et al (2008) Media ion composition controls regulatory and virulence response of Salmonella in spaceflight. PLoS One 3:e3923

Chapter 2
Bacterial Resistance to Hospital Disinfection

Matthew E. Wand

Abstract With the number of infections caused by bacteria that are multidrug resistant (MDR) increasing, it is imperative that these infections can be controlled. Biocides are used to prevent contamination in a variety of topical and hard-surface applications and are an essential component of infection control in hospitals. They are also used for disinfection and preservation of foodstuffs and within the home environment for a range of applications which again includes hard-surface disinfection. Whilst antibiotic resistance has been well studied in a variety of bacteria, there is little known about potential resistance mechanisms to biocides. We have yet to see well-documented outbreaks caused by biocide-resistant bacteria, although this is rarely looked for. Development of resistance to biocides is thought to be more difficult than for antibiotics for several reasons including the high concentration of biocides used in many applications and that biocides act on multiple targets. However, the increased use of biocides has led to speculation of development of resistance to biocides and whether this leads to cross resistance to antibiotics. This chapter will consider the different resistance mechanisms to biocides in bacteria, the possible link between biocide and antibiotic resistance and whether we should be concerned about bacterial biocide resistance.

2.1 Introduction

Biocides have been used for centuries mainly as preservative agents for water (e.g. the use of copper and silver vessels) and foodstuffs as well as the use of vinegar and honey for cleansing wounds. The nineteenth century saw the onset of antiseptic surgery and the introduction of disinfectants into clinical use. Iodine was used as a wound disinfectant, phenol (carbolic acid) in wound dressings and water containing chlorine used in general disinfection and obstetrics. In the early part of the twentieth century, many biocides were introduced including other chlorine-releasing agents (CRAs) and some quaternary ammonium compounds (QACs)

M.E. Wand (✉)
Public Health England, National Infection Service, Technology Development Group, Porton Down SP4 OJG, UK
e-mail: matthew.wand@phe.gov.uk

C.J. Hurst (ed.), *Modeling the Transmission and Prevention of Infectious Disease*,
Advances in Environmental Microbiology 4, DOI 10.1007/978-3-319-60616-3_2

(Hugo 1999). The mid-twentieth century saw the introduction of biguanides (chlorhexidine, alexidine), amphoteric surfactants, bisphenols (triclosan), aldehydes (glutaraldehyde, ortho-phthalaldehyde, succinaldehyde-based products), CRAs (isocyanurates), iodine-releasing agents (iodophors), isothiazolones and peracetic acid (Russell 2002). Whilst biocides have been in use for centuries as disinfectants, the ability of bacteria to survive and adapt to the presence of biocides has only been documented within the last 50 years. Comparison of strains of *Klebsiella pneumoniae* isolated pre (Murray) and post (modern) the widespread introduction of many biocides showed that more modern strains, which have presumably been more exposed to particular biocides, have become less susceptible to biocides such as triclosan, chlorhexidine and benzalkonium chloride (Wand et al. 2015) (Table 2.1). Although more work is needed, this suggests that whilst bacteria are still highly susceptible to in-use concentrations of biocides, they are beginning to adapt to lower concentrations. In hospitals and the environment, many bacteria are now resistant to several in-use antibiotics which places an even greater burden on biocides for infection prevention and control. This means that bacteria which have adapted (become more resistant) to particular biocides will be increasingly problematic for infection control and could lead to outbreaks which are difficult to combat. Still more problematic is whether biocide adaptation leads to increased resistance to antibiotics and other antimicrobial agents.

2.2 Types of Biocides and Mode of Action

2.2.1 Cationic Antimicrobial Agents

The surface of bacterial cells is usually negatively charged which is stabilized by the presence of divalent cations such as Mg^{2+} and Ca^{2+}. For Gram-positive bacteria, this is associated with the teichoic acid and polysaccharide components of the cell wall, for Gram-negative bacteria, the lipopolysaccharide and the cytoplasmic membrane. Therefore, antimicrobial agents which are cationic have a high binding affinity for bacterial cells. Cationic antimicrobials interact initially with the bacterial cell wall by displacing the divalent cations with subsequent interactions with membrane proteins and the lipid bilayer dependent upon the actual biocide.

2.2.2 Quaternary Ammonium Compounds

The QACs are amphoteric surfactants with activity towards specific species of bacteria dependent upon the hydrophobicity and chain length of the *n*-alkyl chain (Tomlinson et al. 1977). Those QACs with a C_{16} hydrophobic tail length have more activity against Gram-negative bacteria than do shorter-chain compounds, possibly

Table 2.1 Comparison of susceptibility to disinfectants between Murray and modern *K. pneumoniae* strains

		Murray		Modern	
		Range	Mode	Range	Mode
Disinfectants	CHX	0.25 to 32	8	8 to 32	16
	BAC	1 to 16	4	8 to 32	16
	HDPCM	1 to 8	2	4 to 32	8
	TRI	0.007 to 0.25	0.06 to 0.125	0.125 to 1	0.5
	H_2O_2 (%)	0.03 to 0.06	0.03	0.03 to 0.06	0.06
	Per acid (%)	0.04 to 0.08	0.08	0.04 to 0.08	0.08
	HSD1 (% WC)	1.56 to 3.125	3.125	1.56 to 3.125	3.125
	EtoH (%)	1.56 to 6.25	6.25	3.125 to 6.25	6.25
	Chlorine (ppm)	250 to 500	500	250 to 500	500
	HSD2 (% WC)	1.56 to 3.125	3.125	1.56 to 3.125	3.125
Antibiotics	AMK	1 to 4	2	4 to >64	8
	GEN	0.5 to 4	2	4 to >64	>64
	PIP	0.25 to >64	0.5	8 to >64	>64
	TZP	0.25 to 8	0.5	1 to >64	>64
	AMX	0.5 to >64	0.5 to 1	64 to >64	>64
	AMC	0.5 to 8	0.5	1 to >64	>64
	CTX	≤0.06 to 0.125	≤0.06	1 to >64	>64
	MEM	≤0.06 to 0.125	≤0.06	≤0.06 to 0.125	≤0.06
	CIP	≤0.06 to 0.125	≤0.06	≤0.5 to 128	≤0.5
	CHL	2 to 4	2	1 to >512	4

Potential adaptation of *K. pneumoniae* isolates through exposure. Several *K. pneumoniae* isolates ($N = 39$) from the Murray Collection (isolated pre-1950) were compared to more modern *K. pneumoniae* isolates (isolated post-2000) ($N = 39$). For certain disinfectants, particularly the antiseptics, e.g. BAC, there was a clear significant shift towards decreased susceptibility between the Murray isolates and the modern isolates. This did not occur for all disinfectants, e.g. Per acid and chlorine. Several antibiotics were tested against the same panel and are shown for comparison. The large increase in resistance to antibiotics such as AMX and CTX is likely due to the presence of antibiotic resistance genes, e.g. bla_{SHV} in the majority of modern isolates

All values are given as mg/L unless otherwise stated (adapted from Wand et al. 2015)

CHX chlorhexidine, *BAC* benzalkonium chloride, *HDPCM* hexadecylpyridinium chloride monohydrate, *TRI* triclosan, H_2O_2 hydrogen peroxide, *Per acid* peracetic acid, *HSD1 (2)* hard-surface disinfectant 1 (2), *EtoH* ethanol, *AMK* amikacin, *GEN* gentamicin, *PIP* piperacillin, *TZP* piperacillin/tazobactam (TZP), *CTX* cefotaxime, *MEM* meropenem, *CIP* ciprofloxacin, *CHL* chloramphenicol, *WC* working concentration

due to the increased affinity of the C_{16} chain with the fatty acid portion of lipid A (Ahlström et al. 1999). Their mode of action is also dependent on the concentration of the biocide. Low concentrations of QAC cause cellular leakage of potassium and hydrogen ions and loss of osmoregulatory capability by binding to anionic sites on the surface of the bacterial cell membrane (Lambert and Hammond 1973). At in use, high-concentration QACs kill bacterial cells by solubilization of the cellular membrane and rapid cell leakage occurring at bactericidal concentrations (Ioannou et al. 2007).

As a group, QACs have been in use since the 1930s, and although some changes in resistance have been noted, these are generally increases in MIC (minimum inhibitory concentration) level which is still far below the in-use concentration level (Gilbert and McBain 2003). Some bacteria such as *Pseudomonas aeruginosa* are relatively resistant to QACs possibly due to the inability of the compounds to penetrate the outer membrane. Examples of QACs include benzalkonium chloride and cetrimide.

2.2.3 Biguanides

Bisbiguanide antiseptics (e.g. chlorhexidine) have a mode of action which is similar to QACs in that the biguanide groupings strongly associate with anionic sites on the cell surface, particularly acidic phospholipids and proteins (Chawner and Gilbert 1989a). Chlorhexidine has both bacteriostatic (inhibits bacterial growth) and bactericidal (kills bacteria) mechanisms of action, depending on its concentration. The chlorhexidine cation forms a bridge between pairs of adjacent phospholipid head groups and displaces the associated divalent cations (Mg^{2+} and Ca^{2+}) (Davies 1973). This results in reduction of membrane fluidity and osmoregulation as well as a change in metabolic capability of the enzymes associated with the cell membrane (Hugo and Longworth 1966). At higher concentrations the interaction between chlorhexidine and the cellular membrane causes the membrane to lose its structural integrity, and the membrane adopts a liquid crystalline state which leads to a rapid loss of cellular contents (Longworth 1971; Chawner and Gilbert 1989b). Again, increased resistance to chlorhexidine has been described, mainly resulting from increased efflux; this is mainly far below in use concentrations, although some bacteria have been isolated from chlorhexidine-containing solutions (Brooks et al. 2004).

For polyhexamethylene biguanides (PHMB), the mode of action is similar to the biguanide antiseptics except that the bridging is not restricted to adjacent phospholipid pairs which results in the formation of a mosaic of individual phospholipid domains. These each have a different phase transition temperature causing the membrane to separate into liquid and fluid crystalline regions. This again results in generalized cellular leakage (Broxton et al. 1983; Gilbert and Moore 2005). Again, PHMB is bacteriostatic at low concentrations, but bactericidal at higher concentrations and increased resistance has been associated with upregulation of Rhs (recombinational hot spots) elements which suggests that PHMB may interact directly with DNA (Allen et al. 2006).

2.2.4 *Peroxygens*

2.2.4.1 Hydrogen Peroxide

Hydrogen peroxide (H_2O_2) works as an oxidant by the production of hydroxyl free radicals which then attack essential cellular components such as membrane lipids, DNA and proteins. Exposed sulphydryl groups and double bonds are particularly targeted (Block 1991). Hydrogen peroxide is active against a wide range of bacteria and has been shown to be sporicidal (Block 1991; Turner 1983) but has greater activity against Gram-positive bacteria compared to Gram-negative bacteria. It is widely used in hospitals and other environments as a whole room decontaminator. The presence of bacterial enzymes called catalases can help to protect against low concentrations of hydrogen peroxide.

2.2.4.2 Peracetic Acid

Its mode of action is thought to be similar to other oxidizing agents, i.e. it functions by denaturation of proteins, disruption of the cell wall permeability and oxidation of sulphydryl and sulphur bonds in enzymes and other proteins (Baldry and Fraser 1998; Block 1991). Again it has a broad spectrum of activity and has been shown to be especially penetrative against bacterial biofilms (Perumal et al. 2014).

2.2.5 *Alcohols*

It is generally thought that alcohols are membrane disruptors and act by denaturation of proteins. Alcohol can destroy the dehydrogenases in *E. coli* (Sykes 1939). They have increased activity in water with proteins being denatured more quickly in the presence of water rather than absolute alcohol (Ali et al. 2001; Morton 1983). Ethanol is able to cause the rapid release of intracellular components and membrane disruption by interaction with the hydrocarbon component of the phospholipid bilayer. Other examples of alcohols include phenylethanol and phenoxyethanol which induce generalized loss of cytoplasmic membrane function (Fitzgerald et al. 1992). Ethanol has also been shown to inhibit DNA, RNA, protein and peptidoglycan synthesis in *E. coli* (Nunn 1975). Alcohols exhibit rapid broad spectrum antimicrobial activity against vegetative bacteria but are not sporicidal. Low concentrations of alcohol are often used in conjunction with other biocides, e.g. chlorhexidine.

2.2.6 Aldehydes

Glutaraldehyde has a broad spectrum of activity against both vegetative bacteria and bacterial spores. For vegetative bacteria glutaraldehyde strongly associates with the outer layers of the cell wall, specifically with unprotonated amines on the cell surface (Bruck 1991) which causes cross-linking of amino groups within proteins and subsequent inhibition of transport processes into the bacterial cell. The action of glutaraldehyde against bacterial spores is dependent upon its concentration. At low concentrations spore germination is inhibited, whereas at high concentrations glutaraldehyde strongly interacts with the outer spore layers (Thomas and Russell 1974a, b).

Formaldehyde is also sporicidal and can penetrate the outer spore layers. It is an extremely reactive chemical and works to inactivate bacteria by alkylating the amino and sulphydryl groups of proteins and ring nitrogen atoms of purine bases (Favero and Bond 1991).

2.2.7 Phenolics

Phenol derivatives or phenolics are derivatives of phenol where a functional group (e.g. alkyl, phenyl, benzyl, halogen) replaces one of the hydrogen atoms on the aromatic ring. Many have improved biocide activity over phenol. Phenolic compounds in high concentrations are gross protoplasmic poisons which penetrate and disrupt the cell wall and precipitating cellular proteins. At low concentrations phenol can induce the progressive leakage of intracellular constituents and inactivation of essential enzyme systems (Prindle 1983).

Bisphenols are hydroxyl-halogenated derivatives of two phenolic groups connected by a variety of bridges (Gump 1977). They have a broad host range although some bacteria, e.g. *P. aeruginosa*, exhibit intrinsic resistance and they only show sporostatic activity. Examples of bisphenols include hexachlorophene and 2,4,4′-trichloro-2′-hydroxydiphenylether (triclosan). Triclosan is unusual for a biocide in that, although it has been shown to have a variety of targets, it primarily appears to target fatty acid synthesis by inhibition of the enzyme enoyl reductase FabI by competitive inhibition of the enzyme's natural substrate (McMurry et al. 1998a; Heath et al. 1998). However, this inhibition process is relatively slow, and the rapid killing of bacteria by high concentrations of triclosan cannot be explained solely by inhibition of fatty acid synthesis (Gomez Escalada et al. 2005). Studies have shown that insertion of triclosan into the bacterial cell membrane will comprise the membrane functional integrity (Villalain et al. 2001; Guillén et al. 2004).

2.2.8 *Halogen-Releasing Agents*

2.2.8.1 Chlorine-Releasing Agents

The exact mechanisms of microbial destruction by free chlorine are still not fully understood. Oxidation of sulphydryl enzymes and amino acids, ring chlorination of amino acids, decreased uptake of nutrients and oxygen, inhibition of protein synthesis, oxidation of respiratory components and decreased ATP production and effects on DNA synthesis have all been shown to be factors resulting from inactivation by chlorine (Dychdala 2001). Low pH is thought to increase activity. Examples of chlorine-releasing agents include sodium hypochlorite, chlorine dioxide and sodium dichloroisocyanurate (NaDCC). Low concentrations of these agents are active against vegetative bacteria, but higher concentrations are necessary to have an effect against mycobacteria or bacterial spores (Rutala et al. 1991; Bloomfield and Arthur 1992; Ungurs et al. 2011). The activity of CRAs is affected by the presence of organic matter, e.g. blood (Coates 1991).

2.2.8.2 Iodophors

Iodophors are a combination of iodine and a solubilizing agent or carrier, which acts as an active 'free' iodine reservoir (Gottardi 1991). Similar to chlorine, iodine acts by rapidly penetrating into bacteria and attacking cysteine and methionine amino acids as well as nucleotides and fatty acids (Kruse 1970; Gottardi 1991). This results in disruption to protein and nucleic acid structure and synthesis. They are generally used as antiseptics but have been used for disinfection of endoscopes and other medical equipment. Dilution is a critical factor with contamination of iodophor solutions reported (Craven et al. 1981; Parrott et al. 1982).

2.2.9 *Others*

There are many other types of biocide which have not been discussed here including silver-containing biocides (silver nitrate, silver sulphadiazine), diamidines which are used for wound disinfection and anilides which show good activity against Gram-positive bacteria but are poor against Gram-negative bacteria. These are reviewed in the article by McDonnell and Russell (1999).

2.3 Resistance to Biocides

Bacterial resistance to biocides is not a new phenomenon having been described over 50 years ago (Russell 2004; Maher et al. 1961). However, there are a now growing number of reports of resistance to biocides, both within clinical and laboratory conditions, suggesting that biocide resistance is increasingly problematic. The response of bacteria to particular biocides, and the nature of their resistance to it, is driven primarily by the nature of the biocide itself and the type of organism. Biocide activity is also influenced by a variety of factors including pH, presence of organic matter, temperature, concentration and contact time. Changes to the environment can result in alterations to biocide susceptibility. The outer cell wall generally plays a critical role in determination of intrinsic bacterial susceptibility (or insusceptibility) to biocides. Bacteria can also develop resistance to particular biocides (acquired) through various mechanisms, which will be discussed below.

2.3.1 Intrinsic Resistance to Biocides

Although biocides usually have a low degree of selectivity in their action against different types of microorganism, they do show varying degrees of activity against different species of bacteria. Some bacterial species are intrinsically resistant to certain biocides. This is defined as the innate (natural) ability of a bacterial species to resist the activity of a particular antimicrobial agent (biocide) through its inherent structural or functional characteristics. This can be due to one or more factors which include (a) the lack of affinity of the biocide for the bacterial target, (b) the inaccessibility of the biocide into the bacterial cell, (c) the extrusion of the biocide by chromosomally encoded active exporters and (d) innate production of enzymes which inactivate the biocide (Fig. 2.1). Production of specific enzymes is rare, and since biocides have multiple targets (one exception being triclosan), lack of affinity is also uncommon.

Bacterial endospores are considered more resistant to biocides than vegetative bacteria (Fig. 2.2). Gram-negative bacteria are intrinsically more resistant to biocides than Gram-positive bacteria due to their outer membrane composition. The exception is mycobacteria which, for vegetative bacteria, are the most resistant, again due to their cell wall composition which limits access of biocides. There are also additional factors, such as whether the bacteria are within a biofilm, which will also affect and generally decrease the effectiveness of biocides.

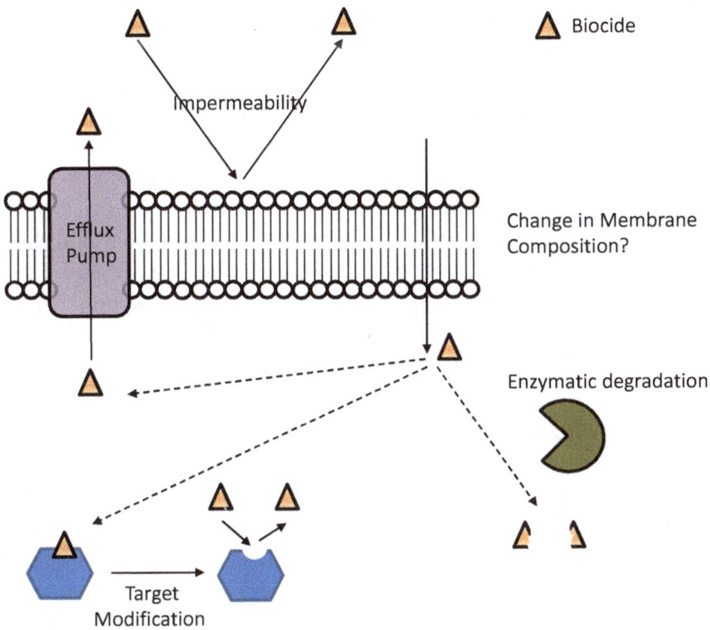

Fig. 2.1 Examples of mechanisms of resistance to biocides found in bacteria. Membrane impermeability and increased bacterial efflux are the most common forms of resistance mechanisms found in bacteria. Target alteration and enzymatic degradation of biocides are limited to a few examples

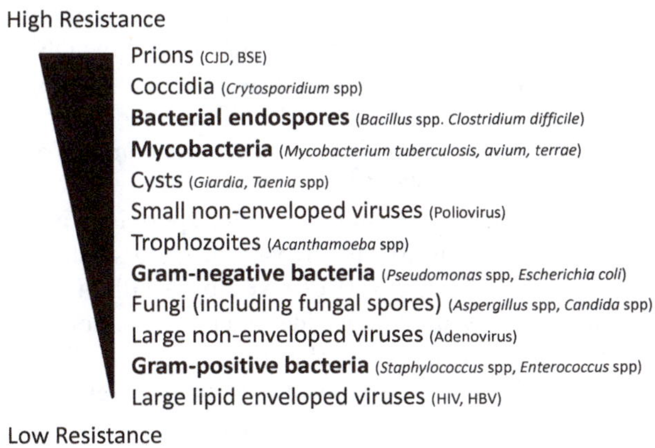

Fig. 2.2 Classification of micro-organisms according to their resistance to biocides. Groups containing bacteria are highlighted in *bold*. Adapted from Maillard (2002)

2.3.1.1 Spore-Forming Bacteria

Many biocides are active against non-sporulating bacteria but have little sporicidal activity; for others sporicidal activity is only achieved at high concentrations. *Clostridium* and *Bacillus* spores are the most resistant of all types of bacteria to biocides (Russell 1990). Examples of biocides which are described as sporicidal include glutaraldehyde and formaldehyde. Resistance of bacterial endospores is mainly due to their structure with factors such as the spore coat and the cortex all providing biocide protection due to inaccessibility of the target site. The biochemical composition of the spore coat varies between species and can even vary between different strains of the same species (Driks 1999; Henriques and Moran 2007). It has been shown to play an essential role in protection against many chemicals including hydrogen peroxide (Young and Setlow 2004) and hypochlorite (Young and Setlow 2003). *Bacillus subtilis* spores which lack several spore coat layers due to mutations in genes which code for a protein essential for outer spore coat formation (*cotE*) and a regulator of coat protein formation (*gerE*) are more susceptible to hypochlorite as compared to wild-type spores produced by this species (Ghosh et al. 2008). Superoxide dismutase (SOD) when present on the spore surface was shown in *Bacillus anthracis* spores to protect against oxidative stress (Cybulski et al. 2009); however, this was not shown to be the case for *B. subtilis* (Casillas-Martinez and Setlow 1997).

Water availability within the spore core and core water content may also play a role in the ability of bacterial spores to resist biocides. *B. subtilis* spores which had a higher core water content appeared to be more susceptible to killing by liquid hydrogen peroxide (Popham et al. 1995). The presence of small acid-soluble spore proteins (SASPs) has been shown to protect nucleic acids within the spore by binding directly to and saturating potential binding sites on the DNA. This prevents access of the DNA to many biocides which are DNA damaging, e.g. hydrogen peroxide (Imlay and Linn 1988; Setlow 2007). Mutants lacking particular types of SASPs in *B. subtilis* correspondingly have been found to be more sensitive to hydrogen peroxide (Setlow et al. 2000).

There have been a number of studies on the sporicidal activity of various biocides and how their activity is affected by various factors including temperature, level of soiling, type of surface, etc. (Reviewed by Maillard 2011). For surface disinfection, the most widely used sporicidal agents are chlorine-based disinfectants, which include hypochlorites (e.g. bleach) and sodium dichloroisocyanurate (NaDCC) (Maillard 2011). The latter appeared to be more effective against *Clostridium difficile* spores on stainless steel rather than PVC surfaces (Block 2004), and the activity of NaDCC is improved when coupled with rigorous surface pre-cleaning using detergent (Ungurs et al. 2011). There is a distinct lack of methodological clarity when testing disinfectants which claim to have sporicidal activity, with different procedures used to evaluate different disinfectants. This has led to the proposed standardized test for testing sporicidal activity of disinfectants against *C. difficile* spores (Fraise et al. 2015).

2.3.1.2 Mycobacteria

The mycobacterial cell wall is more complex than other Gram-positive bacterial cell walls. It has an abundance of high molecular weight lipids, whilst the inner region contains peptidoglycan linked to another polysaccharide polymer, arabinogalactan (Lambert 2002). Anchored to this skeleton are mycolic acids which combine to give the cell a thick waxy coat which acts as an efficient permeability barrier (Brennan and Nikaido 1995). This waxy layer has been shown to contribute towards increased biocide resistance. Spanning the outer layer are porin proteins, and deletion of Msp group of porins decreases the susceptibility of *Mycobacterium smegmatis* to many widely used biocides, e.g. octenidine and polyhexamethylene biguanide (PHMB) (Frenzel et al. 2011). One biocide which is effective against *Mycobacterium* is triclosan which inhibits the action of the enoyl reductase InhA (McMurry et al. 1999; Parikh et al. 2000) though there are reports suggesting that this is not the only site of action (Boshoff et al. 2004). In response to triclosan, *Mycobacterium* significantly differentially expressed several hundreds of genes and these included potential drug detoxification and efflux mechanisms (Betts et al. 2003; Boshoff et al. 2004). Ligand-binding studies have shown which residues are important in triclosan binding to InhA (Cohen et al. 2011). Presumably, mutations at these residues would affect the binding affinity and therefore effectiveness of triclosan against *Mycobacterium*.

2.3.1.3 Other Gram-Positive Bacteria

For bacteria, Gram-positive micro-organisms are highly susceptible to many biocides. Their cell wall is essentially peptidoglycan and teichoic acid which does not act as an effective barrier against the entry of biocides. This is in part due to the ability of high molecular weight substances to readily traverse the cell wall of staphylococci and vegetative *Bacillus* spp. which renders these organisms highly sensitive to QACs and chlorhexidine (Russell 1991, 1995).

The plasticity of the bacterial cell envelope is a well-known phenomenon (Poxton 1993). The cell physiological state will be affected by growth rate and limiting nutrients which will alter the thickness and degree of peptidoglycan cross-linking, altering the cell's biocide sensitivity. Chlorhexidine sensitivity was altered in *Bacillus megaterium* following changes in nutrient availability (Gilbert and Brown 1995). Mucoid strains of *S. aureus*, where the cells are surrounded by a slime layer, are less sensitive to killing by chlorhexidine than non-mucoid strains, and removal of this slime layer rendered the cells more sensitive (Kolawole 1984). This indicates that the slime layer plays a role in protection either as a physical barrier or by absorption of biocide molecules (Kolawole 1984). This is different from experimental work on the Gram-negative species *K. pneumoniae* where there was no difference observed in sensitivity to several biocides between non-mucoid and mucoid isolates (M. Wand, unpublished results).

There are differences in susceptibility within Gram-positive bacteria with *Enterococci* being, in general, less sensitive to biocides than *Staphylococci*. There is, to date, no correlation between the carriage of antibiotic resistance and biocide resistance, e.g. vancomycin-resistant *Enterococcus* (VRE) are not more resistant to biocides than those *Enterococcus* strains which are vancomycin sensitive (VSE) (Sakagami and Kajimura 2002). MRSA (methicillin-resistant *Staphylococcus aureus*) does not appear to have a selective advantage over MSSA (methicillin-sensitive *Staphylococcus aureus*) with respect to biocide resistance (Wootton et al. 2009).

2.3.1.4 Gram-Negative Bacteria

As already mentioned, Gram-negative bacteria are, in general, more resistant to biocides than Gram-positive bacteria due to their cell wall composition. The outer membrane acts as a barrier which helps either to limit or prevent the entry of biocides into the cell. Some cationic biocides are able to damage the outer membrane which may help to promote their own uptake (Hancock 1984). Gram-negative bacteria which are particularly resistant to biocides are *Pseudomonas* species (especially *P. aeruginosa*) due to a number of efflux systems (Morita et al. 2014) and *Burkholderia cepacia*. *B. cepacia* has been readily isolated as a contaminant of disinfectants and other anti-infective solutions (Romero-Gomez et al. 2008; Frank and Schaffner 1976). Biocide testing showed that *B. cepacia* is able to remain viable in several commercial biocide formulations (Rose et al. 2009). *B. cepacia* is also able to biodegrade and inactivate benzyldimethylalkylammonium chloride (benzalkonium chloride) by cleavage of the C-alkyl-N bond (Ahn et al. 2016). Other species highly resistant to disinfectants, especially chlorhexidine, are *Providencia stuartii* and *Proteus* spp. (Strickler and Thomas 1976; Martin 1969; Strickler 1974). This is linked to their hydrophobicity of the cell surface and especially for *Proteus* spp. the membrane lipid content (el Moug et al. 1985; Thomas and Strickler 1979).

2.3.2 Acquired Resistance to Biocides

Increased tolerance to biocides can arise through several strategies including mutation of existing genes, expression of previously silent genes or acquisition of new genetic information by horizontal gene transfer on extrachromosomal elements, e.g. plasmids and transposons. Within the laboratory organisms can be 'adapted' to the presence of particular biocides by continuous exposure (MacGregor and Elliker 1958), whilst residue disinfectant can remain on equipment which could lead to a selective pressure for more biocide-resistant organisms (Cousins 1963). There is now a prevalence to generate biocide-tolerant bacteria by exposure to gradually increasing concentrations of biocide, e.g. chlorhexidine (Furi et al. 2013; Bock et al. 2016).

Biocide resistance, unlike antibiotic resistance, is rarely associated with acquisition of genes, and relatively few 'disinfectant resistance' genes have been identified. For Gram-positive organisms, particularly *S. aureus*, the plasmid-encoded *qacA/B* genes (Gillespie and Skurray 1986; Lyon and Skurray 1987) and for Gram-negative organisms *qacE* and *qacΔE1* (Paulsen et al. 1993) have been linked to disinfectant resistance. These genes are often carried on plasmids which also contain either antibiotic or heavy metal resistance genes, which leads to the debate of whether acquired biocide resistance can also lead to antibiotic resistance.

2.3.2.1 Impermeability

As already described, target sites for biocides are usually situated within the cytoplasm or for Gram-negative bacteria at the cytoplasmic membrane. Changes in the outer membrane, in particular LPS (lipopolysaccharide) and loss of porin proteins, are the principal methods used to change bacterial permeability in response to biocides (Denyer and Maillard 2002). Chlorhexidine-resistant strains of *Pseudomonas stutzeri* have been found which possessed alterations in their OMP (outer membrane proteins) profiles (Tattawasart et al. 2000a) and also had reduced uptake of chlorhexidine associated with decreased cellular amounts of magnesium, calcium and phosphorous (Tattawasart et al. 2000b). Outer membrane fatty acid composition was observed to have changed in *P. aeruginosa* in response to treatment to quaternary ammonium compounds (Guérin-Méchin et al. 2000). Downregulation of porin expression has been noticed in *Salmonella typhimurium* and *E. coli* in response to disinfectant exposure (Karatzas et al. 2008; Zhang et al. 2011). Increased expression of *rarA*, an AraC-type regulator in *K. pneumoniae*, following challenge with disinfectants has resulted in decreased porin expression (De Majumdar et al. 2013).

2.3.2.2 Target Alteration

This method of resistance is uncommon due to the ability of biocides to act on multiple cellular components. The most widely described target alteration in response to a biocide is mutations in the *fabI* gene resulting in increased resistance to triclosan. At low concentrations, triclosan functions to inhibit the action of the enoyl-acyl carrier protein (ACP) reductase (FabI) and therefore prevents fatty acid synthesis (Heath et al. 1998). Mutations in this gene were first described in *E. coli* (Heath et al. 1999) but have since been found in other organisms including *P. aeruginosa* (Hoang and Schweizer 1999), *S. aureus* (Heath et al. 2000), *Acinetobacter baumannii* (Chen et al. 2009) and *Rhodobacter sphaeroides* (Lee et al. 2002). Current *Staphylococcus epidermidis* isolates have also been shown to have mutations in the *fabI* gene or its putative promoter region coupled to increased triclosan resistance when compared to isolates from the 1960s suggesting adaptation to the widespread use of triclosan (Skovgaard et al. 2013). Other studies have shown that high levels of triclosan resistance are not just dependent on mutations in FabI, but other distinct pathways are involved. In *P. aeruginosa* a Δ*fabI* strain retained high levels of resistance to triclosan, but a deletion of another enoyl reductase, *fabV*, resulted in increased susceptibility to triclosan (Zhu et al. 2010). In *Salmonella* substitutions within a gene not directly related to fatty acid synthesis *gryA* also led to a small decrease in triclosan susceptibility (Webber et al. 2008).

2.3.2.3 Efflux Systems

Efflux systems commonly have a wide range of structurally unrelated substrates, and those that are linked to biocide resistance (Table 2.2) often are involved in the efflux of antibiotics. Efflux pumps have a number of different structures and may consist of single proteins which are either ATP driven, e.g. VcaM from *Vibrio cholera* (Huda et al. 2003), HorA from *Lactobacillus brevis* (Sakamoto et al. 2001) and LmrA in *Lactococcus lactis* (van Veen et al. 1999), or are proton-motive force (PMF) driven, e.g. Qac proteins in *S. aureus*. More complex efflux pumps include three-component systems which are composed of an outer membrane and an inner membrane protein and a protein that traverses the periplasm to connect the two membrane proteins. Examples include the AcrAB-TolC transporter in *E. coli* and the MexAB-OprM transporter in *P. aeruginosa*.

Efflux pumps are broadly grouped into five major classes (Putman et al. 2000) (Fig. 2.3): (1) the ATP (adenosine triphosphate)-binding cassette (ABC) family, (2) the major-facilitator superfamily (MFS), (3) the resistance-nodulation-division (RND) family, (4) the small multidrug resistance (SMR) family (a member of the drug/metabolite transporter (DMT) superfamily) and (5) the multidrug and toxic compound extrusion (MATE) family. Most of the multidrug transporters, apart from the ABC transporters which are powered by ATP hydrolysis, utilize the transmembrane H^+ or Na^+ gradient to catalyse drug extrusion.

Insusceptibility to several biocides including benzalkonium chloride, triclosan and chlorhexidine can be attributed to expression of several MATE transporters in a variety of bacteria. These include, for Gram-negative bacteria, PmpM from *P. aeruginosa* (He et al. 2004), NorM from *Neisseria* species (Rouquette-Loughlin et al. 2003) and its homologue VmrA from *Vibrio* species (Chen et al. 2002) and AbeM in *A. baumannii* (Su et al. 2005). From Gram-positive bacteria the most well-characterized MATE transporter is MepA in *S. aureus* (DeMarco et al. 2007) where exposure to sublethal levels of numerous biocides led to the appearance of mutants overexpressing MepA (Huet et al. 2008).

ABC transporters contributing to biocide resistance include EfrAB from *Enterococcus faecalis* (Lee et al. 2003) which have also been detected in *Staphylococcus* and *Bacillus* species (Fernández-Fuentes et al. 2014) where overexpression of this pump led to increased resistance to chlorhexidine and triclosan (Lavilla Lerma et al. 2014).

Table 2.2 Efflux determinants of biocide resistance

Efflux determinant	Type[a]	Biocide[b]	Organism	Reference
Gram-positive				
MepA	MATE	CHX, QAC	*S. aureus*	DeMarco et al. (2007)
EfrAB	ABC	CHX, TRI	*E. faecalis*, *S. aureus*, *Bacillus* spp.	Lee et al. (2003), Fernández-Fuentes et al. (2014)
QacA	MFS	QAC, BG	*S. aureus*, *E. faecalis*	Brown and Skurray (2001), Bischoff et al. (2012)
QacB	MFS	QAC	*S. aureus*, *E. faecalis*	McDonnell and Russell (1999), Bischoff et al. (2012)
QacC/D smr	SMR	QAC	*S. aureus*	Noguchi et al. (1999)
QacEΔ1	SMR	QAC	*S. aureus*, *E. faecalis*	Kazama et al. (1998)
QacG	SMR	QAC	*S. aureus*	Heir et al. (1999a)
QacH	SMR	QAC	*S. aureus*	Heir et al. (1998)
QacJ	SMR	QAC	*S. aureus*	Bjorland et al. (2003)
NorA	MFS	QAC, CTM, CHX	*S. aureus*	Noguchi et al. (1999), DeMarco et al. (2007)
NorB	MFS	QAC, CTM, CHX	*S. aureus*	Truong-Bolduc and Dunman (2005), DeMarco et al. (2007)
MdeA	MFS	QAC, CHX	*S. aureus*	DeMarco et al. (2007)
EmeA	MFS	QAC	*E. faecalis*	Schwaiger et al. (2014)
Mmr	SMR	CTAB	*M. tuberculosis*	Rodrigues et al. (2013)
Gram-negative				
QacE	SMR	QAC	Widespread	Paulsen et al. (1993)
QacEΔ1	SMR	QAC	Widespread	Paulsen et al. (1993)
QacF	SMR	QAC	*Enterobacter* spp., *P. aeruginosa*	Ploy et al. (1998), Jeong et al. (2009)
QacG	SMR	QAC	*P. aeruginosa*	Laraki et al. (1999)
PmpM	MATE	QAC	*P. aeruginosa*	He et al. (2004)
NorM	MATE	BAC	*N. gonorrhoea*, *N. meningitis*	Rouquette-Loughlin et al. (2003)
VmrA	MATE	TPPCI	Vibrio spp.	Chen et al. (2002)
AbeM	MATE	TRI	*A. baumannii*	Su and Chen (2005)
MexAB-OprM	RND	PHN, TRI	*P. aeruginosa*, *P. azelaica*	Chuanchuen et al. (2001), Czechowska et al. (2013)
MexCD-OprJ	RND	BAC, CHX, TRI	*P. aeruginosa*	Morita et al. (2003), Chuanchuen et al. (2001)
MexEF-OprN	RND	TRI	*P. aeruginosa*	Chuanchuen et al. (2001)
SmeDEF	RND	TRI	*S. maltophilia*	Sanchez et al. (2005)
SdeXY	RND	TRI	*S. marcescens*	Chen et al. (2003)
AdeABC	RND	CHX	*A. baumannii*	Rajamohan et al. (2010)
AdeIJK	RND	TRI	*A. baumannii*	Fernando et al. (2014)
AceI	Novel	CHX	*A. baumannii*	Hassan et al. (2013)

(continued)

Table 2.2 (continued)

Efflux determinant	Type[a]	Biocide[b]	Organism	Reference
CmeABC	RND	TRI	*C. jejuni*	Pumbwe et al. (2005)
CmeDEF	RND	TRI	*C. jejuni*	Pumbwe et al. (2005)
AbuO	RND	TRI, CHX, BAC	*A. baumannii*	Srinivasan et al. (2015)
AdeF	MFS	CHX	*A. baumannii*	Hassan et al. (2011)
AbeS	SMR	CHX	*A. baumannii*	Srinivasan et al. (2009)
OqxAB	RND	TRI, CHX, BAC, CTM	*E. coli*	Hansen et al. (2007)
CepA	SMR	CHX	*K. pneumoniae*	Fang et al. (2002)
KpnEF	SMR	BAC, CHX, TRI	*K. pneumoniae*	Srinivasan and Rajamohan (2013)
KpnGH	MFS	BAC, CHX, TRI	*K. pneumoniae*	Srinivasan et al. (2014)
SmvA	MFS	CHX	*K. pneumoniae*	Wand et al. (2017)
AcrAB-TolC	RND	QAC, TRI	*E. coli*, *S. enterica* serovar Typhimurium	Piddock (2006), Webber et al. (2008)
MdtM	MFS	QAC	*E. coli*	Holdsworth and Law (2013)
YhiUV-TolC	RND	BAC	*E. coli*	Nishino and Yamaguchi (2001)
TriABC-OpmH	RND	TRI	*P. aeruginosa*	Mima et al. (2007)
MexJK-OpmH	RND	TRI	*P. aeruginosa*	Chuanchuen et al. (2003)

[a]*MATE* multidrug and toxic compound extrusion (efflux pump type), *ABC* ATP (adenosine triphosphate)-binding cassette, *MFS* major-facilitator superfamily, *SMR* small multidrug resistance, *RND* resistance-nodulation-division, *Novel* not characterized
[b]*BAC* benzalkonium chloride, *BG* biguanides, *CTM* cetrimide, *CHX* chlorhexidine, *QAC* quaternary ammonium compounds, *PHN* phenolics, *TRI* triclosan, *TPPCI* tetraphenylphosphonium chloride, *CTAB* cetyltrimethylammonium bromide (adapted and updated from Poole 2005)

For the RND efflux pumps, MexCD-OprJ in *P. aeruginosa* has been found to be induced by tetraphenylphosphonium chloride (TPPCI) in mutants lacking other Mex pumps (MexAB-OprM, MexEF-OprN and MexXY) (Morita et al. 2001). It is also induced in subinhibitory concentrations of benzalkonium chloride and chlorhexidine gluconate (Morita et al. 2003) in a process that is dependent on the stress response factor, AlgU (Fraud et al. 2008). Other genes affected in subinhibitory concentrations of benzalkonium chloride include mutations in the Mex efflux regulator gene, *nfXB*, which, in turn, led to overexpression of MexCD-OprJ and MexAB-OprM (McCay et al. 2010). MexCD-OprJ is central to the response of *P. aeruginosa* when exposed to chlorhexidine diacetate being, along with another RND efflux pump *oprH-phoPQ*, upregulated (Nde et al. 2009). MexCD-OprJ is also important in the development of chlorhexidine-tolerant subpopulations within *P. aeruginosa* biofilms (Chiang et al. 2012). In *A. baumannii* chlorhexidine tolerance is linked to a broad range of efflux systems including the RND efflux pump

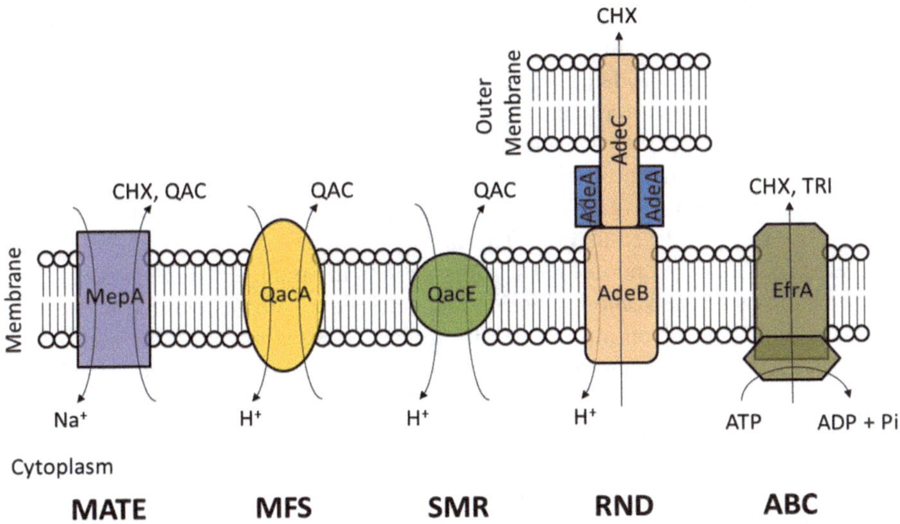

Fig. 2.3 Representatives of the five known families of efflux pumps thought to be involved in bacterial resistance to biocides. The MATE, MFS and SMR families are driven by chemiosmotic energy, with H^+ or Na^+ ions being pumped into the cell whilst pumping biocides out. The RND family spans the inner and outer membranes in Gram-negative bacteria and usually consists of multisubunits. The ABC superfamily utilizes ATP to drive efflux of biocides from the cell. Biocides shown in the diagram are CHX (chlorhexidine), QAC (quaternary ammonium compounds) and TRI (triclosan)

AdeABC (Rajamohan et al. 2010) with exposure to chlorhexidine leading to elevated levels of expression of AdeAB (Hassan et al. 2013). A recently described efflux pump, AceI, from a novel family of bacterial drug efflux transporters was found to be an active chlorhexidine efflux pump (Hassan et al. 2013). This transporter was also found to be present in other bacterial species including *Pseudomonas* sp. and *Burkholderia cenocepacia* (Hassan et al. 2015). Other efflux pumps involved in resistance to chlorhexidine in *A. baumannii* include the MFS efflux pump AedF (Hassan et al. 2011) and an efflux pump, AbeS, belonging to the SMR family (Srinivasan et al. 2009). Although many of these efflux pumps are found on the bacterial chromosome, a plasmid-encoded RND efflux pump, OqxAB, from *E. coli* which confers resistance to a number of biocides, e.g. triclosan, was able to be successfully transferred to other members of the Enterobacteriaceae (Hansen et al. 2007).

Many RND family pumps associated with resistance to antibiotics are also associated with resistance to triclosan including the Mex pumps in *P. aeruginosa* (Chuanchuen et al. 2003), SmeDEF of *Stenotrophomonas maltophilia* (Sanchez et al. 2005), SdeXY of *Serratia marcescens* (Chen et al. 2003), AdeIJK in *A. baumannii* (Fernando et al. 2014) and CmeABC and CmeDEF of *Campylobacter jejuni* (Pumbwe et al. 2005). Enhanced triclosan resistance has been linked to overexpression of AcrAB-TolC and the global regulators of active efflux, MarA and SoxS in *E. coli* and *Salmonella* (McMurry et al. 1998b; Bailey et al. 2009;

Karatzas et al. 2007). Inactivation of AbuO (a TolC homologue) in *A. baumannii* led to decreased survival in subinhibitory concentrations of triclosan, chlorhexidine and benzalkonium chloride (Srinivasan et al. 2015).

Resistance to quaternary ammonium compounds (QACs) has been readily described in Gram-positive bacteria, particularly *S. aureus*. QAC resistance in *S. aureus* is predominantly linked to plasmid-encoded MFS (QacA/B) or SMR family exporters (Smr (QacC/D), QacEΔ1, QacG, QacH, QacJ) with resistance arising from plasmid acquisition. The most prevalent pump is QacA which can mediate resistance to a number of organic cations including lipophilic monovalent cations such as benzalkonium chloride and cetrimide and divalent cations such as chlorhexidine. The seven nucleotide differences between *qacA* and its homologue *qacB* are distributed throughout the gene and result in only one amino acid change (Asp273 in QacA and Ala273 in QacB). However, this change means that QacB offers negligible protection from divalent cations, being primarily associated with resistance to monovalent cations (Paulsen et al. 1996). There is conflicting evidence to suggest that the presence of QacA/B actually plays a role in biocide tolerance. Several studies, in *S. aureus*, have shown that the presence of *qacA* does not always correlate with increased resistance to chlorhexidine; isolates which are *qacA* positive may have comparable chlorhexidine susceptibility levels to isolates which are negative for the presence of *qacA* (Horner et al. 2012). Analysis of over 1600 staphylococcal isolates was unable to show a clear breakpoint between susceptible and non-susceptible populations to benzalkonium chloride and chlorhexidine and the presence of *qacA* and *qacB* genes (Furi et al. 2013). In contrast, there was a clear relationship between resistance to ethidium bromide and the detection of *qacA* and *qacB* (Patel et al. 2010). QacA/B have also been detected in other Gram-positive bacteria, e.g. *Enterococcus faecalis* (Bischoff et al. 2012).

Other efflux pumps contributing to QAC resistance in *S. aureus* are chromosomally encoded and include NorA, NorB and the MFS family efflux protein MdeA (Huang et al. 2004; Noguchi et al. 1999). NorA is a chromosomally encoded MDR (multidrug resistant) efflux pump in *S. aureus* (Patel et al. 2010) and is implicated in resistance to fluoroquinolones including norfloxacin and ciprofloxacin (Yoshida et al. 1990; Ng et al. 1994). Mutations in the promoter of NorA, leading to increased expression, resulted in a slight increase in resistance to ethidium bromide, benzalkonium chloride and chlorhexidine (Furi et al. 2013). Homologues to NorA exist in other Gram-positive bacteria and include EmeA (*Enterococcus faecalis*) (Jonas et al. 2001) with variants in this gene leading to increased resistance to QAC (Schwaiger et al. 2014).

For Gram-negative organisms efflux-mediated biocide resistance is generally chromosomally encoded. The exceptions are *qacE*, *qacEΔ1*, *qacF* and *qacG* which are widely disseminated and are associated with mobile genetic elements. However, again there is conflicting evidence as to whether the presence of the *qacE* and *qacEΔ1* genes correlates with increased QAC resistance. Isolates from a range of Gram-negative organisms showed similar MIC values to benzalkonium chloride whether *qacE* or *qacEΔ1* was detected or not (Kücken et al. 2000). In *K. pneumoniae*, chlorhexidine resistance has been linked to another SMR family efflux pump, *cepA* (Fang et al. 2002), with several resistant strains to chlorhexidine and benzalkonium chloride showing carriage of this gene and *qacEΔ1* (Abuzaid et al. 2012). Homologues to *cepA* have been found in other bacteria, e.g. *P. aeruginosa*, indicating that this mechanism is widely distributed within Gram-negative pathogenic bacteria (Poole 2005). It is again worth noting that there were several isolates which showed elevated levels of antiseptic resistance, in which *cepA* and *qacEΔ1* were not detected, and isolates which were positive for *cepA* had low MIC values to the disinfectants tested (Abuzaid et al. 2012). Another study indicated that reduced susceptibility to chlorhexidine appeared to be independent of the expression of *cepA* and two other efflux pumps, *acrA* and *kdeA* (Naparstek et al. 2012). Therefore, the role, if any, of these genes in antiseptic resistance is unclear. Several *K. pneumoniae* strains which were highly susceptible to chlorhexidine have shown insertions within the *cepA* gene (Wand et al. 2015) which suggests that investigating the genetic sequence of these genes is more important than looking at their presence/absence. Subsequent studies showed that these isolates also lacked the MFS pump SmvA (Wand et al. 2017) which has been implicated in methyl viologen resistance in *S. enterica* (Santiviago et al. 2002). SmvA also has high homology to QacA from *S. aureus* (Santiviago et al. 2001). When strains of *K. pneumoniae* were cultured in sublethal levels of chlorhexidine, increased tolerance was found to be associated with deletion of *smvR* (a Tet repressor acting on *smvA*) and increased expression of *smvA* (Wand et al. 2017).

Other efflux pumps which have been implicated in biocide resistance in *K. pneumoniae* include the SMR-type pump KpnEF (Srinivasan and Rajamohan 2013) and the MFS pump KpnGH (Srinivasan et al. 2014) where the removal of these genes led to a slight increase in susceptibility to benzalkonium chloride, chlorhexidine and triclosan.

2.4 Biofilms

Biofilms are an important source of infection, particularly in clinical environments where an estimated 80% of all infections are linked to biofilms (National Nosocomial Infections Surveillance 1999). Therefore, particularly with surface disinfectants, biocides need to be effective against bacterial biofilms. There is a plethora of literature describing increased antibiotic resistance when bacteria are in a biofilm or when they are attached to surfaces. The same is true for biocides; microbes which are attached to a surface have reduced biocide susceptibility when compared to planktonic cells (Condell et al. 2012; Leung et al. 2012). When bacteria are part of an established biofilm, the biocide resistance is greatly increased which can render certain biocides ineffectual (Smith and Hunter 2008). There have been many studies which have shown that biocides have reduced efficacy against biofilms when compared to planktonic cells, but most are simple observational studies without exploring the reasons for this reduced efficacy. Several studies have also demonstrated that multispecies biofilms are more resistant to biocides than mono-species biofilms (Luppens et al. 2008; Van der Veen and Abee 2010). Within a mixed-species biofilm, *P. aeruginosa* and *K. pneumoniae* were able to survive challenge with above in clinical use concentrations of chlorhexidine (Touzel et al. 2016), and a multispecies biofilm was more resistant to hydrogen peroxide than each individual mono-species biofilm alone (Burmølle et al. 2006). Why multispecies biofilms are more resistant than mono-species biofilms is unclear, but several theories have been hypothesized. Firstly an association or aggregation with more resistant isolates may help protect more susceptible species, e.g. the presence of *Kocuria* sp., which are more resistant to chlorine, was found to protect *Staphylococcus sciuri* from disinfection (Leriche et al. 2003). There also may be a synergistic interaction between different bacterial species within the same biofilm, whether this is production of different enzymes which complement each other to inactivate toxic compounds (Shu et al. 2003) or the production of a more highly viscous matrix due to chemical interactions between different bacterial species polymers (von Canstein et al. 2002). Mechanisms within biofilms which confer reduced susceptibility to antimicrobial agents include reduced penetration (Szomolay et al. 2005). Chlorine dioxide was unable to completely penetrate a mixed-species biofilm, possibly due to interactions between the biocides and components of the biofilm (Jang et al. 2006). The mean penetration time of alkaline hypochlorite into a mixed *P. aeruginosa* and *K. pneumoniae* biofilm was eight times longer than for chlorosulphamate (Stewart et al. 2001). This was potentially due to the greater capacity of alkaline hypochlorite to react with matrix constituents within the biofilm. Cell density and biofilm accumulation invariably play a role in biocide effectiveness with older thicker biofilms being less susceptible than younger less dense biofilms (Stewart 2015).

Other factors influencing biocide effectiveness against biofilms include alteration to growth, modulation of metabolic processes including the stress response and changes in quorum sensing. The expression of *rpoS*, the principal regulator of the general stress response, was over threefold upregulated in a 3-day-old *P. aeruginosa* biofilm when compared to stationary phase planktonic cells (Xu et al. 2001). There are also several experimental factors which have been indicated in biofilm formation which include surface type and growth temperature. An increase in growth temperature has been noted to increase resistance of *S. aureus* biofilm to polyhexamethylene biguanide and QAC disinfectants (Abdallah et al. 2014).

Some biocidal agents are thought to be more effective than others at killing bacteria within biofilms. The oxidizing agents sodium hypochlorite and peroxygens were found to be the most successful at eradication of *P. aeruginosa* and *S. aureus* biofilms (Toté et al. 2010). Sodium hypochlorite was found to be more effective over chlorhexidine at the removal of *E. faecalis* and MRSA in biofilms (Lee et al. 2009; Williamson et al. 2009). Oxidizing agents have multiple targets within the biofilm, whereas other biocides such as chlorhexidine specifically target cell wall components which may explain the high level of efficacy for oxidizing agents against biofilms (Otter et al. 2015). Benzalkonium chloride was shown to have delayed penetration into a *P. aeruginosa* biofilm when compared to peracetic acid (Bridier et al. 2011). Peracetic acid appears to be very effective against bacterial biofilms with multiple strains of *A. baumannii*, *K. pneumoniae* and *P. aeruginosa* having low MBEC (minimum biofilm eradication concentration) against a disinfectant containing peracetic acid (Perumal et al. 2014). Investigation of the ability of certain biocides, e.g. chlorhexidine digluconate and triclosan to actively penetrate and remove oral biofilms, was found to range between 86.8 and 99.5% efficacy after 1 h with chlorhexidine being the most effective (Corbin et al. 2011). Therefore, clearly the ability of a biocide to penetrate biofilms is of paramount importance to eradication of these structures.

Within biofilms, the gene expression profile is thought to be different from planktonic cells. Induction of genes involved in the oxidative stress response has been shown in biofilms in multiple bacteria, e.g. *P. aeruginosa* and *E. coli* (Sauer et al. 2002; Ren et al. 2004). Analysis of gene expression of sessile *A. baumannii* showed changes in expression of genes involved in amino acid and fatty acid metabolism, motility, active transport, DNA methylation, iron acquisition, transcriptional regulation and quorum sensing (Rumbo-Feal et al. 2013). When *B. cenocepacia* biofilms were exposed to chlorhexidine, the expression of several genes including RND and MFS efflux systems, membrane proteins and chaperonins was increased (Coenye et al. 2011). Studies have also shown that subinhibitory concentrations of biocides can lead to upregulation of genes involved in biofilm formation and therefore actively promote biofilm formation (Dong et al. 2012). *Bacillus subtilis* biofilm formation was stimulated in response to sublethal doses of chlorine dioxide by upregulation of the major operons *epsA-epsO* and *yqxM-sipW-tasA* responsible for matrix production (Shemesh et al. 2010). This suggests that biofilm formation is a defensive mechanism employed to help protect the bacteria from the toxic effects of the biocide.

Within biofilms there is thought to be a greater concentration of bacterial persister cells (Lewis 2007; Fauvart et al. 2011). Persister cells are characterized by their ability to survive exposure to antimicrobial agents, yet do not develop resistance to these agents (Lewis 2007) and thus are not true 'mutants'. The persistence of these cells is linked to their metabolic inactivity and slow growth, both of which are characteristic in biofilms. These persister cells are also increasingly likely to survive exposure to a biocide (Simöes et al. 2011) though it is not known if these cells remain metabolic inactive or alter their gene expression profile in response to biocide exposure. When *S. aureus* cells were challenged with triclosan, they were observed to completely shut down the *agr* quorum-sensing system but they also increased the expression of *fabI* (Nielsen et al. 2013).

Other phenotypic resistance mechanisms linked to biocide resistance include swarming motility. Swarming has similar aspects to biofilm communities, and swarm cells from a variety of Gram-negative organisms including *P. aeruginosa* showed increased resistance to biocides, e.g. triclosan (Lai et al. 2009).

2.5 Link to Antibiotic Resistance

Recent evidence has shown that the indiscriminate use of biocides can select for particular bacterial pathogens which have increased resistance to both biocides and antibiotics with many studies showing links between increased biocide and antibiotic resistance (e.g. Schwaiger et al. 2014; Russell et al. 1998). However, often these studies showed no clear mechanism. In addition, there are many other studies which were unable to find evidence for co-resistance to biocides and antibiotics (e.g. Suller and Russel 2000; Loughlin et al. 2002). There is also little evidence to suggest that isolates that are multidrug resistant are also more resistant to biocides (Shinoda et al. 2016) or that the carriage of specific antibiotic resistance genes, e.g. bla_{SHV}, is linked to biocide resistance.

Since biocide usage is less regulated than antibiotic usage and biocides are widely used in a variety of settings, there is a danger that this will lead to outbreaks of pathogens which are both multidrug resistant and resistant to a number of biocides. The Scientific Committee on Emerging and Newly Identified Health Risks (SCENIHR) published a strategy to investigate the antimicrobial resistance effects of biocides (SCENIHR 2010). Within this document there were several recommendations for future research. These covered areas such as the effect of biofilms and exposure to sublethal levels of biocides on biocide and antibiotic resistance.

Horizontal transfer of antibiotic resistance genes and genetic elements between bacteria is common and occurs through a variety of methods including cell to cell conjugation, use of a bacteriophage intermediate or transformation of DNA. Many of these mobile antibiotic resistance genes are found on plasmids. QacA/B genes which are thought to confer increased resistance to benzalkonium chloride were found on multi-resistance plasmids containing *bla* and *tet* resistance genes in *S. aureus* (Heir et al. 1999b; Sidhu et al. 2002). Thus, acquisition of these plasmids will invariably lead to increased resistance to benzalkonium chloride, penicillin and tetracycline. There is also reported genetic linkage between *qac* genes and antibiotic resistance genes (*blaZ*, *aacA-aphD*, *dfrA* and *ble*) on the same plasmids in *staphylococcal* species isolated from clinical and food environments (Sidhu et al. 2001). The higher frequency of antibiotic resistance among benzalkonium chloride-resistant strains indicates that the presence of either resistance determinant selects for the other during selective pressure caused by either antimicrobial therapy or clinical disinfection (Sidhu et al. 2002). The presence of these resistant determinants on plasmids leads to the spread of biocide and antibiotic 'multi-resistance' to other strains or species which are independent of their original carrier (Bjorland et al. 2005).

Upregulation of efflux pumps, as already discussed, is a common resistance mechanism associated with increased antibiotic and biocide tolerance. In *Salmonella* increased triclosan tolerance was associated with increased resistance to ampicillin, tetracycline and kanamycin. This was probably down to overexpression of the *acrAB* efflux pump (Karatzas et al. 2007). Exposure to triclosan also led to cross resistance to antibiotics in *S. maltophilia* (Sanchez et al. 2005) through its binding to the transcriptional repressor SmeT; this in turn causes overexpression of the SmeDEF efflux pump and reduced susceptibility to quinolones (Hernandez et al. 2011). However, another study showed that increased expression of SmeDEF caused by exposure to another biocide, benzalkonium chloride, did not show any chances in susceptibility in *S. maltophilia* to antibiotics. This may be due to the fact that the concentration of benzalkonium chloride required to observe changes in antibiotic susceptibility is lethal (Sanchez et al. 2015).

In *K. pneumoniae* exposure to sublethal levels of chlorhexidine led to mutations in the two-component regulator PhoPQ which in turn led to increased resistance to the last resort antibiotic colistin (Wand et al. 2017). Although these mutations on their own do not appear to increase chlorhexidine tolerance, the fact that they were regularly observed following exposure to chlorhexidine is cause for concern. Similarly, exposure to MIC levels of chlorhexidine led to increased expression of *vanHAX* and *liaXYZ* (involved in vancomycin and daptomycin resistance, respectively) in *Enterococcus faecium* (Bhardwaj et al. 2016). Increased expression of these genes, however, again did not provide any increased protection against chlorhexidine. For *Salmonella* culture in sublethal levels of several commercially available biocides resulted in mutants which had elevated levels of resistance to nalidixic acid, ciprofloxacin, chloramphenicol and tetracycline as well as triclosan but not to the biocides themselves (Webber et al. 2015). Mutations observed with exposure to multiple biocides were in genes *gyrA* (involved in fluoroquinolone resistance), *ramR* (involved in regulation of *acrAB*) and *fabI* (implicated in triclosan resistance) (Webber et al. 2015).

2.6 Problems and Challenges Associated with Testing Biocide Efficacy

Although the number of reports showing an increase in bacterial resistance to biocides is growing and there are outbreaks caused by a failure in disinfection (Gillespie et al. 2007; Gamble et al. 2007; Duarte et al. 2009), there are many factors which should be taken into account with regard to biocide efficacy against bacteria. Often biocide resistance is defined in terms of MIC/MBC values, and whilst an increase in these values is seen, the levels are still far below the in-use concentration. Therefore, are these increased MIC/MBC values clinically relevant? Many studies will show laboratory adaptation to biocides but make little attempt to understand the mechanism of increased resistance. Often bacterial adaptation in the laboratory to biocides occurs over a prolonged period of time; within a clinical environment, most biocides only have a contact time of a few minutes; therefore, is this enough time for bacterial adaptation to occur? The contact time is important since again with susceptibility to biocides being measured by MIC/MBC over a period of 20–24 h, this does not mirror contact time of many biocides.

There is some debate as to whether adaptation of bacteria to biocides using a 'stepwise' method is clinically relevant. Although this method regularly produces bacteria with increased MIC/MBC values to particular biocides, again these are rarely in-use concentration levels (Walsh et al. 2003; Lear et al. 2006). Also, we must consider the question of 'Is this likely to occur within a clinical environment such that adaptation seen in the laboratory would be relevant to an in situ setting?' Exposure to high levels of biocide, performed so as to mirror 'in-use' concentrations, has also produced adaptation, and bacteria can be isolated as a contaminant from biocides, e.g. chlorhexidine (Ko et al. 2015; Brooks et al. 2004). Also exposure to low concentrations of biocide which is indicative of residual concentrations has been shown to induce low-level resistance (Thomas et al. 2000).

Many bacteria will be found as a contaminant on surfaces, and indeed, as already discussed, bacteria either attached to surfaces or in the form of biofilms are more resistant to biocides. The majority of biocide efficacy testing occurs against planktonic bacteria which are invariably more susceptible. Where biofilms are used to test disinfectant efficacy, again the exposure is invariably for 24 h although some studies have done a range of contact times to attempt to mimic biocide use (Perumal et al. 2014). Often the biofilms in the laboratory are grown as mono-species and in rich media. In hospital environments the bacteria may be starved of such nutrients, be in mixed populations or alternatively mixed with organic matter (e.g. blood) (so-called biological soil) which will affect biocide killing. The presence of organic matter has been shown to decrease biocide effectiveness (Condell et al. 2012) with chlorine in particular reacting with organic matter, thus reducing effectiveness (Nou and Luo 2010). Mechanical wiping is often important in the removal of bacterial contamination (Ungurs et al. 2011; Sattar and Maillard 2013), and this is often critical in the removal of biofilms yet is seldom taken into account. Less well understood is the growth of bacteria on surfaces with residual biocide concentrations or the effect of low concentrations of biocide on biofilms. Microbial diversity was reduced following exposure to triclosan but not quaternary ammonium-based formulations (Dunne 2002; McBain et al. 2004). There are also a lack of studies highlighting the link between gain of biocide resistance and biological fitness. It is assumed that following removal of selective pressure, the mutations that arise following biocide adaptation are transient. One study on *S. enterica* serovar Typhimurium SL1344 showed that adaptation to various biocides rendered those isolates 'less fit' (Curiao et al. 2016). Those 'biocide-adapted' cells which are less fit will be outcompeted by a more susceptible population following removal of the selective pressure provided by the biocide. However, triclosan-resistant mutants in *S. typhimurium* had comparable fitness with wild-type sensitive strains (Webber et al. 2008). Another study showed that fitness was impaired in some strains of *K. pneumoniae* but not others following adaptation to chlorhexidine (Wand et al. 2017). This shows that retention of fitness is dependent upon a number of factors, e.g. strain, the individual mutation and particular biocide used. Those mutations which show no impact on fitness are perhaps more clinically relevant especially if they occur in MDR isolates. Perhaps more worryingly is a recent study which links the carriage of a plasmid containing the benzalkonium chloride tolerance efflux pump gene, *emrC* and clinical outcomes of meningitis linked to infections with ST6 *Listeria monocytogenes* (Kremer et al. 2016). This is one of the first studies to link the use of biocides with an increase in disease outcome.

Bacterial resistance to biocides has been well documented in vitro, but concrete evidence of clinical resistance is lacking. There are also issues surrounding the lack of consensus on the methodologies used to study the emergence of bacterial resistance to biocides and also the lack of guidance on the use of biocides within clinical environments. Especially for certain species of bacteria which are multidrug resistant, the use of biocides is critical to combat the spread of these infections. Since antibiotics appear to have failed, it is imperative that biocides do not.

Compliance with Ethical Standards
Conflict of Interest: Matthew E. Wand declares that he has no conflict of interest.

Ethical approval: This chapter does not contain any studies with human participants or animals performed by any of the authors.

References

Abdallah M, Chataigne G, Ferreira-Theret P et al (2014) Effect of growth temperature, surface type and incubation time on the resistance of *Staphylococcus aureus* biofilms to disinfectants. Appl Microbiol Biotechnol 98:2597–2607

Abuzaid A, Hamouda A, Amyes SG (2012) *Klebsiella pneumoniae* susceptibility to biocides and its association with *cepA*, *qac*ΔE and *qacE* efflux pump genes and antibiotic resistance. J Hosp Infect 81:87–91

Ahlström B, Thompson RA, Edebo L (1999) The effect of hydrocarbon chain length, pH and temperature on the binding and bacterial effect of amphiphilic betaine esters on *Salmonella typhimurium*. APMIS 107:318–324

Ahn Y, Kim JM, Kweon O et al (2016) Intrinsic resistance of *Burkholderia cepacia* complex to benzalkonium chloride. mBio 7(6):e01716-16. doi:10.1128/mBio.01716-16

Ali Y, Dolan MJ, Fendler EJ et al (2001) Alcohols. In: Block SS (ed) Disinfection, sterilization, and preservation. Lippincott Williams and Wilkins, Philadelphia, PA, pp 229–504

Allen MJ, White GF, Morby AP (2006) The response of *Escherichia coli* to exposure to the biocide polyhexamethylene biguanide. Microbiology 152:989–1000

Bailey AM, Constantinidou C, Ivens A (2009) Exposure of *Escherichia coli* and *Salmonella enterica* serovar Typhimurium to triclosan induces a species-specific response, including drug detoxification. J Antimicrob Chemother 64:973–985

Baldry MGC, Fraser JAL (1998) Disinfection with peroxygens. Crit Rep Appl Chem 22:91–116

Betts JC, McLaren A, Lennon MG et al (2003) Signature gene expression profiles discriminate between isoniazid-, thiolactomycin-, and triclosan-treated *Mycobacterium tuberculosis*. Antimicrob Agents Chemother 47:2903–2913

Bhardwaj P, Ziegler E, Palmer KL (2016) Chlorhexidine induces VanA type vancomycin resistance genes in Enterococci. Antimicrob Agents Chemother 60:2209–2221

Bjorland J, Steinum T, Sunde M et al (2003) Novel plasmid-borne gene *qacJ* mediates resistance to quaternary ammonium compounds in equine *Staphylococcus aureus*, *Staphylococcus simulans*, and *Staphylococcus intermedius*. Antimicrob Agents Chemother 47:3046–3052

Bjorland J, Steinum T, Kvitle B (2005) Widespread distribution of disinfectant resistance genes among staphylococci of bovine and caprine origin in Norway. J Clin Microbiol 43:4363–4368

Block SS (1991) Peroxygen compounds. In: Block SS (ed) Disinfection, sterilization, and preservation, vol 4. Lea and Febiger, Philadelphia, PA, pp 167–181

Block C (2004) The effect of Perasafe and sodium dichloroisocyanurate (NaDCC) against spores of *Clostridium difficile* and *Bacillus atrophaeus* on stainless steel and polyvinyl chloride surfaces. J Hosp Infect 57:144–148

Bloomfield SF, Arthur M (1992) Interaction of *Bacillus subtilis* spores with sodium hypochlorite, sodium dichloroisocyanurate and chloramine-T. J Appl Bacteriol 72:166–172

Bock LJ, Wand ME, Sutton JM (2016) Varying activity of chlorhexidine-based disinfectants against *Klebsiella pneumoniae* clinical isolates and adapted strains. J Hosp Infect 93:42–48

Boshoff HI, Myers TG, Copp BR et al (2004) The transcriptional responses of *Mycobacterium tuberculosis* to inhibitors of metabolism: novel insights into drug mechanisms of action. J Biol Chem 279:40174–40184

Bischoff M, Bauer J, Preikschat P et al (2012) First detection of the antiseptic resistance gene *qacA/B* in *Enterococcus faecalis*. Microb Drug Resist 18:7–12

Brennan PJ, Nikaido H (1995) The envelope of mycobacteria. Annu Rev Biochem 64:29–63

Bridier A, Dubois-Brissonnet F, Greub G et al (2011) Dynamics of the action of biocides in *Pseudomonas aeruginosa* biofilms. Antimicrob Agents Chemother 55:2648–2654

Brooks SE, Walczak MA, Malcolm S et al (2004) Intrinsic *Klebsiella pneumoniae* contamination of liquid germicidal hand soap containing chlorhexidine. Infect Control Hosp Epidemiol 25:883–885

Brown MH, Skurray RA (2001) Staphylococcal multidrug efflux protein QacA. J Mol Microbiol Biotechnol 3:163–170

Broxton P, Woodcock PM, Gilbert P (1983) A study of the antibacterial activity of some polyhexamethylene biguanides towards *Escherichia coli* ATCC 8739. J Appl Bacteriol 54:345–353

Bruck CW (1991) Role of glutaraldehyde and other liquid chemical sterilants in the processing of new medical devices. In: Morrissey RF, Prokopenko YI (eds) Sterilization of medical products, vol V. Polyscience Publications, Morin Heights, Canada, pp 376–396

Burmølle M, Webb JS, Rao D et al (2006) Enhanced biofilm formation and increased resistance to antimicrobial agents and bacterial invasion are caused by synergistic interactions in multispecies biofilms. Appl Environ Microbiol 72:3916–3923

Casillas-Martinez L, Setlow P (1997) Alkyl hydroperoxide reductase, catalase, MrgA, and superoxide dismutase are not involved in resistance of *Bacillus subtilis* spores to heat or oxidizing agents. J Bacteriol 179:7420–7425

Chawner JA, Gilbert P (1989a) A comparative study of the bactericidal and growth inhibitory activities of the bisbiguanides alexidine and chlorhexidine. J Appl Bacteriol 66:243–252

Chawner JA, Gilbert P (1989b) Adsorption of the bisbiguanides alexidine and chlorhexidine to Escherichia coli cells and to isolated membrane components. Int J Pharm 55:209–215

Chen J, Morita Y, Huda MN et al (2002) VmrA, a member of a novel class of Na(+)-coupled multidrug efflux pumps from *Vibrio parahaemolyticus*. J Bacteriol 184:572–576

Chen J, Kuroda T, Huda MN et al (2003) An RND-type multidrug efflux pump SdeXY from *Serratia marcescens*. J Antimicrob Chemother 52:176–179

Chen Y, Pi B, Zhou H et al (2009) Triclosan resistance in clinical isolates of *Acinetobacter baumannii*. J Med Microbiol 58:1086–1091

Chiang WC, Pamp SJ, Nilsson M et al (2012) The metabolically active subpopulation in *Pseudomonas aeruginosa* biofilms survives exposure to membrane-targeting antimicrobials via distinct molecular mechanisms. FEMS Immunol Med Microbiol 65:245–256

Chuanchuen R, Beinlich K, Hoang TT et al (2001) Cross-resistance between triclosan and antibiotics in *Pseudomonas aeruginosa* is mediated by multidrug efflux pumps: exposure of a susceptible mutant strain to triclosan selects *nfxB* mutants overexpressing MexCD-OprJ. Antimicrob Agents Chemother 45:428–432

Chuanchuen R, Karkhoff-Schweizer RR, Schweizer HP (2003) High-level triclosan resistance in *Pseudomonas aeruginosa* is solely a result of efflux. Am J Infect Control 31:124–127

Coates D (1991) Disinfection of spills of body fluids: how effective is a level of 10,000 ppm available chlorine? J Hosp Infect 18:319–322

Coenye T, Van Acker H, Peeters E et al (2011) Molecular mechanisms of chlorhexidine tolerance in *Burkholderia cenocepacia* biofilms. Antimicrob Agents Chemother 55:1912–1919

Cohen EM, Machado KS, Cohen M et al (2011) Effect of the explicit flexibility of the InhA enzyme from *Mycobacterium tuberculosis* in molecular docking simulations. BMC Genomics 22:S7

Condell O, Iversen C, Cooney S et al (2012) Efficacy of biocides used in the modern food industry to control *Salmonella enterica*, and links between biocide tolerance and resistance to clinically relevant antimicrobial compounds. Appl Environ Microbiol 78:3087–3097

Corbin A, Pitts B, Parker A et al (2011) Antimicrobial penetration and efficacy in an *in vitro* oral biofilm model. Antimicrob Agents Chemother 55:3338–3344

Cousins CM (1963) Methods for the detection of survivors on milk handling equipment with reference to the use of disinfectant inhibitors. J Appl Bact 26:376–382

Craven DE, Moody B, Connolly MG et al (1981) Pseudobacteremia caused by povidone-iodine solution contaminated with *Pseudomonas cepacia*. N Engl J Med 305:621–623

Curiao T, Marchi E, Grandgirard D et al (2016) Multiple adaptive routes of *Salmonella enterica* Typhimurium to biocide and antibiotic exposure. BMC Genomics 17:491

Cybulski RJ Jr, Sanz P, Alem F et al (2009) Four superoxide dismutases contribute to *Bacillus anthracis* virulence and provide spores with redundant protection from oxidative stress. Infect Immun 77:274–285

Czechowska K, Reimmann C, van der Meer JR (2013) Characterization of a MexAB-OprM efflux system necessary for productive metabolism of *Pseudomonas azelaica* HBP1 on 2-hydroxybiphenyl. Front Microbiol 4:203

Davies A (1973) The mode of action of chlorhexidine. J Periodontol 8:68–75

De Majumdar S, Veleba M, Finn S et al (2013) Elucidating the regulon of multidrug resistance regulator RarA in *Klebsiella pneumoniae*. Antimicrob Agents Chemother 57:1603–1609

DeMarco CE, Cushing LA, Frempong-Manso E et al (2007) Efflux-related resistance to norfloxacin, dyes, and biocides in bloodstream isolates of *Staphylococcus aureus*. Antimicrob Agents Chemother 51:3235–3239

Denyer SP, Maillard JY (2002) Cellular impermeability and uptake of biocides and antibiotics in Gram-negative bacteria. J Appl Microbiol 92:35S–45S

Dong L, Tong Z, Linghu D et al (2012) Effects of sub-minimum inhibitory concentrations of antimicrobial agents on *Streptococcus mutans* biofilm formation. Int J Antimicrob Agents 39:390–395

Driks A (1999) Bacillus spore coat. Microbiol Mol Biol Rev 63:1–20

Duarte RS, Lourenço MC, Fonseca Lde S et al (2009) Epidemic of postsurgical infections caused by *Mycobacterium massiliense*. J Clin Microbiol 47:2149–2155

Dunne WM Jr (2002) Bacterial adhesion: seen any good biofilms recently? Clin Microbiol Rev 15:167–193

Dychdala GR (2001) Chlorine and chlorine compounds. In: Block SS (ed) Disinfection, sterilization and preservation. Lippincott Williams and Williams, Philadelphia, PA, pp 135–137

El Moug T, Rogers ST, Furr JR (1985) Antiseptic-induced changes in the cell surface of a chlorhexidine-sensitive and a chlorhexidine-resistant strain of *Providencia stuartii*. J Antimicrob Chemother 16:685–689

Fang CT, Chen HC, Chuang YP et al (2002) Cloning of a cation efflux pump gene associated with chlorhexidine resistance in *Klebsiella pneumoniae*. Antimicrob Agents Chemother 46:2024–2028

Fauvart M, De Groote VN, Michiels J (2011) Role of persister cells in chronic infections: clinical relevance and perspectives on anti-persister therapies. J Med Microbiol 60:699–709

Favero MS, Bond WW (1991) Chemical disinfection of medical and surgical materials. In: Block SS (ed) Disinfection, sterilization and preservation. Lea and Febiger, Philadelphia, PA, pp 617–641

Fernández-Fuentes MA, Abriouel H, Ortega Morente E et al (2014) Genetic determinants of antimicrobial resistance in Gram positive bacteria from organic foods. Int J Food Microbiol 172:49–56

Fernando DM, Xu W, Loewen PC et al (2014) Triclosan can select for an AdeIJK-over expressing mutant of *Acinetobacter baumannii* ATCC 17978 that displays reduced susceptibility to multiple antibiotics. Antimicrob Agents Chemother 58:6424–6431

Fitzgerald KA, Davies A, Russell AD (1992) Sensitivity and resistance of Escherichia coli and Staphylococcus aureus to chlorhexidine. Lett Appl Microbiol 14:33–36

Fraise AP, Wilkinson MA, Bradley CR et al (2015) Development of a sporicidal test method for *Clostridium difficile*. J Hosp Infect 89:2–15

Frank MJ, Schaffner W (1976) Contaminated aqueous benzalkonium chloride. An unnecessary hospital infection hazard. JAMA 236:2418–2419

Fraud S, Campigotto AJ, Chen Z et al (2008) MexCD-OprJ multidrug efflux system of *Pseudomonas aeruginosa*: involvement in chlorhexidine resistance and induction by membrane-damaging agents dependent upon the AlgU stress response sigma factor. Antimicrob Agents Chemother 52:4478–4482

Frenzel E, Schmidt S, Niederweis M et al (2011) Importance for porins for biocide efficacy against *Mycobacterium smegmatis*. Appl Environ Microbiol 77:3068–3073

Furi L, Ciusa ML, Knight D et al (2013) Evaluation of reduced susceptibility to quaternary ammonium compounds and bisquanides in clinical isolates and laboratory-generated mutants of *Staphylococcus aureus*. Antimicrob Agents Chemother 57:3488–3497

Gamble HP, Duckworth GJ, Ridgway GL (2007) Endoscope decontamination incidents in England 2003–2004. J Hosp Infect 67:350–354

Ghosh S, Setlow B, Wahome PG et al (2008) Characterization of spores of *Bacillus subtilis* that lack most coat layers. J Bacteriol 190:6741–6748

Gilbert P, Brown MRW (1995) Some perspectives on preservation and disinfection in the present day. Int Biodeterior Biodegrad 36:219–226

Gilbert P, McBain AJ (2003) An evaluation of the potential impact of the increased use of biocides within consumer products upon the prevalence of antibiotic resistance. Clin Microbiol Rev 16:189–208

Gilbert P, Moore LE (2005) Cationic antiseptics: diversity of action under a common epithet. J Appl Microbiol 99:703–715

Gillespie MT, Skurray RA (1986) Plasmids in multiresistant *Staphylococcus aureus*. Microbiol Sci 3:53–58

Gillespie JL, Arnold KE, Noble-Wang J et al (2007) Outbreak of *Pseudomonas aeruginosa* infections after transrectal ultrasound-guided prostate biopsy. Urology 69:912–914

Gomez Escalada M, Harwood JL, Maillard JY et al (2005) Triclosan inhibition of fatty acid synthesis and its effect on growth of *E. coli* and *P. aeruginosa*. J Antimicrob Chemother 55:879–882

Gottardi W (1991) Iodine and iodine compounds. In: Block SS (ed) Disinfection, sterilization and preservation, vol 4. Lea and Febiger, Philadelphia, PA, pp 152–166

Guérin-Méchin L, Dubois-Brissonnet F, Heyd B et al (2000) Quaternary ammonium compound stresses induce specific variations in fatty acid composition of *Pseudomonas aeruginosa*. Int J Food Microbiol 55:157–159

Guillén J, Bernabeu A, Shapiro S et al (2004) Location and orientation of Triclosan in phospholipid model membranes. Eur Biophys J 33:448–453

Gump WS (1977) The bis-phenols. In: Block SS (ed) Disinfection, sterilization and preservation. Lea and Febiger, Philadelphia, PA, pp 252–281

Hancock REW (1984) Alterations in outer membrane permeability. Annu Rev Microbiol 38:237–264

Hansen LH, Jensen LB, Sørensen HI et al (2007) Substrate specificity of the OqxAB multidrug resistance pump in *Escherichia coli* and selected enteric bacteria. J Antimicrob Chemother 60:145–147
Hassan KA, Brzoska AJ, Wilson NL et al (2011) Roles of DHA2 family transporters in drug resistance and iron homeostasis in *Acinetobacter* spp. J Mol Microbiol Biotechnol 20:116–124
Hassan KA, Jackson SM, Penesyan A et al (2013) Transcriptomic and biochemical analyses identify a family of chlorhexidine efflux proteins. Proc Natl Acad Sci USA 110:20254–20259
Hassan KA, Liu Q, Henderson PJ et al (2015) Homologs of the *Acinetobacter baumannii* AceI transporter represent a new family of bacterial multidrug efflux systems. mBio 6:e01982-14. doi:10.1128/mBio.01982-14
He GX, Kuroda T, Mima T et al (2004) An H(+)-coupled multidrug efflux pump, PmpM, a member of the MATE family of transporters, from *Pseudomonas aeruginosa*. J Bacteriol 186:262–265
Heath RJ, Rubin JR, Holland DR et al (1999) Mechanism of triclosan inhibition of bacterial fatty acid synthesis. J Biol Chem 274:11110–11114
Heath RJ, Yu YT, Shapiro MA et al (1998) Broad spectrum antimicrobial biocides target the FabI component of fatty acid synthesis. J Biol Chem 273:30316–30320
Heath RJ, Li J, Roland GE et al (2000) Inhibition of the *Staphylococcus aureus* NADPH-dependent enoyl-acyl carrier protein reductase by triclosan and hexachlorophene. J Biol Chem 275:4654–4659
Heir E, Sundheim G, Holck AL (1998) The *Staphylococcus qacH* gene product: a new member of the SMR family encoding multidrug resistance. FEMS Microbiol Lett 163:49–56
Heir E, Sundheim G, Holck AL (1999a) The *qacG* gene on plasmid pST94 confers resistance to quaternary ammonium compounds in staphylococci isolated from the food industry. J Appl Microbiol 86:378–388
Heir E, Sundheim G, Holck AL (1999b) Identification and characterisation of quaternary ammonium compound resistant staphylococci from the food industry. Int J Food Microbiol 48:211–219
Henriques AO, Moran CP Jr (2007) Structure, assembly, and function of the spore surface layers. Annu Rev Microbiol 61:555–588
Hernandez A, Ruiz FM, Romero A et al (2011) The binding of triclosan to SmeT, the repressor of the multidrug efflux pump SmeDEF, induces antibiotic resistance in *Stenotrophomonas maltophilia*. PLoS Pathog 7:e1002103. doi:10.1371/journal.ppat.1002103
Hoang TT, Schweizer HP (1999) Characterization of *Pseudomonas aeruginosa* enoyl-acyl carrier protein reductase (FabI): a target for the antimicrobial triclosan and its role in acylated homoserine lactone synthesis. J Bacteriol 181:5489–5497
Holdsworth SR, Law CJ (2013) The major facilitator superfamily transporter MdtM contributes to the intrinsic resistance of *Escherichia coli* to quaternary ammonium compounds. J Antimicrob Chemother 68:831–839
Horner C, Mawer D, Wilcox M (2012) Reduced susceptibility to chlorhexidine in staphylococci: is it increasing and does it matter? J Antimicrob Chemother 67:2547–2559
Huang J, O'Toole PW, Shen W (2004) Novel chromosomally encoded multidrug efflux transporter MdeA in *Staphylococcus aureus*. Antimicrob Agents Chemother 48:909–917
Huda N, Lee E, Chen J et al (2003) Molecular cloning and characterization of an ABC multidrug efflux pump, VcaM, in Non-O1 *Vibrio cholerae*. Antimicrob Agents Chemother 47:2413–2417
Huet AA, Raygada JL, Mendiratta K et al (2008) Multidrug efflux pump overexpression in *Staphylococcus aureus* after single and multiple *in vitro* exposures to biocides and dyes. Microbiology 154:3144–3153
Hugo WB (1999) Disinfection mechanisms. In: Russell AD, Hugo WB, Aycliffe GAJ (eds) Principles and practice of disinfection, preservation and sterilization, 3rd edn. Blackwell Science, Oxford, pp 258–283

Hugo WB, Longworth AR (1966) The effect of chlorhexidine on the electrophoretic mobility, cytoplasmic constituents, dehydrogenase activity and cell walls of *Escherichia coli* and *Staphylococcus aureus*. J Pharm Pharmacol 18:569–578

Imlay JA, Linn S (1988) DNA damage and oxygen radical toxicity. Science 240:1302–1209

Ioannou CJ, Hamlon GW, Denyer SP (2007) Action of disinfectant quaternary ammonium compounds against *Staphylococcus aureus*. Antimicrob Agents Chemother 51:296–306

Jang A, Szabo J, Hosni AA et al (2006) Measurement of chlorine dioxide penetration in dairy process pipe biofilms during disinfection. Appl Microbiol Biotechnol 73:3742–3746

Jeong JH, Skin KS, Lee JW et al (2009) Analysis of novel class integrin containing metallo-beta-lactamase gene VIM-2 in *Pseudomonas aeruginosa*. J Microbiol 47:753–759

Jonas BM, Murray BE, Weinstock GM (2001) Characterization of *emeA*, a NorA homolog and multidrug resistance efflux pump, in *Enterococcus faecalis*. Antimicrob Agents Chemother 45:3574–3579

Karatzas KA, Webber MA, Jorgensen F et al (2007) Prolonged treatment of *Salmonella enterica* serovar Typhimurium with commercial disinfectants selects for multiple antibiotic resistance, increased efflux and reduced invasiveness. J Antimicrob Chemother 60:947–955

Karatzas KA, Randall LP, Webber M et al (2008) Phenotypic and proteomic characterization of multiply antibiotic-resistant variants of *Salmonella enterica* serovar Typhimurium selected following exposure to disinfectants. Appl Environ Microbiol 74:1508–1516

Kazama H, Hamashima H, Sasatsu M et al (1998) Distribution of the antiseptic-resistance gene *qacEΔ1* in gram-positive bacteria. FEMS Microbiol Lett 165:295–299

Ko S, An HS, Bang JH et al (2015) An outbreak of *Burkholderia cepacia* complex pseudobacteremia associated with intrinsically contaminated commercial 0.5% chlorhexidine solution. Am J Infect Control. doi:10.1016/j.ajic.2014.11.010

Kolawole DO (1984) Resistance mechanisms of mucoid-grown *Staphylococcus aureus* to the antibacterial action of some disinfectants and antiseptics. FEMS Microbiol Lett 25:205–209

Kremer PHC, Lees JA, Koopmans MM et al (2016) Benzalkonium tolerance genes and outcome in *Listeria monocytogenes* meningitis. Clin Microbiol Infect. pii:S1198-743X(16)30614-0. doi:10.1016/j.cmi.2016.12.008

Kruse WC (1970) Halogen action on bacteria, viruses and protozoa. In: Proceedings of the National Special Conference on Disinfection. ASCE, Amherst, MA, pp 113–137

Kücken D, Feucht H, Kaulfers P (2000) Association of *qacE* and *qacEDelta1* with multiple resistance to antibiotics and antiseptics in clinical isolates of Gram-negative bacteria. FEMS Microbiol Lett 183:95–98

Lai S, Tremblay J, Déziel E (2009) Swarming motility: a multicellular behaviour conferring antimicrobial resistance. Environ Microbiol 11:126–136

Lambert PA (2002) Cellular impermeability and uptake of biocides and antibiotics in Gram-positive bacteria and mycobacteria. J Appl Microbiol 92:46S–54S

Lambert PA, Hammond SM (1973) Potassium fluxes, first indications of membrane damage in microorganisms. Biochem Biophys Acta 54:796–799

Laraki N, Galleni M, Thamm I et al (1999) Structure of In31, a blaIMP-containing *Pseudomonas aeruginosa* integron phyletically related to In5, which carries an unusual array of gene cassettes. Antimicrob Agents Chemother 43:890–901

Lavilla Lerma L, Benomar N, Valenzuela AS et al (2014) Role of EfrAB efflux pump in biocide tolerance and antibiotic resistance of *Enterococcus faecalis* and *Enterococcus faecium* isolated from fermented foods and the effect of EDTA as EfrAB inhibitor. Food Microbiol 44:249–257

Lear JC, Maillard JY, Dettmar PW (2006) Chloroxylenol- and triclosan-tolerant bacteria from industrial sources – susceptibility to antibiotics and other biocides. Int Biodeterior Biodegrad 57:51–56

Lee IH, Kim EJ, Cho YH et al (2002) Characterization of a novel enoyl-acyl carrier protein reductase of diazaborine-resistant *Rhodobacter sphaeroides* mutant. Biochem Biophys Res Commun 299:621–627

Lee EW, Huda MN, Kuroda T et al (2003) EfrAB, an ABC multidrug efflux pump in *Enterococcus faecalis*. Antimicrob Agents Chemother 47:3733–3738

Lee D, Howlett J, Pratten J et al (2009) Susceptibility of MRSA biofilms to denture-cleansing agents. FEMS Microbiol Lett 291:241–246

Leriche V, Briandet R, Carpentier B (2003) Ecology of mixed biofilms subjected daily to chlorinated alkaline solution: spatial distribution of bacterial species suggests a protective effect of one species to another. Environ Microbiol 5:64–71

Leung CY, Chan YC, Samaranayake LP et al (2012) Biocide resistance of *Candida* and *Escherichia coli* biofilms is associated with higher antioxidative capacities. J Hosp Infect 81:79–86

Lewis K (2007) Persister cells, dormancy and infectious disease. Nat Rev Microbiol 5:48–56

Longworth AR (1971) Chlorhexidine. In: Hugo WB (ed) Inhibition and destruction of the microbial cell. Academic Press, London, pp 95–106

Loughlin MF, Jones MV, Lambert PA (2002) *Pseudomonas aeruginosa* cells adapted to benzalkonium chloride show resistance to other membrane active agents but not to clinically relevant antibiotics. J Antimicrob Chemother 49:631–639

Luppens SBI, Kara D, Bandounas L et al (2008) Effect of *Veillonella parvula* on the antimicrobial resistance and gene expression of *Streptococcus mutans* grown in a dual-species biofilm. Oral Microbiol Immunol 23:193–189

Lyon BR, Skurray R (1987) Antimicrobial resistance of *Staphylococcus aureus*: genetic basis. Microbiol Rev 51:88–134

MacGregor DR, Elliker PR (1958) A comparison of some properties of strains of *Pseudomonas aeruginosa* sensitive and resistant to quaternary ammonium compounds. Can J Microbiol 4:499–503

Maher JT, Rogers MR, Peterson DW (1961) Evaluation of disinfection techniques for, and their effects on, rectal thermocouple catheters. Appl Microbiol 9:273–280

Maillard JY (2002) Bacterial target sites for biocide action. J Appl Microbiol 92(Suppl):16S–27S

Maillard JY (2011) Innate resistance to sporicides and potential failure to decontaminate. J Hosp Infect 77:204–209

Martin TDM (1969) Sensitivity of the genus *Proteus* to chlorhexidine. J Med Microbiol 2:101–108

McBain AJ, Ledder RG, Moore LE et al (2004) Effects of quaternary-ammonium based formulations on bacterial community dynamics and antimicrobial susceptibility. Appl Environ Microbiol 70:3449–3456

McCay PH, Ocampo-Sosa AA, Fleming GT (2010) Effect of subinhibitory concentrations of benzalkonium chloride on the competitiveness of *Pseudomonas aeruginosa* grown in continuous culture. Microbiology 156:30–38

McDonnell G, Russell AD (1999) Antiseptics and disinfectants: activity, action, and resistance. Clin Microbiol Rev 12:147–179

McMurry LM, Oethinger M, Levy SB (1998a) Triclosan targets lipid synthesis. Nature 394:531–532

McMurry LM, Oethinger M, Levy SB (1998b) Overexpression of *marA*, *soxS*, or *acrAB* produces resistance to triclosan in laboratory and clinical strains of *Escherichia coli*. FEMS Microbiol Lett 166:305–309

McMurry LM, McDermott PF, Levy SB (1999) Genetic evidence that InhA of *Mycobacterium smegmatis* is a target for triclosan. Antimicrob Agents Chemother 43:711–713

Mima T, Joshi S, Gomez-Escalada M et al (2007) Identification and characterization of TriABC-OpmH, a triclosan efflux pump of *Pseudomonas aeruginosa* requiring two membrane fusion proteins. J Bacteriol 189:7600–7609

Morita Y, Komori Y, Mima T et al (2001) Construction of a series of mutants lacking all of the four major *mex* operons for multidrug efflux pumps or possessing each one of the operons from *Pseudomonas aeruginosa* PA01: MexCD-OprJ is an inducible pump. FEMS Microbiol Lett 202:139–143

Morita Y, Murata T, Mima T et al (2003) Induction of *mexCD-oprJ* operon for a multidrug efflux pump by disinfectants in wild-type *Pseudomonas aeruginosa* PAO1. J Antimicrob Chemother 51:991–994

Morita Y, Tomida J, Kawamura Y (2014) Responses of *Pseudomonas aeruginosa* to antimicrobials. Front Microbiol 4:422

Morton HE (1983) Alcohols. In: Block SS (ed) Disinfection, sterilization, and preservation. Lea and Febiger, Philadelphia, PA, pp 225–239

Naparstek L, Carmeli Y, Chmelnitsky I et al (2012) Reduced susceptibility to chlorhexidine among extremely-drug resistant strains of *Klebsiella pneumoniae*. J Hosp Infect 81:15–19

Nde CW, Jang HJ, Toghrol F et al (2009) Global transcriptomic response of *Pseudomonas aeruginosa* to chlorhexidine diacetate. Environ Sci Technol 43:8406–8415

Ng EY, Trucksis M, Hooper DC (1994) Quinolone resistance mediated by *norA*: physiologic characterization and relationship to *flqB*, a quinolone resistance locus on the *Staphylococcus aureus* chromosome. Antimicrob Agents Chemother 38:1345–1355

Nielsen LN, Larsen MH, Skovgaard S et al (2013) *Staphylococcus aureus* but not *Listeria monocytogenes* adapt to triclosan and adaptation correlates with increased *fabI* expression and *agr* deficiency. BMC Microbiol 13:177

Nishino K, Yamaguchi A (2001) Analysis of a complete library of putative drug transporter genes in *Escherichia coli*. J Bacteriol 183:5803–5812

NNIS System (1999) National Nosocomial Infections Surveillance (NNIS) System Report, data summary from January 1990–May 1999. A report from the NNIS System. Am J Infect Control 27(6):520–532

Noguchi N, Hase M, Kitta M et al (1999) Antiseptic susceptibility and distribution of antiseptic-resistance genes in methicillin-resistant *Staphylococcus aureus*. FEMS Microbiol Lett 172:247–253

Nou X, Luo Y (2010) Whole-leaf wash improves chlorine efficacy for microbial reduction and prevents pathogen cross-contamination during fresh-cut lettuce processing. J Food Sci 75: M283–M290

Nunn WD (1975) The inhibition of phospholipid synthesis in *Escherichia coli* by phenethyl alcohol. Biochim Biophys Acta 380:403–413

Otter JA, Vickery K, Walker JT et al (2015) Surface-attached cells, biofilms and biocide susceptibility: implications for hospital cleaning and disinfection. J Hosp Infect 89:16–27

Parikh SL, Xiao G, Tonge PJ (2000) Inhibition of InhA, the enoyl reductase from *Mycobacterium tuberculosis*, by triclosan and isoniazid. Biochemistry 39:7645–7650

Parrot PL, Terry PM, Whitworth EN et al (1982) *Pseudomonas aeruginosa* peritonitis associated with contaminated poloxamer-iodine solution. Lancet 2:683–685

Patel D, Kosmidis C, Seo SM et al (2010) Ethidium bromide MIC screening for enhanced efflux pump gene expression or efflux activity in *Staphylococcus aureus*. Antimicrob Agents Chemother 54:5070–5073

Paulsen IT, Littlejohn TG, Rådstöm P et al (1993) The 3' conserved segment of integrons contains a gene associated with multidrug resistance to antiseptics and disinfectants. Antimicrob Agents Chemother 37:761–768

Paulsen IT, Brown MH, Littlejohn TG et al (1996) Multidrug resistance proteins QacA and QacB from *Staphylococcus aureus*: membrane topology and identification of residues in substrate specificity. Proc Natl Acad Sci 93:3630–3635

Perumal PK, Wand ME, Sutton JM et al (2014) Evaluation of the effectiveness of the hydrogen-peroxide based disinfectants on biofilms formed by Gram-negative pathogens. J Hosp Infect 87:227–233

Piddock LJV (2006) Clinically relevant chromosomally encoded multidrug resistance efflux pumps in bacteria. Clin Microbiol Rev 19:382–402

Ploy MC, Courvalin P, Lambert T (1998) Characterization of In40 of *Enterobacter aerogenes* BM2688, a class 1 integron with two new gene cassettes, *cmlA2* and *qacF*. Antimicrob Agents Chemother 42:2557–2563

Poole K (2005) Efflux-mediated antimicrobial resistance. J Antimicrob Chemother 56:20–51
Popham DL, Sengupta S, Setlow P (1995) Heat, hydrogen peroxide, and UV resistance of *Bacillus subtilis* spores with increased core water content and with or without major DNA binding proteins. Appl Environ Microbiol 61:3633–3638
Poxton IR (1993) Prokaryote envelope diversity. J Appl Bacteriol Symp Suppl 70:1S–11S
Prindle RF (1983) Phenolic compounds. In: Block SS (ed) Disinfection, sterilization and preservation. Lea and Febiger, Philadelphia, PA, pp 197–224
Pumbwe L, Randall LP, Woodward MJ et al (2005) Evidence for multiple-antibiotic resistance in *Campylobacter jejuni* not mediated by CmeB or CmeF. Antimicrob Agents Chemother 49:1289–1293
Putman M, van Veen HW, Konings WN (2000) Molecular properties of bacterial multidrug transporters. Microbiol Mol Biol Rev 64:672–693
Rajamohan G, Srinivasan VB, Gebreyes WA (2010) Novel role of *Acinetobacter baumannii* RND efflux transporters in mediating decreased susceptibility to biocides. J Antimicrob Chemother 65:228–232
Ren D, Bedzyk LA, Thomas SM et al (2004) Gene expression in *Escherichia coli* biofilms. Appl Microbiol Biotechnol 64:515–524
Rodrigues L, Villellas C, Bailo R et al (2013) Role of the Mmr efflux pump in drug resistance in *Mycobacterium tuberculosis*. Antimicrob Agents Chemother 57:751–757
Romero-Gómez MP, Quiles-Melero MI, Peña Garcia P et al (2008) Outbreak of *Burkholderia cepacia* bacteraemia caused by contaminated chlorhexidine in a haemodialysis unit. Infect Control Hosp Epidemiol 29:377–378
Rose H, Baldwin A, Dowson CG et al (2009) Biocide susceptibility of the *Burkholderia cepacia* complex. J Antimicrob Chemother 63:502–510
Rouquette-Loughlin C, Dunham SA, Kuhn M et al (2003) The NorM efflux pump of *Neisseria gonorrhoeae* and *Neisseria meningitides* recognizes antimicrobial cationic compounds. J Bacteriol 185:1101–1106
Rumbo-Feal S, Gómez MJ, Gayoso C et al (2013) Whole transcriptome analysis of *Acinetobacter baumannii* assessed by RNA-sequencing reveals different mRNA expression profiles in biofilm compared to planktonic cells. PLoS One 8:e72968
Russell AD (1990) Bacterial spores and chemical sporicidal agents. Clin Microbiol Rev 3:99–119
Russell AD (1991) Mechanisms of bacterial resistance to no-antibiotics: food additives and food and pharmaceutical preservatives. J Appl Bacteriol 71:191–201
Russell AD (1995) Mechanisms of bacterial resistance to biocides. Int Biodeterior Biodegrad 36:247–265
Russell AD (2002) Introduction of biocides into clinical practice and the impact on antibiotic-resistant bacteria. J Appl Microbiol 92:121S–135S
Russell AD (2004) Bacterial adaptation and resistance to antiseptics, disinfectants and preservatives is not a new phenomenon. J Hosp Infect 57:97–104
Russell AD, Tattawasart U, Maillard JY et al (1998) Possible link between bacterial resistance and use of antibiotics and biocides. Antimicrob Agents Chemother 42:2151
Rutala WA, Cole EC, Wannamaker NS et al (1991) Inactivation of *Mycobacterium tuberculosis* and *Mycobacterium bovis* by 14 hospital disinfectants. Am J Med 91:267S–271S
Sakagami Y, Kajimura K (2002) Bactericidal activities of disinfectants against vancomycin-resistant enterococci. J Hosp Infect 50:140–144
Sakamoto K, Margolles A, van Veen HW et al (2001) Hop resistance in the beer spoilage bacterium *Lactobacillus brevis* is mediated by the ATP-binding cassette multidrug transporter HorA. J Bacteriol 183:5371–5375
Sanchez P, Moreno E, Martinez JL (2005) The biocide triclosan selects *Stenotrophomonas maltophilia* mutants that overproduce the SmeDEF multidrug efflux pump. Antimicrob Agents Chemother 49:781–782

Sanchez MB, Decorosi F, Viti C et al (2015) Predictive studies suggest that the risk for the selection of antibiotic resistance by biocides is likely low in *Stenotrophomonas maltophilia*. PLoS One 10:e0132816

Santiviago CA, Toro CS, Bucarey SA et al (2001) A chromosomal region surrounding the *ompD* porin gene marks a genetic difference between *Salmonella typhi* and the majority of Salmonella serovars. Microbiology 147:1897–1907

Santiviago CA, Fuentes JA, Bueno SM et al (2002) The *Salmonella enterica* sv.Typhimurium *smvA*, *yddG* and *ompD* (porin) genes are required for the efficient efflux of methyl viologen. Mol Microbiol 46:687–698

Sattar SA, Maillard JY (2013) The crucial role of wiping in decontamination of high-touch environmental surfaces: review of current status and directions for the future. Am J Infect Control 41:S97–104

Sauer K, Camper AK, Ehrlich GD et al (2002) *Pseudomonas aeruginosa* displays multiple phenotypes during development as a biofilm. J Bacteriol 185:1140–1154

SCENIHR (2010) Assessment of the antibiotic resistance effects of biocides. http://ec.europa.eu/health/ph_risk/committees/04_scenihr/docs/scenihr_o_021.pdf

Schwaiger K, Harms KS, Bischoff M et al (2014) Insusceptibility to disinfectants in bacteria from animals, food and humans – is there a link to antimicrobial resistance? Front Microbiol 5:88

Setlow P (2007) I will survive: DNA protection in bacterial spores. Trends Microbiol 15:172–180

Setlow B, McGinnis KA, Ragkousi K et al (2000) Effects of major spore-specific DNA-binding proteins on *Bacillus subtilis* sporulation and spore properties. J Bacteriol 182:6906–6912

Shemesh M, Kolter R, Losick R (2010) The biocide chlorine dioxide stimulates biofilm formation in *Bacillus subtilis* by activation of the histidine kinase KinC. J Bacteriol 192:6352–6356

Shinoda N, Mitarai S, Suzuki E et al (2016) Disinfectant-susceptibility of multi-drug-resistant *Mycobacterium tuberculosis* isolated in Japan. Antimicrob Resist Infect Control 5:3. doi:10.1186/s13756-016-0102-y

Shu M, Browngardt CM, Chen YYM et al (2003) Role of urease enzymes in stability of a 10-species oral biofilm consortium cultivated in a constant-depth film fermenter. Infect Immun 71:7188–7192

Sidhu MS, Heir E, Sørum H et al (2001) Genetic linkage between resistance to quaternary ammonium compounds and β-lactam antibiotics in food-related *Staphylococcus* spp. Microb Drug Resist 7:363–371

Sidhu MS, Heir E, Leegaard T et al (2002) Frequency of disinfectant resistance genes and genetic linkage with β-lactamase transposon Tn552 among clinical staphylococci. Antimicrob Agents Chemother 46:2797–2803

Simöes LC, Lemos M, Pereira AM et al (2011) Persister cells in a biofilm treated with a biocide. Biofouling 27:403–411

Skovgaard S, Nielsen LN, Larsen MH et al (2013) *Staphylococcus epidermidis* isolated in 1965 are more susceptible to triclosan than current isolates. PLoS One 8:e62197

Smith K, Hunter IS (2008) Efficacy of common hospital biocides with biofilms of multi-drug resistant clinical isolates. J Med Microbiol 57:966–973

Srinivasan VB, Rajamohan G (2013) KpnEF, a new member of the *Klebsiella pneumoniae* cell envelope stress response regulon is an SMR-type efflux pump involved in broad-spectrum antimicrobial resistance. Antimicrob Agents Chemother 57:4449–4462

Srinivasan VB, Rajamohan G, Gebreyes WA (2009) Role of AbeS, a novel efflux pump of the SMR family of transporters, in resistance to antimicrobial agents in *Acinetobacter baumannii*. Antimicrob Agents Chemother 53:5312–5316

Srinivasan VB, Singh BB, Priyadarshi N et al (2014) Role of novel multidrug efflux pump involved in drug resistance in *Klebsiella pneumoniae*. PLoS One 9:e96288

Srinivasan VB, Vaidyanathan V, Rajamohan G (2015) AbuO, a TolC-like outer membrane protein of *Acinetobacter baumannii*, is involved in antimicrobial and oxidative stress resistance. Antimicrob Agents Chemother 59:1236–1245

Stewart PS (2015) Antimicrobial tolerance in biofilms. Microbiol Spectr 3. doi:10.1128/microbiolspec.MB-0010-2014

Stewart PS, Rayner J, Roe F et al (2001) Biofilm penetration and disinfection efficacy of alkaline hypochlorite and chlorosulfamates. J Appl Microbiol 91:525–532

Stickler DJ (1974) Chlorhexidine resistance in *Proteus mirabilis*. J Clin Pathol 27:284–287

Stickler DJ, Thomas B (1976) Sensitivity of Providence to antiseptics and disinfectants. J Clin Pathol 29:815–823

Su XZ, Chen J, Mizushima T et al (2005) AbeM, an H+-coupled *Acinetobacter baumannii* multidrug efflux pump belonging to the MATE family of transporters. Antimicrob Agents Chemother 49:4362–4364

Suller MT, Russell AD (2000) Triclosan and antibiotic resistance in *Staphylococcus aureus*. J Antimicrob Chemother 46:11–18

Sykes G (1939) The influence of germicides on the dehydrogenases of Bact. coli Part 1. The succinic acid dehydrogenase of Bact. coli. J Hyg 39:463–469

Szomolay B, Klapper I, Dockery J et al (2005) Adaptive responses to antimicrobial agents in biofilms. Environ Microbiol 7:1186–1191

Tattawasart U, Maillard JY, Furr JR et al (2000a) Outer membrane changes in *Pseudomonas stutzeri* resistant to chlorhexidine diacetate and cetylpyridinium chloride. Int J Antimicrob Agents 16:233–238

Tattawasart U, Hann AC, Maillard JY et al (2000b) Cytological changes in chlorhexidine-resistant isolates of *Pseudomonas stutzeri*. J Antimicrob Chemother 45:145–152

Thomas S, Russell AD (1974a) Studies on the mechanism of the sporicidal action of glutaraldehyde. J Appl Bacteriol 37:83–92

Thomas S, Russell AD (1974b) Temperature-induced changes in the sporicidal activity and chemical properties of glutaraldehyde. Appl Microbiol 28:331–335

Thomas B, Stickler DJ (1979) Chlorhexidine resistance and the lipids of *Providencia stuartii*. Microbios 24:141–150

Thomas L, Maillard JY, Lambert RJW et al (2000) Development of resistance to chlorhexidine diacetate in *Pseudomonas aeruginosa* and the effect of 'residual' concentration. J Hosp Infect 46:297–303

Tomlinson E, Brown MRW, Davis SS (1977) Effect of colloidal association on the measured activity of alkylbenzyldimethylammonium chlorides against *Pseudomonas aeruginosa*. J Med Chem 20:1277–1282

Toté K, Horemans T, Vanden Berghe D et al (2010) Inhibitory effect of biocides on the viable masses and matrices of *Staphylococcus aureus* and *Pseudomonas aeruginosa* biofilms. Appl Environ Microbiol 76:3135–3142

Touzel RE, Sutton JM, Wand ME (2016) Establishment of a multi-species biofilm model to evaluate chlorhexidine efficacy. J Hosp Infect 92:154–160

Truong-Bolduc QC, Dunman PM, Strahilevitz J et al (2005) MgrA is a multiple regulator of two new efflux pumps in *Staphylococcus aureus*. J Bacteriol 187:2395–2405

Turner FJ (1983) Hydrogen peroxide and other oxidant disinfectants. In: Block SS (ed) Disinfection, sterilization, and preservation, vol 4. Lea and Febiger, Philadelphia, PA, pp 240–250

Ungurs M, Wand M, Vassey M et al (2011) The effectiveness of sodium dichloroisocyanurate against *Clostridium difficile* spores contaminating stainless steel. Am J Infect Control 39:199–205

Van der Veen S, Abee T (2010) Mixed species biofilm's of *Listeria monocytogenes* and *Lactobacillus plantarum* show enhanced resistance to benzalkonium chloride and peracetic acid. Int J Food Microbiol 144:421–431

Van Veen HW, Putman M, Margolles A et al (1999) Structure-function analysis of multidrug transporters in *Lactococcus lactis*. Biochim Biophys Acta 1461:201–206

Villalain J, Mateo CR, Aranda FJ et al (2001) Membranotropic effects of the antibacterial agent triclosan. Arch Biochem Biophys 390:128–136 von Canstein H, Kelly S, Li Y et al (2002) Species diversity improves the efficiency of mercury-reducing biofilms under changing environmental conditions. Appl Environ Microbiol 68:2829–2837

Yoshida H, Bogaki M, Nakamura S et al (1990) Nucleotide sequence and characterization of the *Staphylococcus aureus norA* gene, which confers resistance to quinolones. J Bacteriol 172:6942–6949

Young SB, Setlow P (2003) Mechanisms of killing of *Bacillus subtilis* spores by hypochlorite and chlorine dioxide. J Appl Microbiol 95:54–67

Young SB, Setlow P (2004) Mechanisms of killing of *Bacillus subtilis* spores by Decon and Oxone, two general decontaminants for biological agents. J Appl Microbiol 96:289–301

Walsh SE, Maillard JY, Russell AD et al (2003) Development of bacterial resistance to several biocides and effects on antibiotic susceptibility. J Hosp Infect 55:98–107

Wand ME, Baker KS, Benthall G et al (2015) Characterization of pre-antibiotic era *Klebsiella pneumoniae* isolates with respect to antibiotic/disinfectant susceptibility and virulence in *Galleria mellonella*. Antimicrob Agents Chemother 59:3966–3972

Wand ME, Bock LJ, Bonney LE et al (2017) Mechanisms of increased resistance to chlorhexidine and cross-resistance to colistin following exposure of *Klebsiella pneumoniae* clinical isolates to chlorhexidine. Antimicrob Agents Chemother 61:e01162-16

Webber MA, Randall LP, Cooles S et al (2008) Triclosan resistance in *Salmonella enterica* serovar Typhimurium. J Antimicrob Chemother 62:83–91

Webber MA, Whitehead RN, Mount M et al (2015) Parallel evolutionary pathways to antibiotic resistance selected by biocide exposure. J Antimicrob Chemother 70:2241–2248

Williamson AE, Cardon JW, Drake DR (2009) Antimicrobial susceptibility of monoculture biofilms of a clinical isolate of *Enterococcus faecalis*. J Endod 35:95–97

Wootton M, Walsh TR, Davies EM et al (2009) Evaluation of the effectiveness of common hospital hand disinfectants against methicillin resistant *Staphylococcus aureus*, glycopeptide-intermediate *S. aureus*, and heterogeneous glycopeptide-intermediate *S. aureus*. Infect Control Hosp Epidemiol 30:226–232

Xu KD, Franklin MJ, Park CH et al (2001) Gene expression and protein levels of the stationary phase sigma factor, RpoS, in continuously fed *Pseudomonas aeruginosa* biofilms. FEMS Microbiol Lett 199:67–71

Zhang DF, Li H, Lin XM et al (2011) Characterization of outer membrane proteins of *Escherichia coli* in response to phenol stress. Curr Microbiol 62:777–783

Zhu L, Lin J, Ma J et al (2010) Triclosan resistance of *Pseudomonas aeruginosa* PA01 is due to FabV, a triclosan-resistant enoyl-acyl carrier protein reductase. Antimicrob Agents Chemother 54:689–698

Chapter 3
Disinfection of Microbial Aerosols

Atin Adhikari and Scott Clark

Abstract Disinfection of microbial aerosols gained significant attention among researchers worldwide due to increasing flu pandemics and bioterrorism threats. Ultraviolet germicidal irradiation, HEPA filtration, and photocatalytic oxidation are conventional methods of bioaerosol inactivation, and new approaches developed in recent years include application of cold plasma and plasmacluster ions, microwave irradiation, ion emission, thermal treatment, applications of nanoparticles and nanotubes in filtration media, and application of natural products in filtration media. Although all these methods have shown promising responses for airborne microbial inactivation, they have some inevitable limitations. Most of the methods were tested in the laboratories, and adequate field data are still lacking. Furthermore, most of these methods were never tested for real pathogens and emerging drug-resistant pathogens. Advantages and disadvantages of all these conventional and newly developed approaches have been discussed in this review article. The authors conclude that a perfect solution to inactivate all airborne microorganisms does not exist yet. We can combine different microbial inactivation methods to achieve a more effective disinfection approach depending on the types of suspected microorganisms and indoor environmental conditions.

3.1 Introduction

Microbial aerosols of pathogenic organisms can serve as infective sources either through direct inhalation or through pathways involving the deposition of microbes onto food substances or other surfaces, which eventually contaminate objects entering an individual's body. Besides infection, some microbial aerosols can also cause respiratory allergy, allergic and non-allergic asthma, and toxic reactions

A. Adhikari (✉)
Department of Epidemiology and Environmental Health Sciences, Jiann-Ping Hsu College of Public Health, Georgia Southern University, PO Box: 8015, Statesboro, GA 30460, USA
e-mail: aadhikari@georgiasouthern.edu

S. Clark
Department of Environmental Health, College of Medicine, University of Cincinnati, PO Box: 670056, Cincinnati, OH 45267, USA

C.J. Hurst (ed.), *Modeling the Transmission and Prevention of Infectious Disease*,
Advances in Environmental Microbiology 4, DOI 10.1007/978-3-319-60616-3_3

in human body (Douwes et al. 2003; Edwards et al. 2012). The disinfection of microbial aerosols has gained more attention due to the 2009 H1N1 flu pandemic and the increasing threats of bioterrorism worldwide (Kortepeter and Parker 1999; Henderson 2004; Fraser et al. 2009). This chapter will deal methods to disinfect air that contains microbial aerosols, resulting in a reduction in the airborne concentration of the microbial aerosols and hopefully a decrease in their infectivity and other adverse health effects. Specifically, we will cover the advantages and disadvantages of ultraviolet germicidal irradiation, HEPA filtration, photocatalytic oxidation, and other newly developed technologies for air disinfection and purification from airborne pathogens, such as cold plasma and plasmacluster ions, microwave irradiation, ion emission, thermal treatment, applications of nanoparticles and nanotubes in filtration media, and application of natural products in filtration media. The disinfection of surfaces that may have been contaminated by microbial aerosols and the ultraviolet disinfection of liquids contaminated with microbes that may be a reservoir, supplying microbes to the air, are not covered in this chapter.

A comprehensive plan for controlling microbial aerosols must contain a variety of measures to reduce the potential impact from sources of the microbes, to make conditions less favorable for microbial survival, and to reduce the airborne levels by the use of different engineering controls (e.g., ventilation or the covering of vessels containing microbes). In other words, three levels of intervention are involved: elimination at source, controlling the source, and controlling the exposure. Work practice techniques (e.g., type of clothing worn, removing work areas from locations of highest exposure) are also important to control both the concentration and infectivity of microbial aerosols. For instance, the effectiveness of the ventilation system and the effectiveness of clothing in preventing shed bacteria from releasing into the air could be major influencing factors for the number of airborne bacteria in the operating theater (Gosden et al. 1998). The types of staff clothing worn in an ultraclean operating room with downflow air enclosure has been found to affect bacterial air contamination, being able to reduce contamination from the skin of persons attending the operation to below 1 colony-forming unit (CFU) per cubic meter (Sanzén et al. 1990). In one study cotton gowns were used under two types of operating gowns (exhaust and nonwoven) (Whyte et al. 1976). Exhaust gowns incorporate a vacuum system with an air-exhaust helmet and visor to collect those potentially contaminating aerosols originating from the wearer. The use of the exhaust gowns was found to result in lower airborne bacterial levels than did use of the nonwoven disposable gowns. This observation indicates that a significant reduction in the bacterial count would be best achieved by impervious trousers. In another study involving the wearing of cotton and synthetic scrub gowns under nonwoven operating gowns, the use of the synthetic gowns was found to be associated with airborne contamination levels lower than 1 CFU/m^3, and contrastingly in 9 of 20 operations using cotton scrub suits, the air concentration exceeded 1 CFU/m^3 (Sanzén et al. 1990).

Disinfection of microbial aerosols is generally recommended only as a supplement to other engineering controls. Some reports on the efficiency of microbial aerosol disinfection are expressed as the equivalent effect of a certain number of air changes per hour. Under certain circumstances personal protective equipment such as respirators must also be a part of the overall program to protect human health.

Source control measures have considerable potential for indirectly reducing levels of microbial aerosols. Such measures include limiting the exposure to the air of sources of microbially contaminated materials, reducing the opportunity for microbial growth, preventing the entry of aerosols into the environment of concern, and disinfecting surfaces and liquids contaminated with microbes. This chapter, however, focuses on measures relating directly to controlling the airborne microbes once the aerosolization has occurred.

3.2 Microbial Aerosol Disinfection Methods

There are four general methods to reduce airborne concentrations of microbes: the use of ultraviolet lamps for germicidal irradiation (UVGI), air filtration through high-efficiency particulate filters (HEPA), photocatalytic oxidation, and application of plasma. Chemical disinfections are another potential approach; however, there has been little work done in the area of chemical disinfection, because of the dangers involved in releasing such chemical agents into the air. Microbial disinfection methods may have undesirable side effects, which must be considered. UV radiation, for example, has a number of adverse effects. Short-term UV overexposure can cause erythema and keratoconjunctivitis (Sterenborg et al. 1988). Broad-spectrum UV radiation has been associated with increased risk for squamous and basal cell carcinomas of the skin. Some studies have indicated that UV radiation can increase replication of the human immunodeficiency virus (Zmudzka and Beer 1990). Although technically not a true disinfection mechanism, HEPA filtration has virtually equivalent results and is discussed in this chapter as it can also be effective in removing non-microbial contaminants. Photocatalysis involves speeding up of a photoreaction by the presence of a reducing agent or catalyst (TiO_2, WO_3, ZnS, etc.). The efficiency of photocatalytic activity depends on the ability of the catalyst to create electron-hole pairs, which generate free short-lived radicals and secondary reactions mediated by these radicals, which can inactivate microorganisms. Both fluorescent or UV light could be used for photocatalytic oxidation (PCO). Application of PCO is emerging fast in the HVAC industry, especially in removal of airborne bacteria, by utilizing TiO_2 catalysts and short-wave UV. The potential problem with PCO is generated short-lived radicals could react to form secondary chemical species (e.g., aldehydes, ketones), which may have adverse health effects for the occupants of indoor environments. Nonthermal or cold plasma-based methods were recently investigated for inactivation of airborne microorganisms (Gallagher et al. 2007; Vaze et al. 2010; Park et al. 2011). Plasmacluster ion (PCI) technology is a new technology for reducing the risk of infections in indoor environments. An alternating plasma discharge in an ion generator splits the airborne molecules of water into ions, and the interactions of hydrogen with oxygen ions creates groups of highly reactive OH radicals, which inactivate microorganisms

(Electronics SHARP 2005). Ion emission alone (Seo et al. 2001) and in combination with PCO (Grinshpun et al. 2007) were also used for inactivation of airborne microorganisms. A variety of newer technologies have emerged in recent years, including microwave irradiation, short-term thermal treatment, applications of nanoparticles, nanotubes, and the use of natural products (e.g., essential oils) in filtration media. However, all these methods have advantages and limitations, which are discussed later in this chapter.

3.3 Applications of Microbial Aerosol Disinfection

The disinfection of microbial aerosols has application in a wide variety of situations such as those involving food (processing, distribution, sales, preparation, and consumption), healthcare (surgical wards, dental treatment areas, laboratories, laundries, supply and preparation areas, waiting rooms), waste processing industries (wastewater collection and treatment, composting, and other sludge treatment and distribution), agricultural facilities (poultry and swine confinements), and veterinary facilities.

3.4 Ultraviolet Inactivation

Ultraviolet radiation is defined as the portion of the electromagnetic spectrum with wavelengths between 100 and 400 nm. The use of UV radiation to inactivate microorganisms is currently known as ultraviolet lamps for germicidal irradiation or UVGI. The use of ultraviolet light to inactivate microorganisms has been a topic of research for many years (Wells and Wells 1936; Sharp 1940; Wells 1942, 1955; Riley and O'Grady 1961; Riley et al. 1962, 1976; Stead et al. 1996; Xu et al. 2002; Ko et al. 2002). Many of these studies documented the effectiveness of UVGI for the control of tuberculosis in controlled settings. Several factors are thought responsible for the lack of widespread use of UVGI including the knowledge that subsequent experiments in schools did not duplicate the earlier findings, the use of other treatments showed promise for the control of tuberculosis, and the fact that concerns were raised over the safety of UV radiation (Macher 1993). There has been a renewed interest in the use of UVGI, because of the increased concern for the threat of tuberculosis, especially in compromised populations, and the identification of multidrug-resistant microbial strains.

The United States Centers for Disease Control (CDC) recommends UVGI as a supplement to other tuberculosis control measures in situations where there is the need for the killing or inactivation of tubercle bacteria. Much of the material in this section has been taken from the CDC recommendations (Centers for Disease Control 1994).

3.4.1 Techniques Used

There are three types of applications used for UVGI: duct irradiation, self-contained air-cleaning units, and upper-room air irradiation. In ductwork applications, the UV lamps are placed within ductwork systems. Air from rooms are removed through the ducts and treated with UV radiation before returning the air to the rooms. Incorporating the UV lamps within the ductwork, if properly installed and maintained, can restrict human exposure to only those situations where maintenance operations occur. A number of commercially available units now incorporate UV, such that lamps in self-contained portable systems can be placed in areas where UVGI is needed (These units may also contain HEPA filtration.) In upper-air radiation, shielded lamps are placed in upper areas of rooms, and regular room circulation, such as by convection, is used to ensure that room air circulates and comes into contact with the radiation. The lamps may be either suspended from the ceilings or attached to the walls.

3.4.2 Laboratory Studies

Using an experimental air-conditioning system incorporating both HEPA and UVGI, Nakamura (1987) found that the germicidal efficiency of spiral UV lamps was above 99.99% for *Bacillus subtilis* spores with an exposure time of 0.5 s.

Scarpino et al. (1994) reported an inactivation of more than 99% of *Escherichia coli*, *Pseudomonas fluorescens*, *Serratia marcescens*, and *Micrococcus luteus*, using a lamp-powered UVGI unit containing four lamps.

Salie et al. (1995) evaluated a prototype ceiling fan equipped with an enclosed UV lamp, using three test organisms in a laboratory setting: *E. coli*, *Sarcina lutea*, and *B. subtilis*. The residence time in the UV-equipped fan assembly was 26 ms, and the power intensity of the lamp was 64×10^6 μW/cm^2. For geometric mean upstream air microbial concentrations of 35.1, 1.76, and 2.46×10^6 CFU/m^3, reductions of 73.8, 3.8, and 8.6%, respectively, were observed. Based on a room 6.1 m (20 ft) × 6.1 m (20 ft) × 3.05 m (10 ft) in size, and assuming perfect mixing, and a fan equipped with two such lamps, 30 min would be required for all of the air in the room to pass through the blade assembly a single time.

3.4.3 Field Studies

The effectiveness of UVGI is often measured in terms of equivalent number of room air changes per hour (ACH) required to achieve the same reductions. Riley et al. (1976) found that UVGI was the equivalent of 10 ACH in reducing BCG (*Mycobacterium bovis* strain bacille Calmette-Guérin) aerosolized in a room.

Kethley and Branch (1972) reported UVGI to be equivalent to 39 ACH in reducing *Serratia marcescens* aerosolized in another room. Macher et al. (1992) determined the effectiveness of 254-nm ultraviolet radiation from four 15-W wall-mounted germicidal lamps in inactivating airborne microorganisms in an outpatient waiting room. A reduction of an estimated 14–19% in culturable airborne bacteria was observed.

3.4.4 Effects of Relative Humidity on UVGI

Relative humidity (RH) is an important factor for influencing the UVGI efficiency, although the mechanisms of microbial inactivation by UVGI at different RH levels are largely unknown. Recent studies (Peccia et al. 2001; Xu et al. 2005) showed that with the increase in RH in the environment, the airborne pathogens are more likely to survive the germicidal effects of a UVGI system. Significant decrease in airborne bacterial inactivation rates induced by UV irradiation at RH levels in excess of 50% was reported by Peccia et al. (2001). Xu et al. (2005) evaluated the effects of room ventilation rates, UV effluence rates and distribution, airflow patterns, relative humidity, and photoreactivation on the UVGI efficacy. Their observations indicated that in order to obtain maximum microbial inactivation from a ceiling-/wall-mounted UVGI system, an adequate level of UV radiation of at least 6 W of UV-C per m^3 in the upper zone of the room should be delivered. Furthermore, a good room air mixing and even distributions of UV radiation are required. As stated above, having the room RH around 50% or lower is preferable for successful microbial inactivation by UVGI. RH levels above 75% could significantly (up to 40%) reduce UVGI performance. The influences of RH on UVGI effectiveness were partially implicated to photoreactivation (Peccia and Hernandez 2001), a process whereby DNA damaged by UV light subsequently is repaired by a photolyase enzyme and that repair process requires visible light preferentially from the blue to violet end of the electromagnetic spectrum. The amount of photoreactivation could be increased at a high level of RH providing better protections to airborne pathogens against UVGI.

3.4.5 Adverse Effects

Exposure to ultraviolet radiation can have undesirable effects on humans, and recommendations for worker exposure limits have been developed (National Institute for Occupational Safety and Health 1972). Short-term overexposure to UV radiation can cause erythema (reddening of the skin), photokeratitis (inflammation of the cornea), and conjunctivitis (inflammation of the conjunctiva) (National Institute for Occupational Safety and Health 1972). Broad-spectrum UV radiation has been associated with squamous and basal cell carcinomas of the skin and has led

Table 3.1 Maximum recommended exposure limits for selected values of effective irradiation

Recommended exposure time per day	Effective irradiation level[a] (μW/cm^2)
8 h	0.1
4 h	0.2
2 h	0.4
1 h	0.8
30 min	1.7
15 min	3.3
10 min	5.0
5 min	10.0
1 min	50.0
30 s	100.0

Source: National Institute for Occupational Safety and Health
[a]Relative to the effectiveness of 270 nm, the wavelength of maximum ocular sensitivity

to UV-C (100–290 nm) being classified as "probably carcinogenic to humans" by the International Agency for Research on Cancer (1992). The National Institute for Occupational Safety and Health (NIOSH) recommended exposure limit (REL) (NIOSH 1972) is intended to protect workers from the acute effects of UV exposure. However, those photosensitive may not be protected. The NIOSH REL is wavelength dependent, because different wavelengths have different effects. The recommended exposure times for selected values of effective irradiation is shown in Table 3.1. Properly trained individuals should be engaged to ensure that appropriate safety precautions are in place to protect workers and others from overexposure.

3.5 HEPA Filtration

HEPA filters are air-cleaning devices that have been documented to have a minimum removal efficiency of 99.97% of particles >0.3 μm in diameter. Many pathogenic bioaerosols are of a size range that should make them candidates for removal by HEPA filters. *Aspergillus* spores (1.5–6 μm) have been demonstrated to be removed by HEPA filtration (Opel et al. 1986; Sherentz et al. 1987). Since *Mycobacterium tuberculosis* droplet nuclei are probably 1–5 μm in diameter (Centers for Disease Control 1994), they are likely candidates for removal by HEPA filtration, but this has not yet been demonstrated.

3.5.1 Types of Installation

There are a number of effective ways that HEPA filters can be placed in ventilation systems. They can be used either to cleanse air that is to be recirculated into rooms (Woods 1989), or they can be used before air is exhausted to the outside of the contaminated area where the microbes were generated, or as a combination of both. The HEPA filters can be installed within ductwork or located elsewhere in a room, e.g., mounted on the wall or freestanding. Modular HEPA filtration units are available for each of these types of installation. Many modular units are self-contained with mechanisms to pull air through the unit and recirculate it back into the room. These units have the advantage of being portable and being able to be used to supplement existing systems when needed to control microbial air quality. Disadvantages include the fact that their effective operation may be adversely affected either by the positioning and activities of room occupants or by inappropriate location of the filtration units.

3.5.2 Use of HEPA Filtration for Exhaust Air

Air exhausted from healthcare facilities, laboratories, etc. may need to be disinfected or "cleaned" to prevent pathogen exposure to persons who come into contact with this air either in exterior areas near the exhaust, in other interior areas where the exhausted air may be used, or by direct contact with the contaminated filter. In some instances the filtered air may be exhausted from one area for subsequent use in other presumably uncontaminated indoor areas of a facility. Sometimes the exhausted air is passed to a heat recovery device such as a rotary heat exchanger, often termed a heat wheel, before recirculation. In these cases the HEPA filtration device should be installed upstream of the heat recovery device in order to reduce the potential for contamination.

3.5.3 Use of HEPA Filtration for Recirculation Within a Room

The Centers for Disease Control (1994) suggests two scenarios for use of HEPA filtration within a room. One involves exhausting the air into a duct and passing that air through HEPA filters before returning it to the room. The other involves a ceiling-mounted HEPA unit. The appropriate positioning of these HEPA units with respect to a patient's bed will result in contaminated air being pulled away from the patient and cleaned air moving in the direction of the patient's breathing zone. Portable HEPA filtration units may be useful in some situations, but their uses are subject to several limitations. The effectiveness of the operation of the portable unit may be influenced by configurations of the room, placement of the units, and the fact that the actions of room occupants may compromise the efficiency of the units. They have not been evaluated for their effectiveness in tuberculosis infection control programs (Centers for Disease Control 1994).

A ventilation system capable of removing 95% of all air particles larger than 0.3 μm has been found to be effective in reducing airborne bacteria and dust particles in a calf nursery (Hillman et al. 1992).

3.5.4 Maintenance of HEPA Filters

Careful installation and appropriate routine maintenance are essential if the benefits of HEPA filtration are to be realized (Woods and Rask 1988). Timely replacement of the HEPA filters is necessary, and they should be handled with care during their disposal, since the filters contain the pathogenic microorganisms that were contained in the aerosols removed. The manufacturers' guidelines should be carefully followed. When using HEPA filtration units, placing prefilters upstream of the HEPA filters will greatly extend their HEPA filters' useful life.

3.6 Photocatalytic Oxidation

Recent studies on inactivation of aerosolized bacteria (Grinshpun et al. 2007) by photocatalytic oxidation (PCO; acceleration of a photoreaction by the presence of a catalyst) demonstrate lots of potential of this method for indoor air disinfection. As described above, PCO is a process where a chemical compound is oxidized to simpler radicals using a strong reduction agent (TiO_2, WO_3, ZnS, etc.) in the presence of a light source, which could be either fluorescent or UV light. This method is promising due to low power consumption, long service life, low maintenance requirement, and compatibility with HVAC system. Many investigations were conducted in recent years for scaling up and commercial utilization of PCO. As an example, PCO was previously utilized by Goswami et al. (1997) in a recirculating duct system integrating Degussa P25 TiO_2 and UV-A at intensity of 10 mW/cm^2 (at 350 nm wavelength and 50% RH) to inactivate a Gram-negative bacterium, *Serratia marcescens*. Photocatalysis of 8 h at air velocity of 0.376 m/s in the duct caused 82% inactivation of the bacteria. Another study by Keller et al. (2005) reported a 99.1–99.8% removal of airborne nonpathogenic *E. coli* using the combination of UV-A (380 nm) and TiO_2. Studies of Pal et al. (2005) and Pham and Lee (2014) also demonstrated the feasibility of using TiO_2 photocatalysis for continuous inactivation of airborne microorganisms. Vohra et al. (2006) utilized an advanced silver ion-doped TiO_2 catalyst along with UV-A (10 mW/cm^2) and found complete inactivation of various microbes such as *Bacillus cereus, Staphylococcus aureus, E. coli, Aspergillus niger*, and MS2 bacteriophage in air. Pham and

Lee (2014) also followed a similar approach for inactivating *E. coli* bioaerosols. Cram et al. (2004) examined the efficiency of a PCO unit in a surgical operating theater and other clinical settings for a 24-h period and found a 300% reduction of airborne microorganisms. Despite these promising findings, there are some disadvantages of this method. Under real environmental conditions, only a small portion of the pathogens will be absorbed on the catalyst of a PCO unit and chemically attacked from a single pass system. Furthermore, inactivated pathogens and dust will accumulate on catalytic contact surface, which will reduce the effectiveness of the method as the UV light will not desirably activate the catalyst layer. In addition, short-lived radicals from a PCO unit can react further to form secondary chemical species, such as aldehydes and ketones, which may deteriorate the quality of indoor air.

3.7 Nonthermal Plasma

Plasma is the fourth state of matter, which is an assembly of free charged particles moving at random direction. Plasma consists of photons, electrons, positive and negative ions, and free radicals, which can contribute in microbial inactivation. Nonthermal or cold plasma was recently explored by several researchers for inactivation of airborne microorganisms (Gallagher et al. 2007; Vaze et al. 2010; Park et al. 2011; Sharma et al. 2005). Gallagher et al. found 1.5 and 5.5 log reductions of the airborne *E. coli* cells, respectively, upon exposure to plasma for 10 s and 2 min. Park et al. achieved 89% bioaerosol removal efficiency using cold plasma with an energy dose of 38 J/L. Exposure time is an important issue for microbial inactivation by plasma. For example, Sharma et al. (2005) found one log reduction of *Bacillus atrophaeus (B. subtilis)* residing on agar surface after a plasma exposure period of 1 s, but the inactivation rate increased to a 3-log reduction after 10 min exposure. Possible mechanisms of microbial inactivation by plasma include both cell wall rupture and DNA damages (Sharma et al. 2005). Deng et al. (2006) found both the leakage of the cytoplasm contents and a complete burst of the membrane when *Bacillus subtilis* spores collected on filter surfaces were exposed to plasma treatment for 5 min. A recent study by Liang et al. (2012) demonstrated that nonthermal plasma generated by a dielectric barrier discharge (DBD) system for 0.06–0.12 s inactivated aerosolized *B. subtilis* and *Pseudomonas fluorescens* vegetative cells as well as some common indoor and outdoor bioaerosols. All these promising findings indicate that cold plasma can be successfully utilized for inactivation of airborne pathogens; however, release of ozone can be a concern during the application of this technology in indoor environments.

3.8 Plasmacluster Ion Technology

Plasmacluster technology is a new technology where ion generation has been combined with plasma discharge. This method has been commercially adopted in some air-cleaning devices (Sharp Corporation 2005). The ion generator in this technology uses an alternating plasma discharge between anode and cathode, which split airborne molecules of water into positively charged hydrogen (H^+) and negatively charged oxygen (O_2^-) ions. Highly reactive OH radicals form from the interactions between H^+ and O_2^-, which cause damage to proteins and polysaccharides either in the cell wall or surface structure of microorganisms and inactivate them. Ozone generated from the chemical reactions can be a significant factor for microbial inactivation. However, as stated above for the cold plasma, ozone generated by this technology can be harmful for the occupants of indoor environments, and further investigations are required on potential hazardous effects of this technology. Furthermore, application of this technology can change air distribution patterns in rooms and affect thermal comfort for occupants.

3.9 Microwave Irradiation

Unlike ionizing radiations (e.g., gamma and X-rays) which operate in the frequencies of 50,000 THz, non-ionizing microwave irradiation operates in the frequency range between 300 and 300 GHz. In recent years different doses of microwave irradiations were (Wu and Yao 2010; Zhang et al. 2010) applied for inactivating aerosolized microorganisms. Wu and Yao found that the exposure of *B. subtilis* var. *niger* vegetative cells to microwave irradiation at 2450 MHz for about 1.5 min can cause an inactivation of ~70%. Scanning and transmission electron micrographs showed visible damages in the inactivated bacterial cells. Microwave irradiation coupled with nanofibrous filters was also investigated in inactivating bioaerosols (Zhang et al. 2010). All of these studies indicated that power level and exposure time of microwave irradiation play an important role in the microbial inactivation. There are controversies on the mechanisms of microbial inactivation by microwave irradiation. Previous studies emphasized solely the thermal effects from microwave irradiation on microbial inactivation. However, a recent study by Park et al. (2006) compared the effects of microwave irradiation and external heating and demonstrated that bacterial cells treated by two methods had different DNA quantity and optical density. Although microwave inactivation of airborne biological agents offers a simple, cost-effective, and noninvasive decontamination method, this has not been adequately field tested, and more research is necessary before implementing microwave irradiations for large-scale decontamination in clinical settings.

3.10 Ion Emission

Unipolar and bipolar ion emitters, which meet health standards by not generating ozone above the established safety thresholds, have been incorporated into commercial air purification devices that utilize either bipolar or unipolar ion emission. The bactericidal effect of air ionization on viable (as determined by culturability) microorganisms has also been assessed (Marin et al. 1989; Shargawi et al. 1999; Lee 2001; Seo et al. 2001). In addition to causing charge-related deposition of airborne microorganisms, emitted ions may also exert some biocidal effects. A recent study of Fletcher et al. (2007) demonstrated that exposure to air ions for 15 min inactivated more than 80% of microorganisms on agar plates. Ion emission in combination with PCO has been utilized recently by Grinshpun et al. (2007) for inactivation of airborne bacteria. An inactivation of about 75% of *B. subtilis* spores was obtained by combining unipolar ion emission and PCO. However, the mechanisms of microbial inactivation by ions remain poorly understood. Fletcher et al. (2007) and Kim et al. (2011) suggested that air ions demonstrate bactericidal effects through electroporation of bacterial cell membranes. In addition to the possibility of generating excess ozone, other disadvantages of unipolar air ion emitters include charge accumulation on insulating surfaces, which may cause occasional static problems and accumulation of partially inactivated microorganisms on ceilings and walls.

3.11 Short-Term Thermal Treatment

Exposure of airborne microorganisms to very high temperature can cause denaturation of proteins by breaking the structures of polypeptides in microbial cell surfaces (Madigan and Martinko 2006). Thermal treatment of bioaerosols with electric heating coils is feasible by installing these coils in the existing systems of buildings. Recent studies demonstrated that bioaerosols can be inactivated by dry heat during subsecond exposure time (Lee and Lee 2006; Jung et al. 2009; Grinshpun et al. 2010a). Short-term (~1 s) high-temperature treatment of fungal bioaerosols (Jung et al. 2009) demonstrated damage on spore walls. High-temperature exposure at 400 °C was found to inactivate more than 99.99% of *B. subtilis* spore bioaerosols during short-term exposures (Grinshpun et al. 2010a, b). Despite these promising findings, thermal treatment has a major limitation for inactivating airborne microorganisms because this method is not cost-effective.

3.12 Application of Nanotechnology

Nanoparticles and nanotubes are currently being explored for inactivation of airborne microorganisms. Silver nanoparticles atomized in the air were utilized to control the viability of pathogens aerosolized in a small glass chamber (Lee et al. 2006). More than 99% of the exposed bacteria lost culturability; however, this methodology still needs further investigations because this approach was never investigated in field conditions. Furthermore, aerosolizing silver nanoparticles in air may have some adverse health effects. Loading of nanotubes on filter materials can be useful for inactivating captured microorganisms on filter surfaces. In a recent study, carbon nanotube loading of 1.6 $\mu g/cm^2$ demonstrated 95% physical removal of *B. subtilis* var. *niger* cells on filter surfaces (Guan and Yao 2010). The possible inactivation mechanism is the strong toxicity of carbon nanotubes to bacterial cells. Previous studies demonstrated similar toxicity of carbon nanotubes on *E. coli* (Kang et al. 2009; Brady-Estevez et al. 2010).

Yang et al. (2011) found that a corona discharge system using carbon nanotube electrodes had higher bioaerosol inactivation efficiency than did a corona discharge system using stainless steel electrodes.

3.13 Application of Natural Products

Essential oils used in pharmaceuticals and cosmetics and by the food and beverage industries have a strong biocidal effect, which can be applied in ventilation and air purifier industries for inactivating microorganisms deposited in filter (e.g., HVAC) surfaces. Furthermore, several studies have indicated that antimicrobial effects of essential oils are more pronounced in air compared to liquids (Hammer et al. 1999; Pibiri et al. 2003; Inouye et al. 2003). Recent studies of Pyankov et al. (2008) and Huang et al. (2010) demonstrated that coating of fibrous filters by biologically active tea tree oil (TTO) disinfected bacteria and fungal spores, respectively. Interestingly, the TTO was used for filter efficiency enhancing media as well as a disinfectant for bacterial and fungal aerosols collected on the filter surface. Pyankov et al. (2012) used TTO and eucalyptus oil against the influenza virus captured on filter surfaces and found that both tested oils possessed strong antiviral properties when used as fiber coating materials, capable of inactivating captured microorganisms within 5–10 min of contact on the fiber surface. Applications of these natural products, however, are still going through investigations. Major disadvantage of the use of essential oils is their potential role in hypersensitivity reactions among some occupants (e.g., mint, thyme, oregano, etc.), and some of those oils may exhibit cytotoxic activity among exposed occupants (Inouye et al. 2003).

3.14 Conclusions

As we discussed above, there are many promising technologies for inactivating airborne microorganisms—some are commercially available and some are still being tested in lab investigations; however, all of them have some major to minor limitations. Furthermore, most of these methods were never tested for real pathogens and emerging drug-resistant pathogens. Therefore, a perfect solution to inactivate all airborne microorganisms does not exist yet. We can combine different microbial inactivation methods to achieve a more effective disinfection approach depending on the types of suspected microorganisms and indoor environmental conditions. These new hybrid disinfection approaches should be tested in field conditions, particularly for disinfecting airborne pathogens in large clinical settings.

Compliance with Ethical Standards
Conflict of Interest: Atin Adhikari declares that he has no conflict of interest. Scott Clark declares that he has no conflict of interest.

Ethical approval: This chapter does not contain any studies with human participants or animals performed by any of the authors.

References

Brady-Estévez AS, Schnoor MH, Kang S et al (2010) SWNT-MWNT hybrid filter attains high viral removal and bacterial inactivation. Langmuir 26:19153–19158

Centers for Disease Control (1994) Guidelines for preventing the transmission of *Mycobacterium tuberculosis* in health care facilities. MMWR Recomm Rep 43:1–132

Cram N, Shipman N, Quarles JM (2004) Reducing airborne microbes in the surgical operating theater and other clinical settings: a study utilizing a unique photocatalytic reactor biocide unit. J Clin Eng April–June:79–88

Deng XT, Shi JJ, Kong MG (2006) Physical mechanisms of inactivation of *Bacillus subtilis* spores using cold atmospheric plasmas. IEEE Trans Plasma Sci 34:1310–1316

Douwes J, Thorne P, Pearce N et al (2003) Bioaerosol health effects and exposure assessment: progress and prospects. Ann Occup Hyg 47:187–200

Edwards MR, Bartlett NW, Hussell T et al (2012) The microbiology of asthma. Nat Rev Immunol 10:459–471

Electronics SHARP (2005) Sharp's plasmacluster ions effectively deactivate H5N1 avian influenza virus. Asia Pacific Biotech News 6:469

Fletcher LA, Gaunt LF, Beggs CB et al (2007) Bactericidal action of positive and negative ions in air. BMC Microbiol 7:32

Fraser C, Donnelly CA, Cauchemez S et al (2009) Pandemic potential of a strain of influenza A (H1N1): early findings. Science 324:1557–1561

Gallagher MJ, Gallagher J, Vaze N et al (2007) Rapid inactivation of airborne bacteria using atmospheric pressure dielectric barrier grating discharge. IEEE Trans Plasma Sci 35: 1501–1510

Gosden PE, MacGowan AP, Bannister GC (1998) Importance of air quality and related factors in the prevention of infection in orthopedic implant surgery. J Hosp Infect 39:173–180

Goswami DY, Trivedi DM, Block SS (1997) Photocatalytic disinfection of indoor air. J Sol Energy Eng 119:92–96

Grinshpun SA, Adhikari A, Honda T et al (2007) Control of aerosol contaminants in indoor air: combining the particle concentration reduction with microbial inactivation. Environ Sci Technol 41:606–612
Grinshpun SA, Adhikari A, Li C et al (2010a) Thermal inactivation of airborne viable Bacillus subtilis spores by short-term exposure in axially heated air flow. J Aerosol Sci 41:352–363
Grinshpun SA, Li C, Adhikari A et al (2010b) Method for studying survival of airborne viable microorganisms in combustion environments: development and evaluation. Aerosol Air Qual Res 10:414–424
Guan T, Yao M (2010) Use of carbon nanotube filter in removing bioaerosols. J Aerosol Sci 6:611–620
Hammer KA, Carson CF, Riley TV (1999) Antimicrobial activity of essential oils and other plant extracts. J Appl Microbiol 86:985–990
Henderson D (2004) The threat of aerosolized biological weapons. ASHRAE J 46:50–53
Hillman P, Gebremedhin K, Warner R (1992) Ventilation system to minimize airborne bacteria, dust, humidity, and ammonia in calf nurseries. J Dairy Sci 75:1305–1313
Huang R, Pyankov OV, Yu B et al (2010) Inactivation of fungal spores collected on fibrous filters by *Melaleuca alternifolia* (Tea Tree Oil). Aerosol Sci Technol 44:262–268
Inouye S, Abe S, Yamaguchi H et al (2003) Comparative study of antimicrobial and cytotoxic effects of selected essential oils by gaseous and solution contacts. Int J Aromather 13:33–41
International Agency for Research on Cancer (1992) Monographs on the evaluation of carcinogenic risks to humans: solar and ultraviolet radiation, vol 55. WHO International Agency for Research on Cancer, Lyon
Jung JH, Lee JE, Lee CH et al (2009) Treatment of fungal bioaerosols by a high temperature, short-time process in a continuous flow system. Appl Environ Microbiol 75:2742–2749
Kang S, Mauter MS, Elimelech M (2009) Microbial cytotoxicity of carbon-based nanomaterials: implications for river water and wastewater effluent. Environ Sci Technol 43:2648–2653
Keller V, Keller N, Ledoux MJ et al (2005) Biological agent inactivation in a flowing air stream by photocatalysis. Chem Commun (Camb) June 21(23):2918–2920
Kethley TW, Branch K (1972) Ultraviolet lamps for room air disinfection' effect of sampling location and particle size of bacterial aerosol. Arch Environ Health 25:205–214
Kim YS, Yoon KY, Park JH et al (2011) Application of air ions for bacterial decolonization in air filters contaminated by aerosolized bacteria. Sci Total Environ 409:748–755
Ko G, First MW, Burge HA (2002) The characterization of upper-room ultraviolet germicidal irradiation in inactivating airborne microorganisms. Environ Health Perspect 110:95–101
Kortepeter M, Parker G (1999) Potential biological weapons threats. Emerg Infect Dis 5:523–527
Lee H-K (2001) Electrical sterilization of *Escherichia coli* by electrostatic atomization. J Electrostatics 51–52:71–75
Lee YH, Lee BU (2006) Inactivation of airborne *E. coli* and *B. subtilis* bioaerosols utilizing thermal energy. J Microbiol Biotechnol 16:1684–1689
Lee BU, Yoon K-Y, Bae G-N et al (2006) Airborne silver nanoparticles from an atomizer as an antimicrobial agent against *E. coli* bioaerosols. Proc Healthy Build 2006:345–348
Liang Y, Wu Y, Sun K et al (2012) Rapid inactivation of biological species in the air using atmospheric pressure nonthermal plasma. Environ Sci Technol 46:3360–3368
Macher JM (1993) The use of germicidal lamps to control tuberculosis in healthcare facilities. Infect Control Hosp Epidemiol 14:723–729
Macher JM, Aleantils LE, Chang Y-L et al (1992) Effect of ultra-violet germicidal lamps on airborne microorganisms in an outpatient waiting room. Appl Occup Environ Hyg 7:505–513
Madigan MT, Martinko JM (2006) Brock Biology of microorganisms, 11th edn. Pearson Prentice Hall, Upper Saddle River, NJ, p 52, 673
Marin V, Moretti G, Rassu M (1989) Effects of ionization of the air on some bacterial strains. Annali di Igiene 1:1491–1500
Nakamura H (1987) Sterilization efficiency of ultraviolet irradiation on microbial aerosols under dynamic airflow by experimental air conditioning systems. Bull Tokyo Med Dentistry Univ 34: 25–40

NIOSH (National Institute for Occupational Safety and Health) (1972) Criteria for a recommended standard. Occupational exposure to ultraviolet radiation. Publication No. HSM 73-11009. Department of Health and Human Services, Cincinnati, OH

Opel SM, Asp AA, Cannady PB Jr et al (1986) Efficacy of infection control measures during a nosocomial outbreak of disseminated aspergillosis associated with hospital construction. J Infect Dis 153:634–637

Pal A, Min X, LE Y et al (2005) Photocatalytic inactivation of bioaerosols by TiO_2 coated membrane. Int J Chem Reactor Eng 3:1542–6580

Park DK, Bitton G, Melker R (2006) Microbial inactivation by microwave radiation in the home environment. J Environ Health 69:17–24

Park CW, Byeon JH, Yoon KY et al (2011) Simultaneous removal of odors, airborne particles, and bioaerosols in a municipal composting facility by dielectric barrier discharge. Sep Purif Technol 77:87–93

Peccia J, Hernandez M (2001) Photoreactivation in airborne *Mycobacterium parafortuitum*. Appl Environ Microbiol 67:4225–4232

Peccia J, Werth HM, Miller S et al (2001) Effects of relative humidity on the ultraviolet induced inactivation of airborne bacteria. Aerosol Sci Technol 35:728–740

Pham TD, Lee BK (2014) Effects of Ag doping on the photocatalytic disinfection of *E. coli* in bioaerosol by Ag-TiO_2/GF under visible light. J Colloid Interface Sci 428:24–31

Pibiri MC, Seignez C, Roulet CA (2003) Methods to study the effect of essential oils on microbes present in ventilation systems. CISBAT, p 185–190

Pyankov OV, Agranovski IE, Huang R et al (2008) Removal of biological aerosols by oil coated filters. Clean Soil Air Water 36:609–614

Pyankov OV, Usachev EV, Pyankova O et al (2012) Inactivation of airborne influenza virus by tea tree and eucalyptus oils. Aerosol Sci Technol 46:1295–1302

Riley RL, O'Grady F (1961) Airborne infection. MacMillan, New York

Riley RL, Mills CC, O'Grady F et al (1962) Infectiousness of air from a tuberculosis ward. Am Rev Resp Dis 85:511–525

Riley RL, Knight M, Middlebrook G (1976) Ultraviolet susceptibility of BCG and virulent tubercle bacilli. Am Rev Resp Dis 113:413–418

Salie F, Scarpino P, Clark S et al (1995) Laboratory evaluation of airborne microbial reduction by an ultraviolet light positioned in a modified hollow ceiling fan blade. Am Ind Hyg Assoc J 56: 987–992

Sanzén L, Carlsson AS, Walder M (1990) Air contamination during total hip arthroplasty in an ultraclean air enclosure using different types of staff clothing. J Arthroplasty 6:127–130

Scarpino PV, Stoeckel DM, Jensen PA (1994) Ability of a fan-powered UVGI unit to inactivate selected airborne bacteria (abstract). Conference of the Society of Occupational and Environmental Health, Preventing TB in the Workplace: Principles and Practices for Controlling Transmission, Rockville, Maryland, December 1994

Seo KH, Mitchell BW, Holt PS et al (2001) Bactericidal effects of negative ions on airborne and surface *Salmonella enteriditis* from an artificially generated aerosol. J Food Protec 64:113–116

Shargavi JM, Theaker ED, Drucker DB et al (1999) Sensitivity of *Candida albicans* to negative air ion streams. J Appl Microbiol 87:889–897

Sharma A, Pruden A, Yu Z et al (2005) Bacterial inactivation in open air by the afterglow plume emitted from a grounded hollow slot electrode. Environ Sci Technol 39:339–344

Sharp DG (1940) The effects of ultraviolet light on bacteria suspended in air. J Bact 39:535–547

Sherertz RJ, Belani A, Kramer BS et al (1987) Impact of air filtration on nosocomial *Aspergillus* infections. Am J Med 83:709–718

Stead WW, Yeung C, Hartnett C (1996) Probable role of ultraviolet irradiation in preventing transmission of tuberculosis: a case study. Infect Control Hosp Epidemiol 17:11–13

Sterenborg HJ, van der Putte SC, van der Leun JC (1988) The dose response relationship of tumorigenesis by ultraviolet radiation of 254 nm. Photochem Photobiol 47:245–253

Vaze ND, Gallagher MJ, Park S et al (2010) Inactivation of bacteria in flight by direct exposure to non-thermal plasma. IEEE Trans Plasma Sci 38:3234–3240

Vohra A, Goswami DY, Deshpande DA et al (2006) Enhanced photocatalytic disinfection of indoor air. Appl Catal B: Environ 64:57–65
Wells WF (1942) Radiation disinfection of air. Arch Physic Therapy 23:143–148
Wells WF (1955) Airborne contagion and air hygiene. Harvard University Press, Cambridge, MA
Wells WF, Wells MW (1936) Airborne infection. JAMA 107:1805–1809
Whyte W, Vesley D, Hodgson R (1976) Bacterial dispersion in relation to operating room clothing. J Hyg 76:367–378
Woods JE (1989) Cost avoidance and productivity in owning and operating buildings. J Occup Med 4:753–770
Woods JE, Rask CR (1988) Heating ventilation, air conditioning systems; the enlightening approach to methods of control. In: Kandsin RB (ed) Architectural design and indoor microbial pollution. Oxford University Press, New York, pp 123–163
Wu Y, Yao M (2010) Inactivation of bacteria and fungus aerosols using microwave irradiation. J Aerosol Sci 41:682–693
Xu P, Peccia J, Fabian P et al (2002) Efficacy of ultraviolet germicidal irradiation of upper-room air in inactivating airborne bacterial spore and Mycobacteria in full-scale studies. Atmos Environ 37: 405–419
Xu P, Kujundzic E, Peccia J et al (2005) Impact of environmental factors on efficacy of upper-room air ultraviolet germicidal irradiation for inactivating airborne mycobacteria. Environ Sci Technol 39:9656–9664
Yang S, Huang Y-C, Luo C-H et al (2011) Inactivation efficiency of bioaerosols using carbon nanotube plasma. Clean Soil Air Water 39:201–205
Zhang Q, Damit B, Welch J (2010) Microwave-assisted nanofibrous air filtration for disinfection of bioaerosols. J Aerosol Sci 41:880–888
Zmudzka BZ, Beer J (1990) Activation of human immunodeficiency virus by ultraviolet radiation. Photochem Photobiol 52:1153–1162

Part II
Understanding the Ecology of Infectious Disease

Chapter 4
The Role of the Indigenous Gut Microbiota in Human Health and Disease

Tyler Vunk and Kristin M. Burkholder

Abstract The indigenous intestinal microbiota is a complex community that is crucial for maintaining both gastrointestinal and systemic homeostasis of the host. Although gut microbes have long been recognized for their role in modulating colonization by intestinal pathogens, recent reports have demonstrated that the commensal microbes have additional, far-reaching effects on host physiology and well-being. The purpose of this review is to highlight recent research which demonstrates the role of the human gut microbiota and microbial dysbiosis in common human diseases, including gastrointestinal disorders such as inflammatory bowel syndrome (IBS) and inflammatory bowel disorder (IBD), colorectal cancer, obesity, allergic disease, and mental illnesses such as anxiety and depression. We also discuss recent advances in the field of microbial-derived therapeutics, with a focus on emerging bacterial-based therapies that target gastrointestinal infections, osteoporosis, and cancer.

4.1 Introduction: The Human Indigenous Gut Microbiota

The human intestinal tract houses the richest and most diverse microbial community in the body, consisting of bacteria, fungi, protozoa, and viruses. These intestinal microorganisms represent a very important aspect of how the human microbiome influences the balance between host health and disease, effects which likely occur in combination with those of microbes found elsewhere in the body including the skin and respiratory tract. Our understanding of the complexity and species composition of the gut microbiota has been greatly improved by advances in molecular technologies such as high throughput sequencing, as well as

T. Vunk
Department of Psychology, University of New England, Biddeford, ME, USA

Department of English, University of New England, Biddeford, ME, USA
e-mail: tvunk@une.edu

K.M. Burkholder (✉)
Department of Biology, University of New England, Biddeford, ME, USA
e-mail: kburkholder@une.edu

C.J. Hurst (ed.), *Modeling the Transmission and Prevention of Infectious Disease*, Advances in Environmental Microbiology 4, DOI 10.1007/978-3-319-60616-3_4

developments in nongenomic tools and animal models. It is now estimated that the healthy human gut harbors approximately 1000 different bacterial species alone and that microbial genes outnumber human genes by 100-fold (Qin et al. 2010). The microbiota coevolves with its host, as alterations in gut microbial community structure have significant consequences on human health and can contribute to the etiology of a variety of pathological conditions both within and outside of the gut. Here we discuss current reports of the impact of indigenous microbes on human disease states ranging from intestinal inflammatory disorders, obesity, and allergic inflammatory disease to mental illness, and we highlight the potential utility of microbes as therapeutic agents.

4.2 Association Between the Gut Microbiota and Chronic Gastrointestinal Disorders

4.2.1 Irritable Bowel Syndrome and Irritable Bowel Disorder

Irritable bowel syndrome (IBS) and irritable bowel disorder (IBD) are chronic intestinal inflammatory diseases that afflict a significant proportion of the population. In North America alone, an estimated 10–15% of the population exhibit IBS symptoms (Saito et al. 2002), while incidence of IBD is estimated at over 0.2% of the population (Kappelman et al. 2007). Host factors such as immune dysregulation and altered intestinal barrier function are potential causes of IBS (Xu et al. 2014), but recent reports highlight the additional role of the gut microbiota in IBS development. In particular, patients with IBS or IBD tend to exhibit altered intestinal microbial populations compared to their healthy counterparts, and some evidence suggests that IBS and IBD may be associated with increased colonization by gut pathogens.

IBS is a functional bowel disorder characterized solely by symptom-based diagnostic criteria, and symptoms include abdominal pain, altered bowel habits, and low-grade intestinal inflammation. An altered gut microbiota may contribute to IBS status. Studies using phylogenetic microarrays and qPCR analyses demonstrated marked differences in the gut microbiota between normal patients and those with IBS, in both adult and pediatric populations. Adults with IBS exhibited increased abundance of members of the phylum *Firmicutes* as well as species within the genera *Ruminococcus*, *Clostridium*, and *Dorea*, with concurrent reduction in *Bifidobacterium* and *Faecalibacterium* species (Rajilic-Stojanovic et al. 2011). Similarly, children with IBS had higher levels of the phyla *Firmicutes* and *Proteobacteria*, as well as *Dorea*, *Ruminococcus*, and *Haemophilus parainfluenzae* than their non-IBS counterparts, with decreased abundance within the genus *Bacteroides*. Acute infectious gastroenteritis is another commonly identified IBS risk factor, as a number of bacterial, parasitic, and viral pathogens have been associated with development of IBS and other functional gastrointestinal disorders.

For example, postinfectious IBS has been documented in humans following gastrointestinal infections caused by the bacterial pathogens *Campylobacter jejuni* (Dunlop et al. 2003), *Shigella sonnei* (Ji et al. 2005), *Salmonella enterica* (Mearin et al. 2005), and *Clostridium difficile* (Sethi et al. 2011), as well as the viral pathogen *Norovirus* (Zanini et al. 2012), and parasites *Giardia lamblia* (Dizdar et al. 2007) and *Trichinella britovi* (Soyturk et al. 2007). Although the means by which individual pathogens trigger IBS remain poorly understood and likely differ between pathogens, potential mechanisms include infection-induced cellular changes in the gut mucosa, alterations in intestinal permeability (Spiller et al. 2000), increased production of pro-inflammatory cytokines, and altered expression of pattern recognition receptors, such as toll-like receptor 9 (TLR9) (Villani et al. 2010).

Inflammatory bowel disease (IBD), which encompasses both ulcerative colitis (UC) and Crohn's disease (CD), is a chronic relapsing inflammatory disorder of the gastrointestinal tract. UC and CD are diseases with distinct symptoms; UC is associated with inflammation and ulceration of the colon lining, while CD is a chronic, sometimes patchy inflammation that can occur anywhere along the digestive tract. Although the precise microbial populations involved remain a matter of debate, research suggests that both diseases result from general dysbiosis and impaired diversity of the host gut microbiota (Lepage et al. 2011; Martinez-Medina et al. 2006). For example, loss in abundance and complexity of the phylum *Firmicutes* has been observed in patients with CD, with specific reductions in numbers of *Faecalibacterium prausnitzii*, a major member of the *Firmicutes* and a commensal with anti-inflammatory properties (Manichanh et al. 2006; Sokol et al. 2008). Another report which compared intestinal biopsies and stool samples from healthy individuals and IBD patients revealed altered abundance of the families *Enterobacteriaceae*, *Ruminococcaceae*, and *Leuconostocaceae*, while members of the genus *Clostridium* increased (Morgan et al. 2012). Such studies emphasize the role of the gut microbiota in the manifestation of chronic intestinal inflammation, and further research may aid in development of therapeutics that are either derived from microbes or which target specific gut microbial populations.

4.2.2 Colorectal Cancer

Numerous reports suggest a role for the gut microbiota in the pathogenesis of colorectal cancer (CRC), which is the third most common cancer worldwide. Incidence rates of CRC are greatest in developed countries with Western cultures (Haggar and Boushey 2009). Genetic mechanisms such as DNA damage and genetic instability are important etiologic agents for development of CRC, but recent research indicates that environmental factors such as the intestinal microbiome and its metabolites can also contribute to CRC onset. Although the mechanism by which the microbiota might influence CRC is poorly understood, as with other intestinal disorders, microbial dysbiosis is associated with the disease.

In 2011, three separate groups reported high-resolution maps of the human colonic microbiota from patients with late-stage CRC (Kostic et al. 2012; Marchesi et al. 2011;

Castellarin et al. 2012). In each study the microbiome was evaluated from tissues taken from tumors and surrounding non-tumor sites, and although microbiota composition was similar between tumor and non-tumor sites within individuals, tumor tissue samples did exhibit lower levels of microbial diversity (Chen et al. 2012). In general, mucosa-adherent *Bifidobacterium*, *Faecalibacterium*, and *Blautia* were reduced in CRC patients, whereas *Fusobacterium*, *Porphyromonas*, *Peptostreptococcus*, and *Mogibacterium* were present in higher numbers. In luminal samples, CRC patients were enriched in members of the families *Erysipelotrichaceae*, *Prevotellaceae*, and *Coriobacteriaceae*, which were previously related to host metabolic disorders related to energy production and adiposity (Chen et al. 2012). In addition, another study reported a reduction of butyrate-producing bacteria in feces of CRC patients (Wang et al. 2012), which is notable since butyrate has important anti-oncogenic effects on host tissues (Scharlau et al. 2009).

Microbiota-induced intestinal inflammation is a potential mechanism that links the gut microflora to development of CRC. Indeed, patients with chronic intestinal inflammatory diseases such as IBD have increased risk for development of CRC (Moossavi and Bishehsari 2012). The association between inflammation and cancer was initially noted by Rudolf Virchow who, in 1863, reported presence of leukocytes in neoplastic tissues (Virchow 1989) and by Robert Koch and Louis Pasteur who noted bacterial colonization of tumor sites (Compare and Nardone 2011). There is mounting evidence that microbiota-driven chronic inflammation impacts local immune responses, which can subsequently alter tissue homeostasis. For example, inflammatory release of reactive oxygen and nitrogen species promotes cellular DNA damage. Furthermore, inflammatory cytokines and chemokines, such as TNF-a, IL-6, IL-1, IL-8, nitric oxide, and prostaglandin-2 derivatives, may promote tumorigenesis by driving angiogenesis and repressing immune-mediated tumor elimination (Zhu et al. 2013).

It is likely that host innate immune pattern recognition receptors (PRRs), such as Toll-like receptors (TLRs) and Nod-like receptors (NLRs), are involved in the link between microbial-induced inflammation and cancer development. The TLRs and NLRs monitor extracellular and intracellular compartments for microbial molecules called pathogen-associated molecular patterns (PAMPs) and, once engaged, initiate signaling cascades that lead to inflammatory transcriptional responses. Signaling via TLR pathways is dampened through expression of inhibitory molecules such as signal immunoglobulin IL-1-related receptor (SIGIRR). Notably, *sigirr*-deficient mice are more susceptible to experimental colitis and to colitis-associated CRC induced with procarcinogenic compounds azoxymethane (AOM) and dextran sodium sulfate (DSS), while tissue-specific repletion of SIGIRR eliminated the animals' susceptibility to colitis and tumorigenesis (Xiao et al. 2007). In addition, TLR4 was reported to promote development of CRC in mice subjected to AOM and DSS (Fukata et al. 2007). Together, these findings suggest an important role for microbial-induced inflammation and PRR-mediated signaling in the etiology of CRC.

4.3 Influence of the Indigenous Gut Microbiota on Obesity

Incidence of obesity has risen dramatically in recent decades, particularly in Western cultures. The World Health Organization (WHO) estimated in 2014 that there were over 1.9 billion overweight adults in the world, of which 600 million are considered clinically obese (WHO 2014). Although lifestyle factors including diet and exercise contribute heavily to the current epidemic, an increasing number of reports indicate that the human gut microbiota plays an important role in development of obesity and, to a lesser extent, type II diabetes. An initial hypothesis suggested that increased energy harvest by specific microbial communities contributed to weight gain and obesity (Ley et al. 2005; Turnbaugh et al. 2006). However, more recent studies challenge that assumption and posit a link between obesity and the composition and functionality of gut microbes, suggesting that the microbiota may contribute to obesity through a variety of complex interactions with the host (Clarke et al. 2012).

Studies using obese animal models and DNA sequencing demonstrated that gut microbial composition differs between obese and nonobese individuals. For example, studies which used the *ob/ob* (leptin-deficient with excessive appetite and obese phenotype) mouse model to compare the cecal microbiota composition between *ob/ob*, *ob*/+, and +/+ (lean) mice revealed that the proportion of *Bacteroidetes* was reduced in obese mice compared to lean animals, while abundance of *Firmicutes* was increased (Ley et al. 2005; Turnbaugh et al. 2006). A separate study using the diet-induced obesity (DIO) model, which focuses on obesity arising from consumption of a high-fat (HF) Western diet rather than genetics, also demonstrated a lower proportion of *Bacteroidetes* and an increase in the *Mollicutes* subpopulation of the *Firmicutes* phylum in mice fed with a HF diet compared to those fed a low-fat (LF) diet (Turnbaugh et al. 2008). A more recent study compared the fecal microbiota of lean (+/+), *ob/ob*, as well as +/+ mice fed a HF diet at 7, 11, and 15 weeks of age. This report showed decreased *Bacteroidetes* and increased *Firmicutes* over time in obese animals and those fed with the HF diet compared to lean counterparts, but they also found high levels of *Actinobacteria* in all three groups of mice. The proportion of *Actinobacteria* progressively increased in *ob/ob* and HF-fed mice compared to +/+ animals. These authors also reported decreased levels of *Proteobacteria* in HF-fed mice, and decreased numbers of *Lactococcus* and *Deferribacteria* in ob/ob animals (Murphy et al. 2010), further confirming obesity- and diet-related dysbiosis in community structure.

While the above studies highlighted differences in microbial community composition between obese and lean mice and between mice consuming HF versus LF diets, they did not account for potential disparities between the murine and human gut microbiota. Although the distal intestine of mice and humans contains microbes from similar phylogenetic groups, such as the *Firmicutes*, *Bacteroidetes*, *Actinobacteria*, *Proteobacteria*, *Fusobacteria*, and others, many bacterial genera and species within those phyla are not found in both mice and humans (Ley et al. 2005). To improve upon mouse models used to study the impact of the microbiota on obesity, Turnbaugh et al. (2009b) created a humanized mouse model in which human fecal microbial communities were transplanted into germ-free mice. The use of this humanized mouse model demonstrated that the gut microbiota of individuals consuming a HF Western diet contained a higher proportion of the *Firmicutes* class *Erysipelotrichi* and a lower proportion of *Bacteroidetes* and *Bacillus* than did the microbiota of individuals consuming a LF diet. Together, murine and humanized mouse studies emphasize the differences in microbial communities between obese and lean animals and highlight the association of particular microbial groups with body composition.

Studies using human subjects have confirmed the differences in microbial community structure between obese and lean individuals and also indicate that the gut microbiota may transmit the obesity phenotype to nonobese individuals. In a 2009 study with 154 individuals, Turnbaugh et al. (2009a) examined the intestinal microbiota of monozygotic or dizygotic twin pairs concordant for obesity, and their mothers, and found that gut microbial composition was more similar between related than unrelated individuals. Similar to studies conducted in mouse models, this study also found that samples from obese individuals contained a lower proportion of *Bacteroidetes* and higher proportion of *Actinobacteria* than their lean counterparts. However, unlike the murine studies, there was no significant difference in *Firmicutes* populations between lean and obese humans. In a separate study using twin pairs discordant for obesity, fecal samples were collected from each twin and transplanted into germ-free mice fed a low-fat diet (Ridaura et al. 2013). 16S rRNA analysis of microbial communities from recipient mice revealed that taxonomic features of the donor microbiota were retained within the recipients. Intriguingly, the obese phenotype was transmissible through the gut microbiota; mice transplanted with feces from obese twins gained significantly greater adipose mass than did mice receiving the lean twin's gut community. Differences in adiposity were not associated with differences in feed consumption or with recipient mouse inflammatory responses, suggesting involvement of more complex mechanisms, such as the metabolic influence of the gut microbiota on the host. Indeed, mice harboring the obese donor microbiota exhibited greater expression of microbial metabolic genes, including those involved in amino acid metabolism and in the pentose phosphate pathway. Mass spectrometry-based analysis of amino acids in sera from recipient mice revealed significant increases in the amino acid valine, leucine/isoleucine, phenylalanine, tyrosine, and alanine in recipients of obese microbiota compared to the recipients of lean microbiota. Importantly, previous reports showed elevations in serum concentrations of these amino acids in obese and insulin-resistant humans compared to lean and insulin-sensitive humans (Newgard et al. 2009). Similarly, another study demonstrated that transplantation of intestinal microbiota from lean human donors to individuals with metabolic syndrome increased insulin sensitivity in the recipients (Vrieze et al. 2012). These findings suggest that the indigenous gut microbiota has a significant impact on metabolic function of the host and may impact propensity for weight gain.

Although research on the impact of the gut microbiota on obesity is still in preliminary stages, current findings provide a starting point for future studies aimed at understanding the precise role of specific microbial groups or diets on microbiota-associated body composition and metabolic phenotypes. Great interest centers on potential design of probiotics, prebiotics, or synbiotics (combinations of probiotics and prebiotics) to shift gut microbial populations toward "lean" rather than "obese" community structures. Such microbiome-targeted therapeutics could be manufacturable on a large scale and provide a more practical therapeutic alternative to fecal transplants (Vrieze et al. 2012).

4.4 Allergies and Asthma

4.4.1 Hygiene Hypothesis: Origin and Development

For decades, researchers have considered a potential link between microbial exposure and manifestation of allergic disease in humans. In 1958, David P. Strachan initiated a 23-year-long longitudinal study, investigating incidents of hay fever and eczema among 17,414 British children, in which incidence of hay fever was negatively correlated with both the number of older siblings and total number of siblings raised with allergic children. Findings also indicated a negative correlation between prevalence of eczema during the first year of life and the number of older siblings. Strachan suggested viral infections might provide protection against allergy if infections were acquired prenatally (via an infected mother) or by unhygienic contact with siblings (Strachan 1989). This assertion—that long-lasting effects on immunity could result from exposure to viral infections during childhood—was soon termed the hygiene hypothesis and has since become an integral reference for modern allergy pathology (Kim and Drake-Lee 2003).

Throughout the 1990s, subsequent research expanded upon the scope of Strachan's paper. In support of the hygiene hypothesis, studies from around the world presented associations between decreased allergic disease and childhood infections (Serafini 1997). Researchers suggested allergies manifested via the inflammation derived from cytokine imbalances. The Th1/Th2-shift hypothesis stated that insufficient exposure to viral and bacterial infections during childhood resulted in decreased activation of T-helper 1 (Th1) cells and increased T-helper 2 (Th2) cell activation; this immunological imbalance would induce an overexpression of Th2 inflammatory cytokines, such as IL-4 and IL-5 (promoting IgE maturation and engendering eosinophil recruitment, respectively), while simultaneously decreasing Th1 cytokines (IL-2 and IFN-γ), both of which normally regulated Th2 cytokine expression and, theoretically, deterred the manifestation of allergic diseases (Jirapongsananuruk and Leung 1997).

4.4.2 Gut Microbiota and the Hygiene Hypothesis

A 1998 review by Wold et al. challenged popular interpretations of the hygiene hypothesis (Wold 1998). The group argued against the "infection theory," citing studies in which infections protected neither infants nor children from developing allergies. Equally critical of the Th1/Th2-shift hypothesis, Wold et al. stated that if the rise in Western allergies had been caused by increases in Th2 activation, then incidents of autoimmunity (driven by Th-1 activation) could be expected to rise in nonindustrialized nations. Yet, as increases in both autoimmunity and allergies were only observed in industrialized countries, the authors did not find the model plausible and offered another hypothesis: the hygienic conditions of Western living had altered the gut microbiota during infancy, resulting in an inability to develop normal oral tolerance and, consequently, a proper immune response to antigens (Wold 1998). Indeed, studies of gut microbial composition demonstrated differences in composition between inhabitants of industrialized and nonindustrialized nations (Wold 1998; Noverr and Huffnagle 2004, 2005). Given that skin disorders, food allergies, and asthma are more prevalent in industrialized nations, the observed differences in microbial composition appear worthy of consideration (Holt and Strickland 2009; Leung 2013; Holgerson et al. 2013). Furthermore, Wold et al. argued that the increases in allergies observed by Strachan were actually rooted in disrupted dendritic cell (DC) activity and not viral infections. The researchers hypothesized that the altered gut microbiota could trigger an exaggerated immune response via interactions with macrophages that would have otherwise regulated the antigen-presentation capacity of DCs (Wold 1998). Additional evidence has since supported an association between microbiota, immunoregulation, and the development of DCs (Brandtzaeg 2010). An abnormal accumulation of mature DCs could alter signaling from T-regulatory cells, activating the ignition of Th1, Th2, or Th17 pro-inflammatory responses (Noverr and Huffnagle 2004, 2005; Tulic et al. 2012).

4.4.3 Food Allergies

Recent pathological models of various food allergies continue to focus upon interactions between the gut microbiota and the immune system (Vickery et al. 2011). Whether beneficial or pathogenic to a given host, bacteria possess external structures, referred to as microbe-associated molecular patterns (MAMPs). MAMPs can activate the innate immune system by binding to PRRs which can, ultimately, increase or decrease inflammation depending upon the cytokine responses evoked by a given MAMP. In hosts with dysfunctional food tolerances, MAMPs of microbes normally present within the gut can activate TLRs; this exchange decreases the activity of immunomodulatory T-regulatory cells, triggers Th2 responses, and creates allergic sensitivities to food molecules that would otherwise go unnoticed by the immune system (Brandtzaeg 2010). For example, beta-glucan, which is a staple of the Western diet and found in oats, barley, and on the surface of microbes, might instigate immune responses that trigger allergic reactions (Rylander 2010).

4.4.4 Impact of Probiotics and Microbial Metabolites on Allergic Disease

Research suggests that consumption of probiotics may reduce incidence of allergic disease. Probiotics are live microorganisms found in fermented foods and dairy products, which when consumed in sufficient amounts, have a beneficial effect on the health of the host (Schrezenmeir and de Vrese 2001). Fermented dairy products such as kefir, Maasai milk, and kurut promote intestinal colonization by probiotic species of *Lactobacillus* and *Bifidobacterium.* Several studies have linked consumption of such fermented foods on decreased allergic disease and asthma in children (Noverr and Huffnagle 2004; Butel 2014; Yang et al. 2013). In particular, in families that practice Anthroposophy, a spiritual philosophy in which medical and dietary practices include reduced antibiotic use and daily intake of fermented foods from a young age, children exhibit decreased asthma and allergic disease compared to those raised in non-Anthroposophic communities (Alm et al. 1999).

There is some evidence that microbial metabolic activity can also influence allergy manifestation. For example, polyphenols, which are present in plant-rich diets and are metabolites of indigenous gut microbes, have inhibitory effects on development of allergy and asthma (Marin et al. 2015; Magrone and Jirillo 2012). Additionally, short-chain fatty acids (SCFAs), which are produced during fermentation by gut microbes, have been implicated in amelioration of allergic disease. Microbially produced SCFAs are available as energy sources and influence inflammatory conditions within the host. However, the lack of fiber in the average Western diet changes the raw materials that gut microbiota access, and, consequently, the types of SCFAs circulating in the gut (den Besten et al. 2013). A recent study demonstrated that mice fed high-fiber diets actualized more circulating SCFAs and seemingly developed immunity to allergic lung infection, while subjects on low-fiber regimens exhibited decreased gut SCFA production and increased airway disease (Trompette et al. 2014). In the same study, mice exposed to the SCFA propionate exhibited enhanced production and lung infiltration of DCs, but decreased DC capacity to elicit a Th2 allergic airway inflammatory response, suggesting that propionate production in the gut could have systemic immunomodulatory effects. The SCFA butyrate has similarly been associated with decreased incidence of allergy. A study which examined the gut microbiota of infants with atopic eczema reported that severe cases of eczema inversely correlated with gut microbial diversity and with abundance of butyrate-producing gut microbes (Nylund et al. 2015). Fecal samples obtained from the same infants after eczema symptoms resolved exhibited greater microbial diversity and greater numbers of the butyrate-producing intestinal microbe *Coprococcus eutactus*. Together, these studies suggest that metabolites produced by microbes in the gut may have far-reaching antiallergenic effects.

4.4.5 Impact of Birth Method and Breastfeeding on Manifestation of Allergies and Asthma

Although heredity plays an important role in the development of allergic disease, evidence also supports an association between birth method and mode of infant feeding with the rise in Western allergies. Children born via cesarean section display greater incidences of food allergies, allergic rhinitis, asthma, and atopy compared to those born through the vaginal route (Bager et al. 2008). Researchers have hypothesized that modern delivery techniques could reduce neonatal exposure to microbes and MAMPs normally found in the maternal birth canal and fecal matter, consequently altering both the gut tolerances and microbial compositions of newborns. Further, studies of early human development have demonstrated that manifestations of allergic disease can be preceded by decreases in intestinal symbiont diversity (Yatsunenko et al. 2012; Abrahamsson et al. 2012). Intestinal bifidobacteria and *Enterococcus* populations are decreased in allergic children (Bjorksten et al. 1999, 2001) and increased in children born from vaginal births (Biasucci et al. 2010). In their thorough summary of mucosal immune development, Brugman et al. describe several studies of microbiota composition and birth methods, illustrating how environmental factors further complicate this dichotomy: exposed to more diverse microbial populations, children in developing nations who are delivered by cesarean sections tend to regain more of their lost bacterial colonies than children in industrialized nations, birthed in the same way (Brugman et al. 2015).

After birth, the method by which an infant is fed also has an impact on gut microbiota and manifestation of allergic disease. Gut microbial composition differs between breast-fed and formula-fed infants; breast-fed children amass greater numbers of gut bifidobacteria and oral *Lactobacillus*, whereas the intestinal microbiota of formula-fed babies is enriched in *Clostridia* and *Bacteroides* (Noverr and Huffnagle 2004; Penders et al. 2005), microbes that have been associated with asthma (Vael et al. 2008), eczema (Nylund et al. 2013), and food allergies (Ling et al. 2014). Indeed, several reports suggest a protective effect of breastfeeding against development of asthma and allergic diseases such eczema, food allergies, and allergic rhinitis (Bruno et al. 1993; Saarinen and Kajosaari 1995; Pratt 1984), especially among children with family history of atopic disease (van Odijk et al. 2003). Although specific mechanisms by which the gut microbiota in breastfed versus formula-fed infants might impact asthma or allergy remain ill-defined, it is possible that constituents in breast milk, such as growth factors, cytokines, and sIgA, fortify the beneficial gut microbes and provide protection against atopic disease (Bode 2009; Brandztaeg 2002).

4.5 Influence of Gut Microbiota on the Gut-Brain Axis and Mental Illness

4.5.1 The Gut-Brain Axis

The gut-brain axis (GBA) is a model used to study how gastrointestinal function, cognition, and behavior might affect, and be affected by, the complex neuronal, hormonal, metabolic, and immunological processes that occur in healthy and ill people (Collins et al. 2012). In recent years, researchers have incorporated the indigenous gut microbiota into the GBA paradigm, and microbial dysbiosis has recently been associated with certain mental disorders. As a result, new models for studying these disorders have since surfaced, many of which propose novel therapeutic strategies designed to address gut microbiota abnormalities (Preidis and Versalovic 2009).

The GBA is a bidirectional communication system, comprised of the enteric nervous system (ENS), the brain and spinal cord of the central nervous system (CNS), the autonomic nervous system (ANS), and the hypothalamic pituitary adrenal (HPA) axis (Carabotti et al. 2015). The sympathetic and parasympathetic branches of the ANS drive both afferent and efferent signals to and from the gut. Afferent signals that arise in the gut are transmitted through enteric, spinal, and vagal pathways to the CNS, whereas efferent signals travel from the CNS to the intestine. The HPA governs the organism's adaptive responses to stressors, including those arising from environmental, physiological, and emotional sources (McEwen 2007). When activated by a stressor, the hypothalamus secretes corticotrophin-releasing factor (CRF), which stimulates pituitary release of adrenocorticotropic hormone (ACTH) that subsequently promotes cortisol secretion from the adrenal glands. Cortisol serves as a stress hormone that impacts brain function (Tsigos and Chrousos 2002). Therefore, through the GBA, neural and hormonal lines of communication integrate to promote brain influence on gut activities, such as the function of enteric neurons, smooth muscle cells, immune cells, and intestinal epithelial cells. Of course, these same intestinal structures are also in intimate contact with the diverse gut microbiota, and the potential contribution of this microbiota to GBA communication is now gaining widespread attention.

4.5.2 Communication Between the Gut Microbiota and Host Nervous System

Recent research has demonstrated that the gut microbiota influences communication between the CNS and ENS. The "interkingdom signaling" exchanged between the microbiota and the GBA is bidirectional and occurs via neural, endocrine, immune, and humoral mechanisms (Carabotti et al. 2015). For example, metabolism of foodstuffs ingested by the host can result in microbial production of neuroactive substances, such as acetylcholine (AC) (Stephenson and Rowatt 1947), dopamine (DA), norepinephrine (NE) (Asano et al. 2012), and GABA (Barrett et al. 2012). These microbial-derived neurochemicals influence the host brain by either entering portal circulation or by interacting with the receptors on enteric neurons (Lyte 2014b). As the ENS innervates areas along the entire G.I. tract, symbiont-derived neurochemicals targeting enteric neurons could also control the hormonal secretions of enteroendocrine cells (Peterson and Artis 2014). An early investigation of the rat gut noted the existence of a serotonin precursor in the gastric lumen, following vagal nerve stimulation (Stephens and Tache 1989). Later on, researchers began to favor enterochromaffin (EC) cells in the lamina propria as the "signal transducers" by which microbial neuroactive compounds could activate enteric neurons situated near EC cells, resulting in the release of serotonin, CRF, and other signaling molecules (Rhee et al. 2009). Supporting the EC signal-transducer model, a study involving mice and a human EC cell model reported that the microbially-derived SCFAs butyrate and acetate promoted EC cell production of the neurotransmitter serotonin (Reigstad et al. 2015). Interestingly, microbially produced SCFAs and chemotactic peptides can activate the ENS and even modify gastrointestinal motility (Dass et al. 2007), which, consequently, appears to promote the growth of pathogenic bacteria (Van Felius et al. 2003).

Just as gut symbionts influence host systems, neurochemicals and hormones synthesized by the host nervous system also impact gut bacterial composition and phenotype (Lyte 2014a). For example, change in commensal bacterial composition is likely to occur when stress is present during early development (O'Mahony et al. 2009). Maternal prenatal stress is also associated with altered composition of the neonatal intestinal microbiota; such stress causes reductions in lactic acid bacteria and bifidobacteria, with concurrent increases in proteobacterial groups that contain pathogens such as *Escherichia*, *Serratia*, and *Enterobacter* (Bailey et al. 2004; Zijlmans et al. 2015). Furthermore, animal studies demonstrate that physiological stressors increase host susceptibility to colonization by enteric pathogens (Burkholder et al. 2008; Rostagno 2009) and that host stress can trigger pathogen virulence (Alverdy et al. 2000). Therefore, increased rate of infection due to physiological stressors is likely the result of enhanced host production of neurochemicals, which impact not only the host immune response to infection (Webster Marketon and Glaser 2008) but also the physiology of the pathogens themselves. For example, stress-induced catecholamines enhance growth or virulence expression in bacteria including *E. coli*, *Salmonella*, *Staphylococcus*, and *Pseudomonas aeruginosa* (Neal et al. 2001; Hegde et al. 2009; Verbrugghe et al. 2012). Increasing evidence suggests that enteric bacteria, including *E. coli* and *Salmonella*, sense and respond to host neuroendocrine hormones through microbial adrenergic receptors (Karavolos et al. 2013; Weinstein et al. 2015), indicating that pathogens evolved to exploit the stress response of their hosts.

4.5.3 Influence of the Gut Microbiota on Behavior and Affective Disorders

Emerging evidence demonstrates that, via the GBA, the gut microbiota can elicit behavioral responses within the host and can impact incidence or severity of affective disorders such as anxiety or depression. A recent study demonstrated that germ-free mice exhibited differential production of neurotransmitters and brain metabolites compared to their ex-germ-free counterparts; such disparities in the cerebral metabolome may impact brain homeostasis and influence potential development of anxiety and mood disorders (Matsumoto et al. 2013). In addition, studies showed that adult germ-free mice exhibit reduced anxiety-like actions in classical behavioral tests compared to conventional mice (Neufeld et al. 2011a, b). Similarly, administration of broad-spectrum antibacterial and antifungal agents to adult mice reduced anxiety-like and improved exploratory behaviors (Bercik et al. 2011a). Reconstitution of microbiota in germ-free mice early in life restored some anxiety-like behaviors (Diaz Heijtz et al. 2011; Clarke et al. 2013), but in contrast, when germ-free mice were administered specific pathogen-free microbiota in adulthood, the reduced level of anxiety persisted (Neufeld et al. 2011a, b). Those studies conducted in mice suggest that presence of gut microbiota positively impact manifestation of anxiety phenotypes and that a critical developmental window exists within which microbiota can influence stress-related behaviors (Foster and McVey Neufeld 2013). However, a study conducted in a stress-sensitive rat strain demonstrated that absence of gut microbiota exacerbates the neuroendocrine and behavioral response to acute stressors (Crumeyrolle-Arias et al. 2014). Certainly, discrepancies in the findings of these studies may be caused by differences in rodents and behavioral tests used, but collectively, these reports highlight the crucial influence that gut microbes can have on brain function and host behavior.

Colonization by enteric pathogens may contribute to manifestation of affective disorders with or without induction of inflammation. For example, administration of low levels of the pathogens *Campylobacter jejuni* or *Citrobacter rodentium* triggered anxiety-like behaviors in mice that were independent of immune-neural mechanisms (Lyte et al. 1998, 2006, 2011). Similarly, mice infected with *Trichuris muris*, a close relative of the human parasite *Trichuris trichiura*, exhibited increased anxiety phenotype, concurrent with decreased hippocampal levels of *bdnf* mRNA, which encodes brain-derived neurotrophic factor (BDNF). BDNF engenders neuronal growth and affects both cognition and emotion (Neufeld et al. 2011b) and is believed to modulate depression and anxiety (Suliman et al. 2013). Although inflammation is not always required for microbial-induced alterations in behavior, some studies indicate that it can contribute to development of affective disorders (Bercik et al. 2010). Indeed, a study of approximately 1600 patients with inflammatory gastrointestinal disorders revealed that the majority of the subjects had comorbidities with anxiety or depression (Addolorato et al. 2008). Furthermore, a study which employed a mouse model of colitis demonstrated that dextran sodium sulfate (DSS)-induced intestinal inflammation triggered anxiety-like behaviors (Bercik et al. 2011b).

4.6 Microbial-Based Therapeutics to Prevent or Ameliorate Disease

As our understanding of the complexity and composition of the indigenous gut microbiota has advanced, so has the interest in using intestinal commensals as therapeutic agents. Some bacterial species are being evaluated as potential therapeutics or drug-delivery vectors in their wild-type forms, as well as in genetically modified forms that are safer and better able to target specific diseases. Microbial-based therapeutics exploit host-bacterial interactions and often rely on bacterial-induced immunomodulation of the host. Research has demonstrated that bacterial-based therapeutics can be effective in treating a variety of human diseases that are difficult to treat with traditional medicine, including gastrointestinal infections and disorders, mental illness, osteoporosis, and cancer (Mimee et al. 2016; Toussaint et al. 2013; Hernandez et al. 2016). These novel therapeutics are promising as effective, straightforward, and inexpensive alternatives to conventional medical approaches.

4.6.1 Probiotics to Treat Gastrointestinal Infections

Modern management of infectious disease is hindered by increasing prevalence of antibiotic-resistant pathogens; thus there is dire need for alternative anti-infective strategies, of which probiotic microbes are an attractive option. For decades, researchers have demonstrated that probiotics can prevent or ameliorate gastrointestinal infection in humans (Britton and Versalovic 2008). Although the precise mechanisms for the anti-infective effects of probiotics are still poorly understood, evidence suggests that they work in myriad ways, such as by limiting pathogen access to host tissues, depriving pathogens of nutrients, producing bacteriocins that directly damage the microbial cell envelope, or limiting pathogen production of virulence factors (Bayoumi and Griffiths 2012). For details of the scope and mechanisms of the therapeutic effect of probiotics for gastrointestinal infections as well as systemic diseases, the reader is directed to several excellent reviews (Sarowska et al. 2013; Oelschlaeger 2010; Britton and Versalovic 2008).

Despite overwhelming evidence that probiotics may confer protection against infection, probiotic strains differ from one another, and the beneficial attributes of a single strain often vary between individual hosts and pathogens (Karimi and Pena 2008). Therefore, much current research focuses on the use of "designer" probiotics or those bioengineered to express specific traits that enhance their efficacy and safety. Several studies report anti-infective properties of probiotics engineered to express colonization factors with affinity for host receptors which are otherwise targets for pathogen or toxin binding. For example, a recombinant *Lactobacillus paracasei* expressing the *Listeria monocytogenes* adhesion protein (LAP), a virulence factor that aids in *L. monocytogenes* colonization of and translocation across the gut, was effective at blocking *L. monocytogenes* adhesion, invasion, and transepithelial translocation in a Caco-2 intestinal cell line (Koo et al. 2012). Another promising approach is the generation of "receptor mimic" probiotics, in which probiotic genes encoding surface structures, such as lipopolysaccharide (LPS), are genetically altered so that the recombinant LPS mimics the structure of the host receptor for a given toxin or pathogen adhesin. When orally administered prior to or during infection, the probiotic would outcompete host receptors for interaction with given pathogens and virulence factors. This approach has proven successful in an animal model of Shiga toxigenic *E. coli* (STEC) infection, where nonpathogenic *E. coli* was engineered to express a chimeric LPS with a terminal component identical in structure to the host Shiga toxin receptor. When administered orally to mice, the engineered probiotic conferred 100% protection against a fatal dose of STEC (Paton et al. 2000). Probiotics can also be designed to interfere with the regulation of pathogen virulence. In one study, the probiotic *E. coli* Nissle strain was engineered to express the *Vibrio cholerae* quorum sensing molecule cholera autoinducer 1 (CAI-1), a molecule that represses production of cholera toxin and the toxin-coregulated pilus (Duan and March 2010). When the engineered *E. coli* was administered to infant mice prior to challenge with *V. cholerae*, the probiotic reduced intestinal colonization of *V. cholerae* as well as the binding of cholera toxin to the host epithelium. Together, these studies demonstrate utility of not only wild-type probiotic bacteria against gastrointestinal infections but also the potential utility of recombinant probiotics as targeted anti-infective therapeutics.

4.6.2 *Fecal Microbiota Transplantation as a Treatment for Clostridium difficile Infection*

In contrast to the use of specific microbes as anti-infective therapies, the transfer of whole gut microbial communities in the form of fecal microbiota transplantation (FMT) has gained great attention for its utility in treating gastrointestinal infections that do not respond well to antibiotics, such as infections caused by *Clostridium difficile*. A Gram-positive, sporeforming bacterium, *C. difficile*, produces toxins TcdA, TcdB, and binary toxin, which trigger colonic cell death and inflammation (Voth and Ballard 2005). Intestinal *C. difficile* infections (CDI) cause a range of symptoms, including frequent watery diarrhea or pseudomembranous colitis, which is characterized by diarrhea containing pseudomembranes or sloughed intestinal tissue. Serious cases of CDI result in severe fulminant disease with toxic megacolon and can be fatal.

Interestingly, the gut microbiota of a healthy human provides resistance against *C. difficile* colonization and replication. Indeed, CDIs most often occur after sustained antibiotic regimens, which disrupt the normal microbiota and create a niche for the pathogen to colonize and flourish within the gut. Therefore, *C. difficile* is a major nosocomial pathogen that poses the greatest threat to hospitalized patients and those in chronic care facilities, where patients frequently undergo long-term antibiotic therapy. Host and microbial factors contribute to *C. difficile* invasion of a dysbiotic intestinal community, including altered bile salt metabolism and decreased competition for nutrients in absence of the normal microflora (Britton and Young 2014). The germination of *C. difficile* in the gut is governed by a complex mixture of bile salts, which the bacterium likely senses through a putative protease CspC (Francis et al. 2013). The primary bile acids, cholate and chenodeoxycholate, have disparate effects on *C. difficile* germination; cholate promotes germination, while chenodeoxycholate is a potent inhibitor of the process. In mice with a normal microbiota, cholate is not detectable in the cecum, while chenodeoxycholate is present. However, cecal cholate levels rise in response to antibiotic treatment, indicating that dysbiosis of the gut microbial community shifts relative bile salt concentrations toward an environment that favors *C. difficile* germination (Giel et al. 2010). In addition to maintaining a healthy bile salt balance, the normal gut microbiota outcompetes *C. difficile* for nutrients. Antibiotic therapy reduces diversity within the intestinal microbiota, thereby freeing up previously unavailable nutritional resources. Antibiotics also induce bacterial cell lysis, which releases carbon sources that *C. difficile* can consume (Britton and Young 2014).

Given the resistance that the normal microbiota provide against CDI, and the fact that the standard approach of treating CDI with the broad-spectrum antibiotic vancomycin could ultimately increase patient susceptibility to recurrent infections, the concept of treating CDI via FMT from a healthy donor was an attractive therapeutic strategy. Indeed, FMT has been shown to be a successful means of treating CDI and has proven superior to interventions which used defined, single-strain probiotics. Case reports and small series case studies suggest that recurrent CDI can be cured with a single FMT treatment (Aas et al. 2003; Brandt et al. 2012). Although FMT is a promising and inexpensive procedure that is relatively simple to perform, a number of hurdles need to be overcome before the procedure is adopted into mainstream medicine. For example, donor selection is a lengthy and expensive process that requires donors to be screened by health history and physical exam and depends on monetary compensation. Further, the specific route of administration, whether by nasogastric or nasoduodenal tube, endoscopy, or retention enema, is dictated by preferences of the attending healthcare professionals and is backed by little data on safety or efficacy (Khoruts et al. 2015). However, as none of these hurdles is insurmountable, it is feasible that FMT may be adopted as a routine and cost-effective treatment for serious and recurrent CDI.

4.6.3 Microbial-Derived Therapies for Mental Illness and Other Neuropathologies

Studies conducted in animals and humans demonstrate that probiotic consumption may ameliorate anxiety-like behaviors. For example, administration of *Bifidobacterium longum* reduced anxiety-like symptoms and normalized hippocampal BDNF levels in mice infected with the noninvasive parasite *T. muris* (Bercik et al. 2010). In addition, a probiotic containing *Lactobacillus helveticus* and *B. longum* reduced anxiety behaviors in rats and decreased behavior disorder traits in humans as determined by patient response to questionnaires designed to assess anxiety, depression, stress, and coping mechanisms (Messaoudi et al. 2011). A recent report also suggested that in individuals with neuroticism, consumption of fermented foods containing probiotics has a protective effect against exhibition of social anxiety (Hilimire et al. 2015). Probiotic therapies may also have utility for reducing severity of other mental disorders, such as obsessive-compulsive disorder (via *Lactobacillus rhamnosus*) (Kantak et al. 2014), autism (via *Bacillus fragilis*) (Hsiao et al. 2013), and depression (via reduced inflammatory cytokines and increased serotonin production) (Desbonnet et al. 2008).

FMT is another prospective therapy for behavioral and neurological disorders. Studies in mice have exploited the innate differences in behavioral phenotype between Balb/c and NIH Swiss mouse strains; Balb/c mice are timid and exhibit more anxiety-like behavior, while NIH Swiss mice exhibit less timidity and are more gregarious. Denaturing gradient gel electrophoresis (DGGE) profiling of intestinal contents revealed differing microbiome profiles of the two mouse strains (Bercik et al. 2011a). In FMT studies where germ-free NIH Swiss mice were colonized with Balb/c intestinal flora, the posttransplant NIH Swiss mice expressed greater anxiety-like traits and exhibited a reduction in hippocampal levels of BDNF. Conversely, when germ-free Balb/c mice were transplanted with NIH Swiss flora, the Balb/c mice exhibited less anxiety-like behavior and an increase in BDNF production (Bercik et al. 2011a). Although there is mounting evidence for the impact of FMT on brain function in mice, most evidence in humans remains limited to individual case or case series reports and anecdotal observations. In a case series report of three multiple sclerosis (MS) patients receiving FMT from a healthy donor as a treatment for chronic constipation, the FMT was associated with sustained neurological improvement that lasted up to 15 years (Borody et al. 2011). Although the mechanism by which the gut microbiota might impact MS symptoms remains unclear, a recent study which used 16S rRNA sequencing to compare gut microbial communities in MS patients with that of healthy controls reported differences in microbial community structure and inflammatory gene expression between the two groups. Patients with MS exhibited increased levels of intestinal *Methanobrevibacter* and *Akkermansia* species and decreased *Butyricimonas* populations compared to the control group. The altered microbial community structure in MS patients was correlated with increased expression of genes implicated in pro-inflammatory signaling via IFN and IL-2 pathways (Jangi et al. 2016). Similarly, reports of patients with Parkinson's disease showed improved neurologic symptoms following either antimicrobial therapy or FMT from a non-Parkinson's donor, although the mechanism is still unknown (Borody and Campbell 2012; Anathawamy 2011). While additional research is necessary to elucidate the mechanisms by which individual species of the gut microbiota may elicit change in human behavior and neuropathy, the potential effects of microbial interventions upon pathologies of the GBA are nonetheless promising.

4.6.4 Probiotics for Prevention and Treatment of Osteoporosis

Recent studies indicate that orally administered probiotics have systemic effects on the host and that can impact bone health and reduce incidence or severity of osteoporosis. The bone is a dynamic tissue that undergoes continuous remodeling throughout life, as dictated by actions of bone-building osteoblasts and bone-resorbing osteoclasts. Until recently, bone remodeling was believed to be primarily influenced by age, nutrition, and physiological states such as menopause or diabetes. However, recent studies suggest a previously unrecognized role for signaling between the gut microbiota, particularly species of the probiotic genus *Lactobacillus*, and bone in regulating bone remodeling (Ohlsson et al. 2014; Britton et al. 2014; McCabe et al. 2013).

Menopausal women are particularly susceptible to osteoporosis, as the hormone changes during menopause alter the development and activity of osteoblasts and osteoclasts, shifting the balance toward osteoclast activity and bone resorption. Britton et al. (2014) demonstrated that administration of the probiotic *Lactobacillus reuteri* to female ovariectomized mice, a common menopausal-associated osteoporosis model, could suppress osteoclast development and reverse bone resorption. Mice fed *L. reuteri* three times per week for 4 weeks exhibited decreased levels of osteoclast markers Trap5 and RANKL and increased bone density and strength compared to mice not supplemented with the probiotic. In premenopausal women, estrogen blocks TNFα synthesis by CD4 T cells, but during menopause, decreased estrogen leads to enhanced TNFα production and subsequent TNFα-induced activation of osteoclast development. In a separate study, the authors showed that *L. reuteri* suppressed TNFα production by producing histamine, which serves an immunoregulatory signal that modulates PKA and ERK inflammatory signaling

(Thomas et al. 2012). Together, these studies indicate that *L. reuteri* may be a useful and inexpensive approach for reducing menopause-associated bone loss.

Type 1 diabetes (T1D) is a disease characterized by hypoinsulinemia, hyperglycemia, and inflammation, and over time T1D patients are susceptible to complications such as osteoporosis. In these individuals, bone loss occurs regardless of sex and is associated with marked suppression in bone formation by osteoblasts (Schwartz and Sellmeyer 2007). The hyperglycemia and inflammation associated with T1D triggers increased TNFα production, which contributes to osteoblast apoptosis, and affects the proliferation and differentiation of osteoblast precursor cells (Coe et al. 2011). A recent study demonstrated that, in a diabetic mouse model, oral administration of *L. reuteri* restored osteoblast function and prevented T1D-induced bone loss (Zhang et al. 2015). The authors further showed that T1D bone tissue was deficient in Wnt10b, a signaling protein previously shown to promote osteoblast differentiation (Kubota et al. 2009) and which was linked to regulation of bone formation and bone density in nondiabetic mice (Bennett et al. 2007). Notably, Wnt10b expression was fully restored in T1D mice by *L. reuteri* treatment (Zhang et al. 2015), suggesting that one mechanism by which *L. reuteri* prevents T1D-driven bone loss may be modulation of Wnt10b production.

4.6.5 Bacteria-Based Cancer Vaccines and Treatments

Our improved understanding of host-microbial interactions has created an opportunity for exploiting these interactions to develop anticancer immunotherapies or treatments that have a direct cytotoxic effect on tumors. In particular, well-characterized intracellular bacteria such as *Salmonella* and *Listeria monocytogenes* offer the most promise as potential vectors for anticancer therapeutics. Here we will briefly highlight a few of the most pertinent advances in the area of bacteria-based vaccine vectors and cancer treatments. However, since the field of microbial-based cancer treatments is rapidly expanding, for a more thorough review of the subject, we direct the reader to several good reviews (Toussaint et al. 2013; Wood and Paterson 2014; Patyar et al. 2010).

By far the greatest developments in microbial-based cancer vaccines have resulted from studies using *L. monocytogenes* as a vaccine delivery vector. In particular, *L. monocytogenes* is especially suited as a vaccine vector because its molecular interactions with the host during intracellular infection are well elucidated and because it induces efficient CD4 and CD8 T-cell-mediated immunity by infecting a variety of somatic cells, including antigen-presenting cells like dendritic cells and macrophages. Most *L. monocytogenes*-based vaccines use recombinant *L. monocytogenes* expressing cancer-specific antigens which, when expressed inside of host cells, activate the immune system to target tumor cells for destruction. Dr. Daniel Portnoy and his collaborators at Aduro Biotech in Berkeley, California, have developed *Listeria*-based vaccines against pancreatic, ovarian, and prostate cancer, along with mesothelioma and non-small cell lung carcinomas (NSCLC).

One candidate vaccine in particular, CRS-207, shows special promise for treating pancreatic cancer. The CRS-207 vaccine consists of an attenuated form of *L. monocytogenes* which expresses mesothelin, a marker of pancreatic cancers, malignant melanomas, some ovarian cancers and NSCLCs, and which elicits a strong CD8 T-cell response in patients. In addition, CRS-207 efficiently delivers mesothelin antigen to MCH class I and MHC class II antigen-processing pathways, and in human studies, CRS-207 induced a mesothelin-specific CD8 T-cell response and was associated with improved disease-free state and overall survival of pancreatic cancer patients (Lutz et al. 2011). In a recent phase II clinical trial in humans with metastatic pancreatic cancer, dual administration of CRS-207 along with GVAX, another cancer vaccine, extended survival of patients with pancreatic cancer by 56% (Le et al. 2015). These studies demonstrate the potential power of *Listeria*-based vaccines as novel therapeutics against cancers with typically poor prognosis.

Intracellular bacteria such as *Salmonella* and *L. monocytogenes* can also be used to infect and exert direct cytotoxic effects on cancer cells. *Salmonella* and *L. monocytogenes* are particularly useful for these applications, as these bacteria are motile and can actively migrate into tumor tissue, overcoming the penetration limits of common chemotherapeutics. A promising candidate cancer therapy developed by Dr. Neil Forbes and colleagues is a *Salmonella* strain engineered to express *Staphylococcus aureus* hemolysin (*Salmonella*-SAH), a toxin that damages cell membranes and induces rapid cell death of tumor cells. In a study using mice transplanted with mammary carcinoma cells, injection of *Salmonella*-SAH into tumors induced necrosis of tumor cells and resulted in reduced tumor volume (St Jean et al. 2014).

Listeria- and *Salmonella*-based cancer treatments are especially attractive as novel therapeutics because they are less likely to select for treatment-resistant cancers than traditional chemotherapy drugs. Since bacterial-derived treatments usually involve complex mechanisms of action that include induction of the host immune response as well as eliciting direct cytotoxicity toward cancer cells, it may be more difficult for cancer cells to develop resistance (Patyar et al. 2010). Emerging evidence suggests that these microbial therapies may positively impact treatment outcomes, improving quality of life and survivability of patients suffering from an otherwise incurable disease.

4.7 Future Perspectives: The Benefits of Mining the Indigenous Microbiota

In this era of rapidly improving technology, our ability to mine the indigenous human microbiota will continue to shape our understanding of the role that microbial commensals play in health and disease. Perhaps the greatest advances yet to be made revolve around the development of clinically suitable microbial-based therapeutics to treat human pathologies and infections that are either difficult to treat by traditional medicine or which have evolved drug tolerance. In particular, the unsustainability of traditional antibiotic therapy due to the increasing antibiotic resistance crisis warrants the development of safe and effective microbial-based anti-infectives. Certainly, the opportunity for pharmacologic use of microbes depends on further study of the interactions of commensal microbes with the host, as well as the mechanisms underlying anti-inflammatory and anti-infective properties of specific indigenous microorganisms. In addition, while the ability of these microorganisms to colonize the host may contribute to their therapeutic efficacy, it does raise concern over how such therapies can be controlled and about potential spread of recombinant DNA from therapeutic microbes to the indigenous host microbiota (Mimee et al. 2016). Indeed, further work to address these concerns is necessary for the optimal exploitation of the gut microbiota to treat both intestinal and systemic diseases.

Compliance with Ethical Standards
Conflict of Interest: T. Vunk declares that he has no conflict of interest. K. Burkholder declares that she has no conflict of interest.

Ethical approval: This chapter does not contain any studies with human participants or animals performed by any of the authors.

References

Aas J, Gessert CE, Bakken JS (2003) Recurrent Clostridium difficile colitis: case series involving 18 patients treated with donor stool administered via a nasogastric tube. Clin Infect Dis 36 (5):580–585. doi:10.1086/367657

Abrahamsson TR, Jakobsson HE, Andersson AF, Bjorksten B, Engstrand L, Jenmalm MC (2012) Low diversity of the gut microbiota in infants with atopic eczema. J Allergy Clin Immunol 129 (2):434–440, 440.e1–2. doi:10.1016/j.jaci.2011.10.025

Addolorato G, Mirijello A, D'Angelo C, Leggio L, Ferrulli A, Abenavoli L, Vonghia L, Cardone S, Leso V, Cossari A, Capristo E, Gasbarrini G (2008) State and trait anxiety and depression in patients affected by gastrointestinal diseases: psychometric evaluation of 1641 patients referred to an internal medicine outpatient setting. Int J Clin Pract 62(7):1063–1069. doi:10.1111/j.1742-1241.2008.01763.x

Alm JS, Swartz J, Lilja G, Scheynius A, Pershagen G (1999) Atopy in children of families with an anthroposophic lifestyle. Lancet 353(9163):1485–1488. doi:10.1016/S0140-6736(98)09344-1

Alverdy J, Holbrook C, Rocha F, Seiden L, RL W, Musch M, Chang E, Ohman D, Suh S (2000) Gut-derived sepsis occurs when the right pathogen with the right virulence genes meets the right host: evidence for in vivo virulence expression in Pseudomonas aeruginosa. Ann Surg 232 (4):480–489

Anathawamy A (2011) Faecal transplant eases symptoms of Parkinson's disease. New Scientist (2796):8–9

Asano Y, Hiramoto T, Nishino R, Aiba Y, Kimura T, Yoshihara K, Koga Y, Sudo N (2012) Critical role of gut microbiota in the production of biologically active, free catecholamines in the gut lumen of mice. Am J Physiol Gastrointest Liver Physiol 303(11):G1288–G1295. doi:10.1152/ajpgi.00341.2012

Bager P, Wohlfahrt J, Westergaard T (2008) Caesarean delivery and risk of atopy and allergic disease: meta-analyses. Clin Exp Allergy 38(4):634–642. doi:10.1111/j.1365-2222.2008.02939.x

Bailey MT, Lubach GR, Coe CL (2004) Prenatal stress alters bacterial colonization of the gut in infant monkeys. J Pediatr Gastroenterol Nutr 38(4):414–421

Barrett E, Ross RP, O'Toole PW, Fitzgerald GF, Stanton C (2012) gamma-Aminobutyric acid production by culturable bacteria from the human intestine. J Appl Microbiol 113(2):411–417. doi:10.1111/j.1365-2672.2012.05344.x

Bayoumi MA, Griffiths MW (2012) In vitro inhibition of expression of virulence genes responsible for colonization and systemic spread of enteric pathogens using Bifidobacterium bifidum secreted molecules. Int J Food Microbiol 156(3):255–263. doi:10.1016/j.ijfoodmicro.2012.03.034

Bennett CN, Ouyang H, Ma YL, Zeng Q, Gerin I, Sousa KM, Lane TF, Krishnan V, Hankenson KD, MacDougald OA (2007) Wnt10b increases postnatal bone formation by enhancing osteoblast differentiation. J Bone Miner Res 22(12):1924–1932. doi:10.1359/jbmr.070810

Bercik P, Denou E, Collins J, Jackson W, Lu J, Jury J, Deng Y, Blennerhassett P, Macri J, McCoy KD, Verdu EF, Collins SM (2011a) The intestinal microbiota affect central levels of brain-derived neurotropic factor and behavior in mice. Gastroenterology 141 (2):599–609, 609.e1–3. doi:10.1053/j.gastro.2011.04.052

Bercik P, Park AJ, Sinclair D, Khoshdel A, Lu J, Huang X, Deng Y, Blennerhassett PA, Fahnestock M, Moine D, Berger B, Huizinga JD, Kunze W, McLean PG, Bergonzelli GE, Collins SM, Verdu EF (2011b) The anxiolytic effect of Bifidobacterium longum NCC3001 involves vagal pathways for gut-brain communication. Neurogastroenterol Motil 23 (12):1132–1139. doi:10.1111/j.1365-2982.2011.01796.x

Bercik P, Verdu EF, Foster JA, Macri J, Potter M, Huang X, Malinowski P, Jackson W, Blennerhassett P, Neufeld KA, Lu J, Khan WI, Corthesy-Theulaz I, Cherbut C, Bergonzelli GE, Collins SM (2010) Chronic gastrointestinal inflammation induces anxiety-like behavior and alters central nervous system biochemistry in mice. Gastroenterology 139 (6):2102–2112.e1. doi:10.1053/j.gastro.2010.06.063

Biasucci G, Rubini M, Riboni S, Morelli L, Bessi E, Retetangos C (2010) Mode of delivery affects the bacterial community in the newborn gut. Early Hum Dev 86(Suppl 1):13–15. doi:10.1016/j.earlhumdev.2010.01.004

Bjorksten B, Naaber P, Sepp E, Mikelsaar M (1999) The intestinal microflora in allergic Estonian and Swedish 2-year-old children. Clin Exp Allergy 29(3):342–346

Bjorksten B, Sepp E, Julge K, Voor T, Mikelsaar M (2001) Allergy development and the intestinal microflora during the first year of life. J Allergy Clin Immunol 108(4):516–520. doi:10.1067/mai.2001.118130

Bode L (2009) Human milk oligosaccharides: prebiotics and beyond. Nutr Rev 67(Suppl 2):S183–S191. doi:10.1111/j.1753-4887.2009.00239.x

Borody TJ, Campbell J (2012) Fecal microbiota transplantation techniques, applications, and issues. Gastroenterol Clin N Am 41(4):781–803. doi:10.1016/j.gtc.2012.08.008

Borody TJ, Leis S, Campbell J, Torres M, Nowak A (2011) Fecal microbiota transplantation (FMT) in multiple sclerosis (MS). Am J Gastroenterol 106

Brandt LJ, Aroniadis OC, Mellow M, Kanatzar A, Kelly C, Park T, Stollman N, Rohlke F, Surawicz C (2012) Long-term follow-up of colonoscopic fecal microbiota transplant for recurrent Clostridium difficile infection. Am J Gastroenterol 107(7):1079–1087. doi:10.1038/ajg.2012.60

Brandtzaeg P (2010) Food allergy: separating the science from the mythology. Nat Rev Gastroenterol Hepatol 7(7):380–400. doi:10.1038/nrgastro.2010.80

Brandztaeg P (2002) Role of local immunity and breastfeeding in mucosal homeostasis and defence against infections. In: Calder P, Field C, Gill H (eds) Nutrition and immune function, vol 1. CABI Publishing, Oxon, pp 273–320

Britton RA, Irwin R, Quach D, Schaefer L, Zhang J, Lee T, Parameswaran N, McCabe LR (2014) Probiotic L. reuteri treatment prevents bone loss in a menopausal ovariectomized mouse model. J Cell Physiol 229(11):1822–1830. doi:10.1002/jcp.24636

Britton RA, Versalovic J (2008) Probiotics and gastrointestinal infections. Interdiscip Perspect Infect Dis 2008:290769. doi:10.1155/2008/290769

Britton RA, Young VB (2014) Role of the intestinal microbiota in resistance to colonization by Clostridium difficile. Gastroenterology 146(6):1547–1553. doi:10.1053/j.gastro.2014.01.059

Brugman S, Perdijk O, van Neerven RJ, Savelkoul HF (2015) Mucosal immune development in early life: setting the stage. Arch Immunol Ther Exp (Warsz) 63(4):251–268. doi:10.1007/s00005-015-0329-y

Bruno G, Milita O, Ferrara M, Nisini R, Cantani A, Businco L (1993) Prevention of atopic diseases in high risk babies (long-term follow-up). Allergy Proc 14(3):181–186; discussion 186–187

Burkholder KM, Thompson KL, Einstein ME, Applegate TJ, Patterson JA (2008) Influence of stressors on normal intestinal microbiota, intestinal morphology, and susceptibility to Salmonella enteritidis colonization in broilers. Poult Sci 87(9):1734–1741. doi:10.3382/ps.2008-00107

Butel MJ (2014) Probiotics, gut microbiota and health. Med Mal Infect 44(1):1–8. doi:10.1016/j.medmal.2013.10.002

Carabotti M, Scirocco A, Maselli MA, Severi C (2015) The gut-brain axis: interactions between enteric microbiota, central and enteric nervous systems. Ann Gastroenterol 28(2):203–209

Castellarin M, Warren RL, Freeman JD, Dreolini L, Krzywinski M, Strauss J, Barnes R, Watson P, Allen-Vercoe E, Moore RA, Holt RA (2012) Fusobacterium nucleatum infection is prevalent in human colorectal carcinoma. Genome Res 22(2):299–306. doi:10.1101/gr.126516.111

Chen W, Liu F, Ling Z, Tong X, Xiang C (2012) Human intestinal lumen and mucosa-associated microbiota in patients with colorectal cancer. PLoS One 7(6):e39743. doi:10.1371/journal.pone.0039743

Clarke G, Grenham S, Scully P, Fitzgerald P, Moloney RD, Shanahan F, Dinan TG, Cryan JF (2013) The microbiome-gut-brain axis during early life regulates the hippocampal serotonergic system in a sex-dependent manner. Mol Psychiatry 18(6):666–673. doi:10.1038/mp.2012.77

Clarke SF, Murphy EF, Nilaweera K, Ross PR, Shanahan F, O'Toole PW, Cotter PD (2012) The gut microbiota and its relationship to diet and obesity: new insights. Gut Microbes 3 (3):186–202. doi:10.4161/gmic.20168

Coe LM, Irwin R, Lippner D, McCabe LR (2011) The bone marrow microenvironment contributes to type I diabetes induced osteoblast death. J Cell Physiol 226(2):477–483. doi:10.1002/jcp.22357

Collins SM, Surette M, Bercik P (2012) The interplay between the intestinal microbiota and the brain. Nat Rev Microbiol 10(11):735–742. doi:10.1038/nrmicro2876

Compare D, Nardone G (2011) Contribution of gut microbiota to colonic and extracolonic cancer development. Dig Dis 29(6):554–561. doi:10.1159/000332967

Crumeyrolle-Arias M, Jaglin M, Bruneau A, Vancassel S, Cardona A, Dauge V, Naudon L, Rabot S (2014) Absence of the gut microbiota enhances anxiety-like behavior and neuroendocrine response to acute stress in rats. Psychoneuroendocrinology 42:207–217. doi:10.1016/j.psyneuen.2014.01.014

Dass NB, John AK, Bassil AK, Crumbley CW, Shehee WR, Maurio FP, Moore GB, Taylor CM, Sanger GJ (2007) The relationship between the effects of short-chain fatty acids on intestinal motility in vitro and GPR43 receptor activation. Neurogastroenterol Motil 19 (1):66–74. doi:10.1111/j.1365-2982.2006.00853.x

den Besten G, van Eunen K, Groen AK, Venema K, Reijngoud DJ, Bakker BM (2013) The role of short-chain fatty acids in the interplay between diet, gut microbiota, and host energy metabolism. J Lipid Res 54(9):2325–2340. doi:10.1194/jlr.R036012

Desbonnet L, Garrett L, Clarke G, Bienenstock J, Dinan TG (2008) The probiotic Bifidobacteria infantis: an assessment of potential antidepressant properties in the rat. J Psychiatr Res 43 (2):164–174. doi:10.1016/j.jpsychires.2008.03.009

Diaz Heijtz R, Wang S, Anuar F, Qian Y, Bjorkholm B, Samuelsson A, Hibberd ML, Forssberg H, Pettersson S (2011) Normal gut microbiota modulates brain development and behavior. Proc Natl Acad Sci USA 108(7):3047–3052. doi:10.1073/pnas.1010529108

Dizdar V, Gilja OH, Hausken T (2007) Increased visceral sensitivity in Giardia-induced postinfectious irritable bowel syndrome and functional dyspepsia. Effect of the 5HT3-antagonist ondansetron. Neurogastroenterol Motil 19(12):977–982. doi:10.1111/j.1365-2982.2007.00988.x

Duan F, March JC (2010) Engineered bacterial communication prevents Vibrio cholerae virulence in an infant mouse model. Proc Natl Acad Sci USA 107(25):11260–11264. doi:10.1073/pnas.1001294107

Dunlop SP, Jenkins D, Neal KR, Spiller RC (2003) Relative importance of enterochromaffin cell hyperplasia, anxiety, and depression in postinfectious IBS. Gastroenterology 125 (6):1651–1659

Foster JA, McVey Neufeld KA (2013) Gut-brain axis: how the microbiome influences anxiety and depression. Trends Neurosci 36(5):305–312. doi:10.1016/j.tins.2013.01.005

Francis MB, Allen CA, Shrestha R, Sorg JA (2013) Bile acid recognition by the Clostridium difficile germinant receptor, CspC, is important for establishing infection. PLoS Pathog 9(5): e1003356. doi:10.1371/journal.ppat.1003356

Fukata M, Chen A, Vamadevan AS, Cohen J, Breglio K, Krishnareddy S, Hsu D, Xu R, Harpaz N, Dannenberg AJ, Subbaramaiah K, Cooper HS, Itzkowitz SH, Abreu MT (2007) Toll-like receptor-4 promotes the development of colitis-associated colorectal tumors. Gastroenterology 133(6):1869–1881. doi:10.1053/j.gastro.2007.09.008

Giel JL, Sorg JA, Sonenshein AL, Zhu J (2010) Metabolism of bile salts in mice influences spore germination in Clostridium difficile. PLoS One 5(1):e8740. doi:10.1371/journal.pone.0008740

Haggar FA, Boushey RP (2009) Colorectal cancer epidemiology: incidence, mortality, survival, and risk factors. Clin Colon Rectal Surg 22(4):191–197. doi:10.1055/s-0029-1242458

Hegde M, Wood TK, Jayaraman A (2009) The neuroendocrine hormone norepinephrine increases Pseudomonas aeruginosa PA14 virulence through the las quorum-sensing pathway. Appl Microbiol Biotechnol 84(4):763–776. doi:10.1007/s00253-009-2045-1

Hernandez CJ, Guss JD, Luna M, Goldring SR (2016) Links between the microbiome and bone. J Bone Miner Res 31(9):1638–1646. doi:10.1002/jbmr.2887

Hilimire MR, DeVylder JE, Forestell CA (2015) Fermented foods, neuroticism, and social anxiety: an interaction model. Psychiatry Res 228(2):203–208. doi:10.1016/j.psychres.2015.04.023

Holgerson PL, Vestman NR, Claesson R, Ohman C, Domellof M, Tanner AC, Hernell O, Johansson I (2013) Oral microbial profile discriminates breast-fed from formula-fed infants. J Pediatr Gastroenterol Nutr 56(2):127–136. doi:10.1097/MPG.0b013e31826f2bc6

Holt PG, Strickland DH (2009) Soothing signals: transplacental transmission of resistance to asthma and allergy. J Exp Med 206(13):2861–2864. doi:10.1084/jem.20092469

Hsiao EY, McBride SW, Hsien S, Sharon G, Hyde ER, McCue T, Codelli JA, Chow J, Reisman SE, Petrosino JF, Patterson PH, Mazmanian SK (2013) Microbiota modulate behavioral and physiological abnormalities associated with neurodevelopmental disorders. Cell 155 (7):1451–1463. doi:10.1016/j.cell.2013.11.024

Jangi S, Gandhi R, Cox LM, Li N, von Glehn F, Yan R, Patel B, Mazzola MA, Liu S, Glanz BL, Cook S, Tankou S, Stuart F, Melo K, Nejad P, Smith K, Topcuolu BD, Holden J, Kivisakk P, Chitnis T, De Jager PL, Quintana FJ, Gerber GK, Bry L, Weiner HL (2016) Alterations of the human gut microbiome in multiple sclerosis. Nat Commun 7:12015. doi:10.1038/ncomms12015

Ji S, Park H, Lee D, Song YK, Choi JP, Lee SI (2005) Post-infectious irritable bowel syndrome in patients with Shigella infection. J Gastroenterol Hepatol 20(3):381–386. doi:10.1111/j.1440-1746.2005.03574.x

Jirapongsananuruk O, Leung DY (1997) Clinical applications of cytokines: new directions in the therapy of atopic diseases. Ann Allergy Asthma Immunol 79(1):5–16; quiz 19-20. doi:10.1016/S1081-1206(10)63078-5

Kantak PA, Bobrow DN, Nyby JG (2014) Obsessive-compulsive-like behaviors in house mice are attenuated by a probiotic (Lactobacillus rhamnosus GG). Behav Pharmacol 25(1):71–79. doi:10.1097/FBP.0000000000000013

Kappelman MD, Rifas-Shiman SL, Kleinman K, Ollendorf D, Bousvaros A, Grand RJ, Finkelstein JA (2007) The prevalence and geographic distribution of Crohn's disease and ulcerative colitis in the United States. Clin Gastroenterol Hepatol 5(12):1424–1429. doi:10.1016/j.cgh.2007.07.012

Karavolos MH, Winzer K, Williams P, Khan CM (2013) Pathogen espionage: multiple bacterial adrenergic sensors eavesdrop on host communication systems. Mol Microbiol 87(3):455–465. doi:10.1111/mmi.12110

Karimi O, Pena AS (2008) Indications and challenges of probiotics, prebiotics, and synbiotics in the management of arthralgias and spondyloarthropathies in inflammatory bowel disease. J Clin Gastroenterol 42(Suppl 3 Pt 1):S136–S141. doi:10.1097/MCG.0b013e3181662455

Khoruts A, Sadowsky MJ, Hamilton MJ (2015) Development of fecal microbiota transplantation suitable for mainstream medicine. Clin Gastroenterol Hepatol 13(2):246–250. doi:10.1016/j.cgh.2014.11.014

Kim DS, Drake-Lee AB (2003) Infection, allergy and the hygiene hypothesis: historical perspective. J Laryngol Otol 117(12):946–950. doi:10.1258/002221503322683812

Koo OK, Amalaradjou MA, Bhunia AK (2012) Recombinant probiotic expressing Listeria adhesion protein attenuates Listeria monocytogenes virulence in vitro. PLoS One 7(1): e29277. doi:10.1371/journal.pone.0029277

Kostic AD, Gevers D, Pedamallu CS, Michaud M, Duke F, Earl AM, Ojesina AI, Jung J, Bass AJ, Tabernero J, Baselga J, Liu C, Shivdasani RA, Ogino S, Birren BW, Huttenhower C, Garrett WS, Meyerson M (2012) Genomic analysis identifies association of Fusobacterium with colorectal carcinoma. Genome Res 22(2):292–298. doi:10.1101/gr.126573.111

Kubota T, Michigami T, Ozono K (2009) Wnt signaling in bone metabolism. J Bone Miner Metab 27(3):265–271. doi:10.1007/s00774-009-0064-8

Le DT, Wang-Gillam A, Picozzi V, Greten TF, Crocenzi T, Springett G, Morse M, Zeh H, Cohen D, Fine RL, Onners B, Uram JN, Laheru DA, Lutz ER, Solt S, Murphy AL, Skoble J, Lemmens E, Grous J, Dubensky T Jr, Brockstedt DG, Jaffee EM (2015) Safety and survival with GVAX pancreas prime and Listeria Monocytogenes-expressing mesothelin (CRS-207) boost vaccines for metastatic pancreatic cancer. J Clin Oncol 33 (12):1325-1333. doi:10.1200/JCO.2014.57.4244

Lepage P, Hasler R, Spehlmann ME, Rehman A, Zvirbliene A, Begun A, Ott S, Kupcinskas L, Dore J, Raedler A, Schreiber S (2011) Twin study indicates loss of interaction between microbiota and mucosa of patients with ulcerative colitis. Gastroenterology 141(1):227–236. doi:10.1053/j.gastro.2011.04.011

Leung DY (2013) New insights into atopic dermatitis: role of skin barrier and immune dysregulation. Allergol Int 62(2):151–161. doi:10.2332/allergolint.13-RAI-0564

Ley RE, Backhed F, Turnbaugh P, Lozupone CA, Knight RD, Gordon JI (2005) Obesity alters gut microbial ecology. Proc Natl Acad Sci USA 102(31):11070–11075. doi:10.1073/pnas.0504978102

Ling Z, Li Z, Liu X, Cheng Y, Luo Y, Tong X, Yuan L, Wang Y, Sun J, Li L, Xiang C (2014) Altered fecal microbiota composition associated with food allergy in infants. Appl Environ Microbiol 80(8):2546–2554. doi:10.1128/AEM.00003-14

Lutz E, Yeo CJ, Lillemoe KD, Biedrzycki B, Kobrin B, Herman J, Sugar E, Piantadosi S, Cameron JL, Solt S, Onners B, Tartakovsky I, Choi M, Sharma R, Illei PB, Hruban RH, Abrams RA, Le D, Jaffee E, Laheru D (2011) A lethally irradiated allogeneic granulocyte-macrophage colony stimulating factor-secreting tumor vaccine for pancreatic adenocarcinoma. A phase II trial of safety, efficacy, and immune activation. Ann Surg 253 (2):328–335. doi:10.1097/SLA.0b013e3181fd271c

Lyte M (2014a) Microbial endocrinology and the microbiota-gut-brain axis. Microbial endocrinology: the microbiota-gut-brain axis in health and disease. Springer, New York

Lyte M (2014b) Microbial endocrinology: host-microbiota neuroendocrine interactions influencing brain and behavior. Gut Microbes 5(3):381–389. doi:10.4161/gmic.28682

Lyte M, Li W, Opitz N, Gaykema RP, Goehler LE (2006) Induction of anxiety-like behavior in mice during the initial stages of infection with the agent of murine colonic hyperplasia Citrobacter rodentium. Physiol Behav 89(3):350–357. doi:10.1016/j.physbeh.2006.06.019

Lyte M, Varcoe JJ, Bailey MT (1998) Anxiogenic effect of subclinical bacterial infection in mice in the absence of overt immune activation. Physiol Behav 65(1):63–68

Lyte M, Vulchanova L, Brown DR (2011) Stress at the intestinal surface: catecholamines and mucosa-bacteria interactions. Cell Tissue Res 343(1):23–32. doi:10.1007/s00441-010-1050-0

Magrone T, Jirillo E (2012) Influence of polyphenols on allergic immune reactions: mechanisms of action. Proc Nutr Soc 71(2):316–321. doi:10.1017/S0029665112000109

Manichanh C, Rigottier-Gois L, Bonnaud E, Gloux K, Pelletier E, Frangeul L, Nalin R, Jarrin C, Chardon P, Marteau P, Roca J, Dore J (2006) Reduced diversity of faecal microbiota in Crohn's disease revealed by a metagenomic approach. Gut 55(2):205–211. doi:10.1136/gut.2005.073817

Marchesi JR, Dutilh BE, Hall N, Peters WH, Roelofs R, Boleij A, Tjalsma H (2011) Towards the human colorectal cancer microbiome. PLoS One 6(5):e20447. doi:10.1371/journal.pone.0020447

Marin L, Miguelez EM, Villar CJ, Lombo F (2015) Bioavailability of dietary polyphenols and gut microbiota metabolism: antimicrobial properties. Biomed Res Int 2015:905215. doi:10.1155/2015/905215

Martinez-Medina M, Aldeguer X, Gonzalez-Huix F, Acero D, Garcia-Gil LJ (2006) Abnormal microbiota composition in the ileocolonic mucosa of Crohn's disease patients as revealed by polymerase chain reaction-denaturing gradient gel electrophoresis. Inflamm Bowel Dis 12 (12):1136–1145. doi:10.1097/01.mib.0000235828.09305.0c

Matsumoto M, Kibe R, Ooga T, Aiba Y, Sawaki E, Koga Y, Benno Y (2013) Cerebral low-molecular metabolites influenced by intestinal microbiota: a pilot study. Front Syst Neurosci 7:9. doi:10.3389/fnsys.2013.00009

McCabe LR, Irwin R, Schaefer L, Britton RA (2013) Probiotic use decreases intestinal inflammation and increases bone density in healthy male but not female mice. J Cell Physiol 228 (8):1793–1798. doi:10.1002/jcp.24340

McEwen BS (2007) Physiology and neurobiology of stress and adaptation: central role of the brain. Physiol Rev 87(3):873–904. doi:10.1152/physrev.00041.2006

Mearin F, Perez-Oliveras M, Perello A, Vinyet J, Ibanez A, Coderch J, Perona M (2005) Dyspepsia and irritable bowel syndrome after a Salmonella gastroenteritis outbreak: one-year follow-up cohort study. Gastroenterology 129(1):98–104

Messaoudi M, Violle N, Bisson JF, Desor D, Javelot H, Rougeot C (2011) Beneficial psychological effects of a probiotic formulation (Lactobacillus helveticus R0052 and Bifidobacterium longum R0175) in healthy human volunteers. Gut Microbes 2(4):256–261. doi:10.4161/gmic.2.4.16108

Mimee M, Citorik RJ, Lu TK (2016) Microbiome therapeutics—Advances and challenges. Adv Drug Deliv Rev 105(Pt A):44–54. doi:10.1016/j.addr.2016.04.032

Moossavi S, Bishehsari F (2012) Inflammation in sporadic colorectal cancer. Arch Iran Med 15 (3):166–170. doi:012153/AIM.0012

Morgan XC, Tickle TL, Sokol H, Gevers D, Devaney KL, Ward DV, Reyes JA, Shah SA, LeLeiko N, Snapper SB, Bousvaros A, Korzenik J, Sands BE, Xavier RJ, Huttenhower C (2012) Dysfunction of the intestinal microbiome in inflammatory bowel disease and treatment. Genome Biol 13(9):R79. doi:10.1186/gb-2012-13-9-r79

Murphy EF, Cotter PD, Healy S, Marques TM, O'Sullivan O, Fouhy F, Clarke SF, O'Toole PW, Quigley EM, Stanton C, Ross PR, O'Doherty RM, Shanahan F (2010) Composition and energy harvesting capacity of the gut microbiota: relationship to diet, obesity and time in mouse models. Gut 59(12):1635–1642. doi:10.1136/gut.2010.215665

Neal CP, Freestone PP, Maggs AF, Haigh RD, Williams PH, Lyte M (2001) Catecholamine inotropes as growth factors for Staphylococcus epidermidis and other coagulase-negative staphylococci. FEMS Microbiol Lett 194(2):163–169

Neufeld KA, Kang N, Bienenstock J, Foster JA (2011a) Effects of intestinal microbiota on anxiety-like behavior. Commun Integr Biol 4(4):492–494. doi:10.4161/cib.4.4.15702

Neufeld KM, Kang N, Bienenstock J, Foster JA (2011b) Reduced anxiety-like behavior and central neurochemical change in germ-free mice. Neurogastroenterol Motil 23 (3):255–264, e119. doi:10.1111/j.1365-2982.2010.01620.x

Newgard CB, An J, Bain JR, Muehlbauer MJ, Stevens RD, Lien LF, Haqq AM, Shah SH, Arlotto M, Slentz CA, Rochon J, Gallup D, Ilkayeva O, Wenner BR, Yancy WS Jr, Eisenson H, Musante G, Surwit RS, Millington DS, Butler MD, Svetkey LP (2009) A branched-chain amino acid-related metabolic signature that differentiates obese and lean humans and contributes to insulin resistance. Cell Metab 9(4):311–326. doi:10.1016/j.cmet.2009.02.002

Noverr MC, Huffnagle GB (2004) Does the microbiota regulate immune responses outside the gut? Trends Microbiol 12(12):562–568. doi:10.1016/j.tim.2004.10.008

Noverr MC, Huffnagle GB (2005) The 'microflora hypothesis' of allergic diseases. Clin Exp Allergy 35(12):1511–1520. doi:10.1111/j.1365-2222.2005.02379.x

Nylund L, Nermes M, Isolauri E, Salminen S, de Vos WM, Satokari R (2015) Severity of atopic disease inversely correlates with intestinal microbiota diversity and butyrate-producing bacteria. Allergy 70(2):241–244. doi:10.1111/all.12549

Nylund L, Satokari R, Nikkila J, Rajilic-Stojanovic M, Kalliomaki M, Isolauri E, Salminen S, de Vos WM (2013) Microarray analysis reveals marked intestinal microbiota aberrancy in infants having eczema compared to healthy children in at-risk for atopic disease. BMC Microbiol 13:12. doi:10.1186/1471-2180-13-12

Oelschlaeger TA (2010) Mechanisms of probiotic actions—A review. Int J Med Microbiol 300 (1):57–62. doi:10.1016/j.ijmm.2009.08.005

Ohlsson C, Engdahl C, Fak F, Andersson A, Windahl SH, Farman HH, Moverare-Skrtic S, Islander U, Sjogren K (2014) Probiotics protect mice from ovariectomy-induced cortical bone loss. PLoS One 9(3):e92368. doi:10.1371/journal.pone.0092368

O'Mahony SM, Marchesi JR, Scully P, Codling C, Ceolho AM, Quigley EM, Cryan JF, Dinan TG (2009) Early life stress alters behavior, immunity, and microbiota in rats: implications for irritable bowel syndrome and psychiatric illnesses. Biol Psychiatry 65(3):263–267. doi:10.1016/j.biopsych.2008.06.026

Paton AW, Morona R, Paton JC (2000) A new biological agent for treatment of Shiga toxigenic Escherichia coli infections and dysentery in humans. Nat Med 6(3):265–270. doi:10.1038/73111

Patyar S, Joshi R, Byrav DS, Prakash A, Medhi B, Das BK (2010) Bacteria in cancer therapy: a novel experimental strategy. J Biomed Sci 17(1):21. doi:10.1186/1423-0127-17-21

Penders J, Vink C, Driessen C, London N, Thijs C, Stobberingh EE (2005) Quantification of Bifidobacterium spp., Escherichia coli and Clostridium difficile in faecal samples of breast-fed and formula-fed infants by real-time PCR. FEMS Microbiol Lett 243(1):141–147. doi:10.1016/j.femsle.2004.11.052

Peterson LW, Artis D (2014) Intestinal epithelial cells: regulators of barrier function and immune homeostasis. Nat Rev Immunol 14(3):141–153. doi:10.1038/nri3608

Pratt HF (1984) Breastfeeding and eczema. Early Hum Dev 9(3):283–290

Preidis GA, Versalovic J (2009) Targeting the human microbiome with antibiotics, probiotics, and prebiotics: gastroenterology enters the metagenomics era. Gastroenterology 136(6):2015–2031

Qin J, Li R, Raes J, Arumugam M, Burgdorf KS, Manichanh C, Nielsen T, Pons N, Levenez F, Yamada T, Mende DR, Li J, Xu J, Li S, Li D, Cao J, Wang B, Liang H, Zheng H, Xie Y, Tap J, Lepage P, Bertalan M, Batto JM, Hansen T, Le Paslier D, Linneberg A, Nielsen HB, Pelletier E, Renault P, Sicheritz-Ponten T, Turner K, Zhu H, Yu C, Li S, Jian M, Zhou Y, Li Y, Zhang X, Li S, Qin N, Yang H, Wang J, Brunak S, Dore J, Guarner F, Kristiansen K, Pedersen O, Parkhill J, Weissenbach J, Meta HITC, Bork P, Ehrlich SD, Wang J (2010) A human gut microbial gene catalogue established by metagenomic sequencing. Nature 464 (7285):59–65. doi:10.1038/nature08821

Rajilic-Stojanovic M, Biagi E, Heilig HG, Kajander K, Kekkonen RA, Tims S, de Vos WM (2011) Global and deep molecular analysis of microbiota signatures in fecal samples from patients

with irritable bowel syndrome. Gastroenterology 141(5):1792–1801. doi:10.1053/j.gastro.2011.07.043
Reigstad CS, Salmonson CE, Rainey JF III, Szurszewski JH, Linden DR, Sonnenburg JL, Farrugia G, Kashyap PC (2015) Gut microbes promote colonic serotonin production through an effect of short-chain fatty acids on enterochromaffin cells. FASEB J 29(4):1395–1403. doi:10.1096/fj.14-259598
Rhee SH, Pothoulakis C, Mayer EA (2009) Principles and clinical implications of the brain-gut-enteric microbiota axis. Nat Rev Gastroenterol Hepatol 6(5):306–314. doi:10.1038/nrgastro.2009.35
Ridaura VK, Faith JJ, Rey FE, Cheng J, Duncan AE, Kau AL, Griffin NW, Lombard V, Henrissat B, Bain JR, Muehlbauer MJ, Ilkayeva O, Semenkovich CF, Funai K, Hayashi DK, Lyle BJ, Martini MC, Ursell LK, Clemente JC, Van Treuren W, Walters WA, Knight R, Newgard CB, Heath AC, Gordon JI (2013) Gut microbiota from twins discordant for obesity modulate metabolism in mice. Science 341(6150):1241214. doi:10.1126/science.1241214
Rostagno MH (2009) Can stress in farm animals increase food safety risk? Foodborne Pathog Dis 6 (7):767–776. doi:10.1089/fpd.2009.0315
Rylander R (2010) Organic dust induced pulmonary disease—the role of mould derived beta-glucan. Ann Agric Environ Med 17(1):9–13
Saarinen UM, Kajosaari M (1995) Breastfeeding as prophylaxis against atopic disease: prospective follow-up study until 17 years old. Lancet 346(8982):1065–1069
Saito YA, Schoenfeld P, Locke GR III (2002) The epidemiology of irritable bowel syndrome in North America: a systematic review. Am J Gastroenterol 97(8):1910–1915. doi:10.1111/j.1572-0241.2002.05913.x
Sarowska J, Choroszy-Krol I, Regulska-Ilow B, Frej-Madrzak M, Jama-Kmiecik A (2013) The therapeutic effect of probiotic bacteria on gastrointestinal diseases. Adv Clin Exp Med 22 (5):759–766
Scharlau D, Borowicki A, Habermann N, Hofmann T, Klenow S, Miene C, Munjal U, Stein K, Glei M (2009) Mechanisms of primary cancer prevention by butyrate and other products formed during gut flora-mediated fermentation of dietary fibre. Mutat Res 682(1):39–53. doi:10.1016/j.mrrev.2009.04.001
Schrezenmeir J, de Vrese M (2001) Probiotics, prebiotics, and synbiotics – approaching a definition. Am J Clin Nutr 73(2 Suppl):361S–364S
Schwartz AV, Sellmeyer DE (2007) Diabetes, fracture, and bone fragility. Curr Osteoporos Rep 5 (3):105–111
Serafini U (1997) Do infections protect against asthma and atopy? Allergy 52(9):955–957
Sethi S, Garey KW, Arora V, Ghantoji S, Rowan P, Smolensky M, DuPont HL (2011) Increased rate of irritable bowel syndrome and functional gastrointestinal disorders after Clostridium difficile infection. J Hosp Infect 77(2):172–173. doi:10.1016/j.jhin.2010.09.024
Sokol H, Pigneur B, Watterlot L, Lakhdari O, Bermudez-Humaran LG, Gratadoux JJ, Blugeon S, Bridonneau C, Furet JP, Corthier G, Grangette C, Vasquez N, Pochart P, Trugnan G, Thomas G, Blottiere HM, Dore J, Marteau P, Seksik P, Langella P (2008) Faecalibacterium prausnitzii is an anti-inflammatory commensal bacterium identified by gut microbiota analysis of Crohn disease patients. Proc Natl Acad Sci USA 105(43):16731–16736. doi:10.1073/pnas.0804812105
Soyturk M, Akpinar H, Gurler O, Pozio E, Sari I, Akar S, Akarsu M, Birlik M, Onen F, Akkoc N (2007) Irritable bowel syndrome in persons who acquired trichinellosis. Am J Gastroenterol 102(5):1064–1069. doi:10.1111/j.1572-0241.2007.01084.x
Spiller RC, Jenkins D, Thornley JP, Hebden JM, Wright T, Skinner M, Neal KR (2000) Increased rectal mucosal enteroendocrine cells, T lymphocytes, and increased gut permeability following acute Campylobacter enteritis and in post-dysenteric irritable bowel syndrome. Gut 47 (6):804–811

St Jean AT, Swofford CA, Panteli JT, Brentzel ZJ, Forbes NS (2014) Bacterial delivery of Staphylococcus aureus alpha-hemolysin causes regression and necrosis in murine tumors. Mol Ther 22(7):1266–1274. doi:10.1038/mt.2014.36

Stephens RL, Tache Y (1989) Intracisternal injection of a TRH analogue stimulates gastric luminal serotonin release in rats. Am J Phys 256(2 Pt 1):G377–G383

Stephenson M, Rowatt E (1947) The production of acetylcholine by a strain of Lactobacillus plantarum. J Gen Microbiol 1(3):279–298

Strachan DP (1989) Hay fever, hygiene, and household size. BMJ 299(6710):1259–1260

Suliman S, Hemmings SM, Seedat S (2013) Brain-Derived Neurotrophic Factor (BDNF) protein levels in anxiety disorders: systematic review and meta-regression analysis. Front Integr Neurosci 7:55. doi:10.3389/fnint.2013.00055

Thomas CM, Hong T, van Pijkeren JP, Hemarajata P, Trinh DV, Hu W, Britton RA, Kalkum M, Versalovic J (2012) Histamine derived from probiotic Lactobacillus reuteri suppresses TNF via modulation of PKA and ERK signaling. PLoS One 7(2):e31951. doi:10.1371/journal.pone.0031951

Toussaint B, Chauchet X, Wang Y, Polack B, Le Gouellec A (2013) Live-attenuated bacteria as a cancer vaccine vector. Expert Rev Vaccines 12(10):1139–1154. doi:10.1586/14760584.2013.836914

Trompette A, Gollwitzer ES, Yadava K, Sichelstiel AK, Sprenger N, Ngom-Bru C, Blanchard C, Junt T, Nicod LP, Harris NL, Marsland BJ (2014) Gut microbiota metabolism of dietary fiber influences allergic airway disease and hematopoiesis. Nat Med 20(2):159–166. doi:10.1038/nm.3444

Tsigos C, Chrousos GP (2002) Hypothalamic-pituitary-adrenal axis, neuroendocrine factors and stress. J Psychosom Res 53(4):865–871

Tulic MK, Andrews D, Crook ML, Charles A, Tourigny MR, Moqbel R, Prescott SL (2012) Changes in thymic regulatory T-cell maturation from birth to puberty: differences in atopic children. J Allergy Clin Immunol 129 (1):199-206 e191-194. doi:10.1016/j.jaci.2011.10.016

Turnbaugh PJ, Backhed F, Fulton L, Gordon JI (2008) Diet-induced obesity is linked to marked but reversible alterations in the mouse distal gut microbiome. Cell Host Microbe 3(4):213–223. doi:10.1016/j.chom.2008.02.015

Turnbaugh PJ, Hamady M, Yatsunenko T, Cantarel BL, Duncan A, Ley RE, Sogin ML, Jones WJ, Roe BA, Affourtit JP, Egholm M, Henrissat B, Heath AC, Knight R, Gordon JI (2009a) A core gut microbiome in obese and lean twins. Nature 457(7228):480–484. doi:10.1038/nature07540

Turnbaugh PJ, Ley RE, Mahowald MA, Magrini V, Mardis ER, Gordon JI (2006) An obesity-associated gut microbiome with increased capacity for energy harvest. Nature 444 (7122):1027–1031. doi:10.1038/nature05414

Turnbaugh PJ, Ridaura VK, Faith JJ, Rey FE, Knight R, Gordon JI (2009b) The effect of diet on the human gut microbiome: a metagenomic analysis in humanized gnotobiotic mice. Sci Transl Med 1(6):6ra14. doi:10.1126/scitranslmed.3000322

Vael C, Nelen V, Verhulst SL, Goossens H, Desager KN (2008) Early intestinal Bacteroides fragilis colonisation and development of asthma. BMC Pulm Med 8:19. doi:10.1186/1471-2466-8-19

Van Felius ID, Akkermans LM, Bosscha K, Verheem A, Harmsen W, Visser MR, Gooszen HG (2003) Interdigestive small bowel motility and duodenal bacterial overgrowth in experimental acute pancreatitis. Neurogastroenterol Motil 15(3):267–276

van Odijk J, Kull I, Borres MP, Brandtzaeg P, Edberg U, Hanson LA, Host A, Kuitunen M, Olsen SF, Skerfving S, Sundell J, Wille S (2003) Breastfeeding and allergic disease: a multidisciplinary review of the literature (1966-2001) on the mode of early feeding in infancy and its impact on later atopic manifestations. Allergy 58(9):833–843

Verbrugghe E, Boyen F, Gaastra W, Bekhuis L, Leyman B, Van Parys A, Haesebrouck F, Pasmans F (2012) The complex interplay between stress and bacterial infections in animals. Vet Microbiol 155(2–4):115–127. doi:10.1016/j.vetmic.2011.09.012

Vickery BP, Chin S, Burks AW (2011) Pathophysiology of food allergy. Pediatr Clin N Am 58 (2):363–376., ix-x. doi:10.1016/j.pcl.2011.02.012

Villani AC, Lemire M, Thabane M, Belisle A, Geneau G, Garg AX, Clark WF, Moayyedi P, Collins SM, Franchimont D, Marshall JK (2010) Genetic risk factors for post-infectious irritable bowel syndrome following a waterborne outbreak of gastroenteritis. Gastroenterology 138(4):1502–1513. doi:10.1053/j.gastro.2009.12.049

Virchow R (1989) Cellular pathology. As based upon physiological and pathological histology. Lecture XVI—Atheromatous affection of arteries. 1858. Nutr Rev 47(1):23–25

Voth DE, Ballard JD (2005) Clostridium difficile toxins: mechanism of action and role in disease. Clin Microbiol Rev 18(2):247–263. doi:10.1128/CMR.18.2.247-263.2005

Vrieze A, Van Nood E, Holleman F, Salojarvi J, Kootte RS, Bartelsman JF, Dallinga-Thie GM, Ackermans MT, Serlie MJ, Oozeer R, Derrien M, Druesne A, Van Hylckama Vlieg JE, Bloks VW, Groen AK, Heilig HG, Zoetendal EG, Stroes ES, de Vos WM, Hoekstra JB, Nieuwdorp M (2012) Transfer of intestinal microbiota from lean donors increases insulin sensitivity in individuals with metabolic syndrome. Gastroenterology 143(4):913–916. e917. doi:10.1053/j.gastro.2012.06.031

Wang T, Cai G, Qiu Y, Fei N, Zhang M, Pang X, Jia W, Cai S, Zhao L (2012) Structural segregation of gut microbiota between colorectal cancer patients and healthy volunteers. ISME J 6(2):320–329. doi:10.1038/ismej.2011.109

Webster Marketon JI, Glaser R (2008) Stress hormones and immune function. Cell Immunol 252 (1–2):16–26. doi:10.1016/j.cellimm.2007.09.006

Weinstein LI, Revuelta A, Pando R (2015) Catecholamines and acetylcholine are key regulators of the interaction between microbes and the immune system. Ann N Y Acad Sci:1–13. doi:10.1111/nyas.12792

WHO (2014) Fact sheet: obesity and overweight. http://www.who.int/mediacentre/factsheets/fs311/en/. Accessed 9 July 2015

Wold AE (1998) The hygiene hypothesis revised: is the rising frequency of allergy due to changes in the intestinal flora? Allergy 53(46 Suppl):20–25

Wood LM, Paterson Y (2014) Attenuated Listeria monocytogenes: a powerful and versatile vector for the future of tumor immunotherapy. Front Cell Infect Microbiol 4:51. doi:10.3389/fcimb.2014.00051

Xiao H, Gulen MF, Qin J, Yao J, Bulek K, Kish D, Altuntas CZ, Wald D, Ma C, Zhou H, Tuohy VK, Fairchild RL, de la Motte C, Cua D, Vallance BA, Li X (2007) The Toll-interleukin-1 receptor member SIGIRR regulates colonic epithelial homeostasis, inflammation, and tumorigenesis. Immunity 26(4):461–475. doi:10.1016/j.immuni.2007.02.012

Xu XR, Liu CQ, Feng BS, Liu ZJ (2014) Dysregulation of mucosal immune response in pathogenesis of inflammatory bowel disease. World J Gastroenterol 20(12):3255–3264. doi:10.3748/wjg.v20.i12.3255

Yang G, Liu ZQ, Yang PC (2013) Treatment of allergic rhinitis with probiotics: an alternative approach. N Am J Med Sci 5(8):465–468. doi:10.4103/1947-2714.117299

Yatsunenko T, Rey FE, Manary MJ, Trehan I, Dominguez-Bello MG, Contreras M, Magris M, Hidalgo G, Baldassano RN, Anokhin AP, Heath AC, Warner B, Reeder J, Kuczynski J, Caporaso JG, Lozupone CA, Lauber C, Clemente JC, Knights D, Knight R, Gordon JI (2012) Human gut microbiome viewed across age and geography. Nature 486 (7402):222–227. doi:10.1038/nature11053

Zanini B, Ricci C, Bandera F, Caselani F, Magni A, Laronga AM, Lanzini A, San Felice del Benaco Study Investigators (2012) Incidence of post-infectious irritable bowel syndrome and functional intestinal disorders following a water-borne viral gastroenteritis outbreak. Am J Gastroenterol 107(6):891–899. doi:10.1038/ajg.2012.102

Zhang J, Motyl KJ, Irwin R, MacDougald OA, Britton RA, McCabe LR (2015) Loss of bone and Wnt10b expression in male type 1 diabetic mice is blocked by the probiotic L. reuteri. Endocrinology. 156(9): 3169–3182. doi:10.1210/EN.2015-1308

Zhu Q, Gao R, Wu W, Qin H (2013) The role of gut microbiota in the pathogenesis of colorectal cancer. Tumour Biol 34(3):1285–1300. doi:10.1007/s13277-013-0684-4

Zijlmans MA, Korpela K, Riksen-Walraven JM, de Vos WM, de Weerth C (2015) Maternal prenatal stress is associated with the infant intestinal microbiota. Psychoneuroendocrinology 53:233–245. doi:10.1016/j.psyneuen.2015.01.006

Chapter 5
Buruli Ulcer: Case Study of a Neglected Tropical Disease

M. Eric Benbow, Rachel Simmonds, Richard W. Merritt, and Heather R. Jordan

Abstract Neglected tropical diseases affect almost all human communities in rural areas of mostly developing nations. They have staggering negative effects on human health and local, regional, and national economies through mortality and morbidity. These diseases are neglected in many large-scale disease management and control programs and therefore do not recieve the research and funding attention of diseases with higher pharmaceutical potential. One such disease that epitomozies this situation is Buruli ulcer disease, also known as *Mycobacterium ulcerans* infection. This necrotizing skin disease results in severe and lasting morbidity that primarily affects children in rural regions of Africa and other tropical and subtropical regions. It is caused by a mycobacterium related to other pathogens that are the agents for two other diseases, leprosy and tuberculosis; however, this pathogen secretes myolactone which is a cytotoxic molecule that is both necrotizing and immunodepressive and is unique within its phylogeny. As a neglected tropical disease, research and funding has generally been sporadic and diffuse among countries and agencies, limiting scientific gains in better understanding some basic disease system tenants such as the mode of transmission and where the pathogen grows and replicates in the environment. These limitations compounded

M. Eric Benbow (✉)
Department of Entomology, Michigan State University, East Lansing, MI 48824, USA

Department of Osteopathic Medical Specialties, Michigan State University, East Lansing, MI 48824, USA
e-mail: benbow@msu.edu

R. Simmonds
Faculty of Health and Medical Sciences, Department of Microbial and Cellular Sciences, University of Surrey, Guildford, UK
e-mail: rachel.simmonds@surrey.ac.uk

R.W. Merritt
Department of Entomology, Michigan State University, East Lansing, MI 48824, USA
e-mail: merrittr@msu.edu

H.R. Jordan
Department of Biological Sciences, Mississippi State University, Mississippi State, MS, USA
e-mail: jordan@biology.msstate.edu

C.J. Hurst (ed.), *Modeling the Transmission and Prevention of Infectious Disease*, Advances in Environmental Microbiology 4, DOI 10.1007/978-3-319-60616-3_5

with the fact that it focally affects rural and poor populations have made the control of Buruli ulcer disease challenging. Further, disease emergence and reemergence is thought to be associated with landscape modifications such as deforestation, dam construction, farming, and mining, coupling this disease with degraded environmental conditions that may faciliate either the emergence or sustainability of other water-related diseases of the rural poor. This chapter generally reviews Buruli ulcer disease within the context of neglected tropical diseases in a way that integrates the research that occurs at the molecular and cellular level of pathogen and host investigation with broader ecosystem factors that include other biological interactions (e.g., food webs) considered to be important to elucidating transmission of the pathogen, all of which must be assessed in combination to achieve successful future disease management activities.

5.1 Introduction

5.1.1 Neglected Tropical Diseases

Neglected tropical diseases (NTDs) represent infectious diseases that primarily affect rural communities of developing countries in tropical regions of the world (Moncayo and Yanine 2007; Feasey et al. 2010). While not necessarily considered "research neglected," these diseases have historically been overlooked by funders and policy makers, in part because of insufficient market incentives for the pharmaceutical industry, and so these diseases have remained neglected in the sense that there are few drug-based efforts to control and manage them (Moncayo and Yanine 2007). Even when there are pharmaceutical means and cost-effective tools for better control, often these practices either are not widely available or are too expensive to distribute or implement (Feasey et al. 2010).

The global burden of NTDs is difficult to estimate, but some sources suggest that 1/6 of the Earth's population is affected by one or more of these diseases—over a billion human beings worldwide suffer from NTDs (Hotez et al. 2009), and the World Health Organization estimates that another million is at risk (WHO 2006). The WHO also reports that 100% of low-income countries are affected by five of these diseases simultaneously (WHO 2006), suggesting that the NTD burden is compounding in the poorest parts of the world.

Of a total of 13 core NTDs (Feasey et al. 2010), there are several that represent the predominate global burden on health in developing nations (Table 5.1), with staggering estimates of disease-adjusted life years (DALYs) and widespread socioeconomic effects at local, national, regional, and continent-wide scales (Conteh et al. 2010; WHO 2004a; Mathers et al. 2007). As a calculation, DALY = YLD + YLL, where YLL is years of life lost and YLD is years lived with disability. The burden of one DALY can be considered 1 year of healthy life lost due to death, illness, or disability. Thus, while many NTDs do not afford large death rates

Table 5.1 Some of the most common non-malaria NTDs with associated prevalence, number of deaths, and disease-adjusted life years (DALY)

Disease	Prevalence (×1000)	# Deaths (×1000)	DALY (×1000)
Lymphatic filariasis	60,524	0	5777
Leishmaniasis	3665	51	2090
Schistosomiasis	248,248	15	1702
Trypanosomiasis	200	48	1525
Chagas disease	10,137	14	667
Dengue hemorrhagic fever	75	19	616
Trichuriasis	26,624	900	1006
Hookworm	59,999	3	973
Onchocerciasis	2296	0	484
Leprosy	903	6	198
Buruli ulcer[a]	6	0	Not reported

The DALY represents the years of healthy life lost due to either death, or illness, or disability. Values given for all variables should be multiplied by 1000. Table modified from Table 3 of Mathers et al. (2007)

[a]Data taken from www.who.int/buruli/en/

compared to infectious diseases such as malaria, collectively, they have significant impacts on morbidity and economic development potential in the areas and countries where many overlap in occurence. These developing countries suffer from historic and persistent economic disadvantages that faciliate the sustained burden, emergence, and reemergence of NTDs globally.

Many NTDs are associated with inadequate access to safe water, poorly developed sanitation, and lack of stable housing. The pathogens responsible for NTDs vary taxonomically and can be found in a wide range of habitats often correspondingly associated with poor water quality conditions and either poor or nonexistent sanitation services, making pathogen control and management efforts to be difficult in rural areas where NTDs are widespread. The disease agents include vectorborne viruses, protozoa, filarial worms, helminth worms, and bacteria, including nontuberculosis mycobacteria (Feasey et al. 2010). These diseases represent a range of modes of transmission, including vectorborne (e.g., Chagas disease and dengue fever), airborne (e.g., tuberculosis), and environmental exposure through ingestion (e.g., trichuriasis) and skin penetration (e.g., schistosomiasis, helminths, and some mycobacteria). For this chapter we will focus on mycobacterial diseases and first introduce those representing known transmission pathways, such as tuberculosis and leprosy. We will then fully expand on a less common mycobacterial infection where the mode of transmission has been rigorously investigated, yet remains inconclusive. We will provide a review of the general aspects of this disease and a historical and ongoing context for how the study of this NTD has been approached in a way to demonstrate the pathways for understanding a rapidly emerging and relatively unknown NTD.

5.1.2 Mycobacterial Diseases

The genus *Mycobacterium* includes a diverse group of species, ubiquitous in nature, that have been found in soil, dust, rocks, bioaerosols, water, and both human and animal hosts (Falkinham III 2009; Lakhanpal et al. 2011; Comas et al. 2013; Coscolla et al. 2013; Coscolla and Gagneux 2014). Because of their relatively slow growth and cellular impermeability, mycobacteria are good competitors in nutrient-poor environments. Polluted sites may be ideal habitats because these microbes can metabolize a variety of major groundwater hydrocarbon pollutants (Heitkamp et al. 1988; Burback and Perry 1993) and attach within biofilms to particulates where pollutants are concentrated (Stelmack et al. 1999). Mycobacterial ability to thrive within biofilms is in part due to the microbe-rich outer membrane of mycolic acids that contributes to both their hydrophobicity and impermeability (Recht and Kolter 2001; Hall-Stoodley et al. 2006; Yamazaki et al. 2006a, b; Feazel et al. 2009; Cook et al. 2010; Ojha et al. 2010; Bosio et al. 2012; Islam et al. 2012). Members within this genus are taxonomically divided by their growth rate. Rapid growers are identified by their ability to show visible colonies within 7 days, while slow growers require longer periods for visible growth observation (Stahl and Urbance 1990). Due to many of these environmental fitness advantages, many mycobacterial species are successful pathogens of humans and animals leading to high morbidity and mortality worldwide.

5.1.2.1 Mycobacterium Tuberculosis Complex (MTBC)

Tuberculosis (TB) is caused by members of the mycobacterium tuberculosis complex (MTBC). Members of this complex include human pathogens such as *M. tuberculosis* and *M. africanum* (members of the MTB complex, MTBC) as well as the zoonotic pathogens, *M. bovis*, *M. pinnipedii*, *M. microti*, *M. caprae*, and *M. canettii* (van Soolingen et al. 1994; Brosch et al. 2002; Huard et al. 2006; Alexander et al. 2010). Tuberculosis remains one of the most prolific human diseases worldwide. Among several risk factors of these MTBC species as facultative host pathogens include the ease of their transmissibility either from person-to-person or animal-to-person via aerosol droplets and their capability of persisting in human hosts for many years where they parasitize macrophages and successfully inhibit immune responses such as phagosome-lysosome fusion (Vergne et al. 2004a, b; Unknown 2015).

A 2015 report by the World Health Organization showed that TB is present in at least 202 countries and territories where approximately 9.6 million people developed TB in 2014, with 1.5 million deaths (WHO 2015). Of those deaths, 400,000 were among HIV positive patients (WHO 2015). These data, however, do show a declining mortality rate and an improved surveillance and disease burden as compared to previous years (WHO 2015). While promising, multidrug resistance and poor treatment outcomes remain of major concern.

5.1.2.2 *Mycobacterium leprae*

Mycobacterium leprae is the etiological agent of leprosy, a chronic infectious disease that has been reported for thousands of years. *Mycobacterium leprae* is believed to be spread from person-to-person by aerosol dropolets; however prevalence is low in part due to natural resistance to infection (Lastoria and Abreu 2014b). Clinical manifestations depend upon the immune response of the host and are preceded by a long incubation period of approximately 5 years, where progression to disease may take as long as 20 years (Lastoria and Abreu 2014b). The disease mainly affects the skin, the peripheral nerves, mucosa of the upper respiratory tract, and also the eyes (Lastoria and Abreu 2014b), which, if left untreated, can lead to permanent damage.

Besides humans, natural *M. leprae* infection has been reported in soil, water, chimps, apes, and the armadillo (Turankar et al. 2012; Lastoria and Abreu 2014b). It is thought that the armadillo is a good host due to its low body temperature, and armadillos have become the source of *M. leprae* for laboratory studies including their use for bacterial propogation and the development of vaccine candidates (Sharma et al. 2013).

Historically, fear and stigma have been associated with leprosy, and patients were once isolated to leper colonies as a strategy to prevent spread of the disease. Promin, a sulfone drug, was introduced in the 1940s to treat leprosy, and its effectiveness led to decreased patient isolation (Lastoria and Abreu 2014a). The WHO began recommending multidrug therapy including a combination of dapsone, rifampicin, and clofazimine in 1981 for the treatment of leprosy, and along with early diagnosis, multidrug therapy has drastically reduced the number of cases (Lastoria and Abreu 2014a; Unknown 2014). In fact, leprosy has been eliminated from 119 countries out of 122 countries where the disease was once considered as a public health problem (Unknown 2014).

5.1.2.3 Other Nontuberculosis Mycobacterial (NTM) Infections

Over the past several decades, NTM have increasingly emerged as important pathogens of humans and animals. Many human infections are a result of the hospital settings. Contaminated instruments or solutions have been associated with disease caused by rapidly growing mycobacteria (*M. abscessus* and *M. fortuitum*) and also slow-growing species (*M. avium* complex, *M. gordonae*, and *M. xenopi*) (Kressel and Kidd 2001; Phillips and von Reyn 2001). Those individuals who are most at risk include patients who either are immunosuppressed or have been administered a long course of antimicrobial therapy (Swanson 1998; Reilly and McGowan 2004).

Mycobacterium marinum is the most common NTB mycobacterium that causes opportunistic infections in humans (Collins et al. 1985; Rallis and Koumantaki-Mathioudaki 2007). *Mycobacterium marinum* is categorized as a slow-growing mycobacterium and a natural pathogen of ectotherms but was recognized as the causative agent of human disease in 1951 among swimmers (Norden and Linell 1951). Due to the organisms' restricted growth temperature of 30–32 °C, most infections are cutaneous, though deep tissue infections and osteomyelitis may also occur (Hurst et al. 1987; Lewis et al. 2003; Tran et al. 2008; Jacobs et al. 2012; Al-Anazi et al. 2014; Kamel et al. 2014; Nguyen et al. 2015). Infection is generally associated either with contact with fish or *M. marinum*-contaminated environments, and increased risk is usually associated with fish-related occupations or hobbies (Norden and Linell 1951; Linell and Norden 1954; Collins et al. 1985; Huminer et al. 1986; Kullavanijaya et al. 1993). Infection by this microbial species can be treated either with a wide range of antibiotics or application of heat. In the following sections, we will turn our attention to *M. ulcerans*, a close genetic relative of *M. marinum*, and the causative agent of Buruli ulcer, a NTD leading to high morbidity in infected individuals.

5.2 *Mycobacterium ulcerans* Infection: Buruli Ulcer Disease

One model of NTD, *M. ulcerans* infection, has received intense research interest over the last 20 years and is the third most frequently reported mycobacterial disease in humans (WHO 2000a), and it is a unique infection type from the mycobacterial complex that otherwise broadly is responsible for tuberculosis, leprosy, and fish and amphibian mycobacteriosis. Thus, this NTD represents an association between the evolution and ecology of a mycobacterial phylogeny that has produced multiple microbial species which present different forms of disease among different host species. We consider that this evolutionary history of mycobacterial diseases provides a provocative opportunity to explore the biology and ecology of a highly neglected tropical disease, providing insight into a model disease system that serves to represent other NTDs associated with poor water quality conditions and limited access of hosts to water treatment resources. The disease is known as Buruli ulcer.

Buruli ulcer disease (BU) is also known by its infectious agent as *Mycobacterium ulcerans* infection and has been reported from over 30 countries (Merritt et al. 2010; Röltgen and Pluschke 2015). Buruli ulcer is considered an emerging disease that presents in the form of chronic and necrotizing skin lesions, and while the death rate associated with BU is low, there is severe morbidity that places extreme social and economic hardship on families with infected members that require long-term treatment and care (Asiedu and Etuaful 1998). Highest rates of this disease have been reported from areas of West Africa that are low-lying such as floodplains, typically with complex topography, and most prominently in rural villages within a landscape matrix of agriculture (Merritt et al. 2005, 2010; Röltgen and Pluschke 2015). The route of transmission has been intensively investigated, but it is not entirely certain how humans acquire *M. ulcerans* leading to infection (Merritt et al. 2010).

5.2.1 History

Some of the first cases of BU were described as early as 1897 by Sir Albert Cook from African patients (Horsburgh and Meyers 1997; WHO 2000a). In the early twentieth century, there were also several reports by a missionary physician in the Congo of patients with skin lesions heavily infected with acid-fast bacilli, presumably mycobacteria (WHO 2000a). However, a clinical verification of BU was not published until the late 1930–1940s in Australia, where the disease was called Bairnsdale ulcer (MacCallum et al. 1948). During these early documentations of potential BU in Africa and Australia, the symptoms inspired other names for the disease including Searles ulcer, Kumusi ulcer, and even the "mysterious disease," the latter still being used in more popular descriptions of this necrotozing skin lesion disease, referring to elusive transmission mechanisms. The name Buruli ulcer was finally adopted by the Uganda Buruli Group providing extensive clinicopathological investigations during an outbreak of cases in the 1960s and 1970s which occurred in the Buruli district near Lake Kyoga, Uganda (Uganda Buruli Group 1971; WHO 2000a; Merritt et al. 2005). During this time there were also cases occuring throughout other parts of Africa that since have been reviewed and summarized elswhere (Clancey 1964; Janssens 1972; Meyers et al. 1974a). After these early reports, there was a rapid increase in reporting of BU disease indicating new and expanding emergence patterns in Africa, Australia, Papua New Guinea, and elsewhere.

In 1988, the new reports and perceived emergence of BU from around the world prompted the World Health Organization (WHO) to establish the Global Buruli Ulcer Initiative (GBUI). A resolution by the 57th World Health Assembly in 2004 initiated and has since continued to encourage member states to participate in the GBUI to intensify research on this NTD (WHO 2004b, c). Since the signing of that resolution, the numbers of research studies that have been conducted on BU have rapidly increased as has the interdisciplinary nature of the reseach that has been undertaken (Merritt et al. 2010; Röltgen and Pluschke 2015). The research on BU over the last 15 years is in many ways representative of the geographic distribution over which the disease has been reported.

5.2.2 Global Distribution and Burden

Historically, cases of BU have been reported from 33 countries on the continents of Africa, Australia, Asia, and South America, with most of the sustained reporting coming from regions of West Africa (Fig. 5.1). While there have been incidences of cases associated with international travel (McGann et al. 2009; Thomas et al. 2014), there have been no reports of BU cases being acquired either in Europe or North America. In any one year, cases of BU typically are reported from between 12 and 16 countries with 10 of these countries having shown relatively constant numbers of

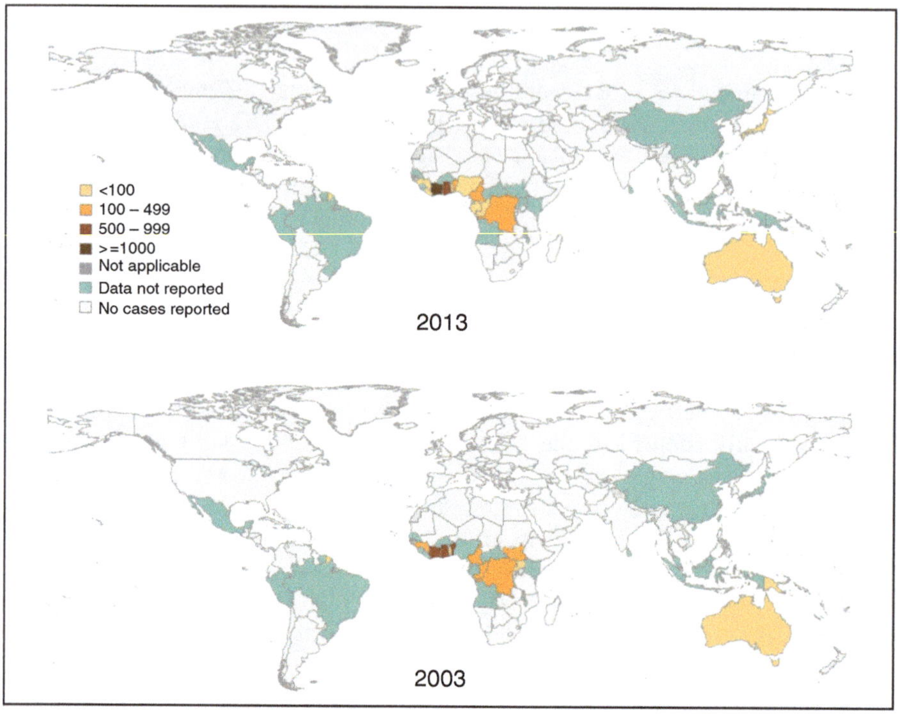

Fig. 5.1 Global distribution of total BU disease cases in 2003 and 2013. Data were downloaded and mapped using the WHO Global Health Observatory Data Repository on 11 February 2015 (http://www.who.int/gho/neglected_diseases/buruli_ulcer/en/)

cases over the last 10 years (Table 5.2). From 2003 to 2013, nearly 40% of the cases reported to the WHO were from Côte d'Ivoire. When additionally considering cases of BU reported in Ghana and Benin, these three countries have represented 75% of the BU cases worldwide over the last 10 years, indicating that while widespread, the global burden of BU falls on the developing nations of West Africa.

On a local scale, the incidence of BU cases geographically is quite focal and even among villages in the same area within a larger geographic region (e.g., <100 km^2) (van Ravensway et al. 2012; Williamson et al. 2012b). For instance, in Benin most BU cases are found in the southern region of the country primarily clustered in the Couffo, Mono, Zou and Atlantique districts; however, there are many villages north of those clusters that report cases in widely separated villages or communes (Sopoh et al. 2007, 2011; Wagner et al. 2008a, b). Even within the southern clusters, villages closer together do not necessarily always report cases of BU. For instance, a study (Williamson et al. 2012b) of both endemic and nonendemic villages of Benin that had undergone active case surveillance within 20 km of each other showed differential prevalence, suggesting that even at very local scales, not all villages are at risk for BU. The reasons for this are unclear and are currently the focus of many research teams. In regions where there is persistent endemicity, as in

Table 5.2 The number of Buruli ulcer cases reported to the WHO worldwide over 10 years from 2003 to 2013

Country	2003	2004	2005	2006	2007	2008	2009	2010	2011	2012	2013	10-year total
Côte d'Ivoire	768	1153	1564	1872	2191	2242	2679	2533	1659	1386	1039	19,086
Ghana	737	1157	1005	1096	668	986	853	1048	971	632	550	9703
Benin	722	925	1045	1195	1203	897	674	572	492	365	378	8468
Cameroon	223	914	265	271	230	312	323	287	256	160	133	3374
Democratic Republic of the Congo	119	487	51	74	340	260	172	136	209	284	214	2346
Togo	38	800	317	40	141	95	52	67	52	51	37	1690
Congo	180	235	53	370	99	126	147	107	56	38	6	1417
Guinea	157	146	208	279		80	61	24	59	82	96	1192
Australia	14	34	47	72	61	40	35	42	143	105	74	667
Gabon		43	91	54	32	53	41	65	59	45	59	542
South Sudan	360	4	24	38	8	3	5	4				446
Uganda	10	7	72	5	31	24	3					152
Papua New Guinea	18	31			26	24	8	5	8			120
Nigeria				9			24	7	4	40	23	107
Japan		1	1	1	3	2	5	9	10	4	10	46
Liberia										21	8	29
Sierra Leone						1			28			29
Central African Republic						3						3
Equatorial Guinea			3									3

Data were downloaded from the WHO Global Health Observatory Data Repository on 11 February 2015 (http://www.who.int/gho/neglected_diseases/buruli_ulcer/en/)

Côte d'Ivoire, up to 16% of rural villages may be burdened by the disease (Marston et al. 1995).

Australia is also consistently reporting BU but from very different populations. Between 1940 and 1975, approximately 40–45 cases of BU were reported in Australia mainly from two general locations: (1) the temperate regions of Southeastern Australia (Bratschi et al) and (2) the tropical regions of Northeastern Australia (Queensland) (Radford 1975). New emergences of BU infections appeared in the 1980s and 1990s near Melbourne, Victoria. Prior to the 1980s, the region around Melbourne had no incidence of BU and was separated from the endemic region of Bairnsdale by approximately 150–200 km (Johnson et al. 1996). Cases in the Melbourne area geographically remain focal but have been routinely reported and show substantial annual variation over time (Fyfe et al. 2010; Lavender et al. 2011; Carson et al. 2014; Huang and Johnson 2014). This annual variation in magnitude (Table 5.2), but consistent case detection, at the country level suggestions that while highly focal within a country or region, there are generally consistent factors that maintain the disease in those human populations historically endemic for BU. Intense investigation over several scales of inquiry has ranged from the entire country surveys to within a single village or town: these investigations are reviewed below.

5.2.3 Mycobacterium ulcerans *Microbiology and Genetics*

Mycobacterium ulcerans (MU) is a slow-growing, acid-fast bacterium that produces a toxic macrolide called mycolactone, and the toxin acts by causing mammalian cells to arrest in G1 of the cell replicative cycle after which the cells die by apoptosis (George et al. 2000). Mycolactone diffuses through infected tissue resulting in pathology, which extends further than the site of bacterial colonization (George et al. 1999, 2000). Genes for mycolactone synthesis are encoded on a large plasmid (Stinear et al. 2005). The structure of mycolactone consists of a 12-membered macrolide core structure and fatty acid side chains (George et al. 1999). The core structure is conserved in all isolates of MU; however, variation in the side chain produces different congeners of mycolactone with differential virulence (Mve-Obiang et al. 2003). Acquisition of the plasmid and mycolactone production are believed to be responsible for the severity of disease as the toxin is responsible for tissue necrosis in patients (Adusumilli et al. 2005; En et al. 2008; Sarfo et al. 2010b; Gama et al. 2014). Previous studies suggested that mycolactone production was restricted to MU (Daniel et al. 2004). However, species of mycolactone-producing mycobacteria (MPMs) have been isolated that produce mycolactones with unique side chains. Mycolactone and its mechanisms of virulence will be discussed in detail in a later section.

In infected humans, MU is found primarily in necrotic cutaneous tissues devoid of inflammatory cells (Adusumilli et al. 2005). Though recent studies have shown phases of both intracellular and extracellular survival and reproduction

(Coutanceau et al. 2005; Torrado et al. 2007b), the infection is primarily extracellular in contrast to other mycobacterial infections. Cross-reactivity with mycobacterial antigens from environmental mycobacteria and *M. tuberculosis* makes the creation of serological tests for MU difficult (Diaz et al. 2006; Pidot et al. 2010; Yeboah-Manu et al. 2012).

Mycobacterium ulcerans grows optimally at 30 °C with a doubling time ranging from 25 to 84 h under laboratory conditions on standard mycobacterial media (Mve-Obiang et al. 2003). Because of the organisms's slow growth, it is difficult to maintain samples for culture in areas where laboratory resources are limited. Infection rarely leads to systemic disease due to its temperature-restricted growth (Pszolla et al. 2003; Pommelet et al. 2014).

Mycobacterium ulcerans is very closely related genetically to the fish pathogen *M. marinum* that causes a granulomatous skin infection in humans (Stinear et al. 2007). *Mycobacterium ulcerans* shares 99% DNA identity to *M. marinum*, although MU was found to have a reduced genome (Raghunathan et al. 2005). In fact, MU differs in only one dimorphic position in the 16s rRNA gene, with no difference in the 16s–23s internal transcribed spacer region (Roth et al. 1998). Sequencing of the MU plasmid found that 60% of the plasmid is composed of three polyketide synthase genes and three accessory genes that are responsible for mycolactone production (Stinear et al. 2005). There are also 26 copies of insertion sequences including IS*2404* and IS*2606* that are also found in high number on the chromosome. Comparative genomic studies showing similarities of all MPM, including MU to *M. marinum*, as well as proliferation of IS*2404* and IS*2606* suggest recent evolution from a *M. marinum*-like progenitor, and genome reduction of *M. ulcerans* suggests possible adaptations to some particular lifestyle and environment (Stinear et al. 2007). This hypothesis has been reinforced by the identification of either deleted or inactivated genes required for pigment biosynthesis and anaerobiosis suggesting the microbe has been adapting to a dark and aerobic environment (Roltgen et al. 2012).

Recently, whole genome comparisons have allowed the identification of single nucleotide polymorphisms (SNPs) that allow high-resolution and strain differentiation into haplotypes (Kaser et al. 2009a; Roltgen et al. 2010). Data from SNP typing among strains isolated in endemic regions has shown a focal distribution and geographical restriction of MU haplotypes, providing insight into transmission (Kaser et al. 2009a; Roltgen et al. 2010). Studies of sequence polymorphisms have also shown that continuous MU evolution has led to at least two distinct phylogenetic lineages (Kaser et al. 2009a, b). Members of the classical lineage include strains from human and other animal isolates from Africa and Australia (Doig et al. 2012). Data using a universal clock rate suggest that the classical lineage diverged approximately 400,000 years ago from the ancestral lineage (Qi et al. 2009). Members of the ancestral lineage include human isolates from South America, Asia, and Mexico and globally distributed fish and frog isolates (Doig et al. 2012) that are genetically closer to *M. marinum* and less virulent than those strains considered as representing the classical lineage (Kaser et al. 2007; Doig et al. 2012). Recent whole genome studies of MU isolates from ten endemic villages reveal many different genotypes across a region and support the mobility of the pathogen and potential reservoirs (Ablordey et al. 2015).

5.2.4 Diagnosis and Surveillance

5.2.4.1 Diagnosis

Currently there is no point of care test for the diagnosis of BU; however, four tests have been employed for clinical diagnosis: direct smear and microscopy for the determination of the presence of acid-fast bacilli, histopathological examination, polymerase chain reaction targeting MU DNA, and bacterial culture (Beissner et al. 2010). The WHO recommends that any two of these laboratory tests be positive in order to assess BU diagnosis and confirmation (Beissner et al. 2010; Sakyi et al. 2016).

Each diagnostic approach has advantages and limitations depending upon the sample and laboratory setting (Table 5.3). Direct smear of ulcers or nodules for acid-fast bacilli detection and histopathological examination are widely used diagnostics in rural areas where resources are limited and are typically accurate when a patient presents from a BU endemic region. Though rapid results are possible, contamination is often an issue when samples are taken in rural areas, and training is required to stain and detect organisms. Additionally, histopathology requires invasive sampling to obtain a biopsy, and samples must be preserved in formalin, which often is of limited supply in rural endemic areas. Neither of these methods can be used to identify viable organisms, nor are they useful for differentiation between strains and they cannot distinguish between recurrence and reinfection. Culturing MU is the most specific method of diagnosis; however this method has low sensitivity and results often take 6–8 weeks (Phillips et al. 2005). Additionally, aseptic technique is necessary to avoid contamination by faster- growing organisms.

Laboratory diagnosis is increasingly relying upon PCR targeting *IS*2404, a genetic target found in many copies on the MU chromosome and plasmid responsible for mycolactone production. This method has high sensitivity and specificity and presently is considered the ideal standard. Currently, the WHO recommends PCR confirmation of at least 50% of samples from suspected patients prior to administering chemotherapeutic agents (Beissner et al. 2010). Due to the extended presence of mycobacterial DNA under antimycobacterial treatment, however, PCR is not suitable for monitoring treatment success as culture is considered the only valid confirmatory test for the detection of viable bacilli (Beissner et al. 2010) (Table 5.3). Depending upon the genetic target, PCR can also be used to distinguish between MU strains, allowing for molecular epidemiology to track chains of transmission with the possibility of differentiating between recurrence and reinfection (Lavender et al. 2008; Williamson et al. 2008, 2012b). Importantly, PCR is a useful tool for diagnosis only when adequate resources are available, and it is

Table 5.3 Advantages and disadvantages of laboratory methods for BU diagnosis

Diagnostic method	Advantages	Disadvantages
Direct smear for acid-fast bacilli	– Rapid – Low-cost reagents – Can be used on swab samples and fine-needle aspirates – Mycobacterial specific	– Low sensitivity (40%) – Cannot distinguish between mycobacteria species – Training required – Does not distinguish between viable and nonviable organisms – Cannot distinguish between *M. ulcerans* strains
Histopathology	– Rapid – Sensitive (60–90%)	– Expensive – Does not distinguish between viable and nonviable organisms – Training required – Requires invasive sampling – Cannot distinguish between *M. ulcerans* strains
Polymerase chain reaction	– Sensitive and specific (90–100%) – Distinguishes between strains depending upon target – Can be used on swab samples and fine-needle aspirates	– Expensive – Requires strict quality control – Requires a sophisticated laboratory – Requires highly trained personnel – Requires isolation of DNA – Cannot distinguish between viable and nonviable organisms
Culture	– Specific (100%) – Distinguishes between viable and nonviable organisms – Can be used on swab samples and fine-needle aspirates	– Low sensitivity (30–50%) – Training required – Requires dedicated laboratory – Slow results (6–8 weeks)

important to note that many laboratories presently servicing affected populations, particularly in West Africa, presently lack the necessary resources for PCR testing such as adequate power supply and necessary equipment. Furthermore, many laboratories in West Africa are located where samples must be shipped from great distances.

Loop-mediated isothermal amplification (Yeboah-Manu et al. 2012) is a method showing great potential for field use. The assay is highly specific due to the recognition of target DNA by several independent sequences and is reported to have sensitivity and specificity equivalent to that of PCR (Ablordey et al. 2012; de Souza et al. 2012). A recent study found the LAMP method could be used to rapidly identify MU in clinical specimens (Ablordey et al. 2012; de Souza et al. 2012). Though promising, this method is not yet field ready, however as obtaining purified DNA, as well as generating isothermal conditions, remains a major challenge (Ablordey et al. 2012).

The broad antigenic cross-reactivity among mycobacterial species is also a challenge for developing a point of care diagnostic. Several MU-specific antigens have been identified, though studies were unable to distinguish between patients and exposed controls (Diaz et al. 2006; Pidot et al. 2010; Yeboah-Manu et al. 2012).

Recent work by Dreyer et al. has identified a MU protein, MUL_3720, as a promising target for the development of a diagnostic antigen capture assay as MUL_3720 demonstrates no cross-reactivity with other pathogenic mycobacteria normally found in BU endemic regions. Furthermore, a recent study was able to generate mycolactone-specific immune sera and monoclonal antibodies, showing new promise for diagnostics development (Dangy et al. 2016).

5.2.4.2 Surveillance

Strategies for BU control emphasize early detection. Many countries rely upon passive case detection where cases are reported upon presentation to a hospital. Often, this method leads to underreporting as there is a high stigma associated with this disease, and hospitals are often at great distances. Furthermore, this has been quite challenging due to the vast number of other skin infections that may exhibit symptoms similar to those known for BU. Buruli ulcer also remains a non-notifiable disease in at least two-dozen countries (Unknown 2002, 2005).

Studies identifying factors contributing to delayed presentation indicate that awareness of BU is generally good in endemic areas (Renzaho et al. 2007; Webb et al. 2009). However, there is still a high level of stigma associated with BU as many perceive transmission, treatment, and disease pathology to be associated with sorcery (Renzaho et al. 2007; Ackumey et al. 2012a, b). Some patients either seek medical attention from traditional healers or attempt to treat the disease themselves, and most of these instances are not reported to the ministries of health (Ackumey et al. 2012a, b; Akoachere et al. 2016). Despite these obstacles, BU control programs that include the implementing of public awareness campaigns have proven successful in decreasing stigma associated with the disease (Ackumey et al. 2012a, b). Furthermore, these health promotion programs have disseminated information and a better understanding of the efficacy and availablility of antibiotic and other treatment therapies, as well as increased awareness of the financial burden concerns which result from disease-related losses in societal productivity (Ackumey et al. 2011a, b; Abass et al. 2015).

5.2.5 Pathology

The first sign of skin infection with MU is often a palpable indurated nodule under the skin, although often the pathology is much more advanced by the time treatment is sought (Fig. 5.2). These nodules are thought to take some time to develop, and estimates of the incubation period between the infection event and nodule appearance range up to 6 months (Silva et al. 2009). The nodule consists of a necrotic core surrounded by still healthy dermal tissue. However, at this stage the MU has already undergone sufficient replication so that bacilli are readily detectable (Phillips et al. 2005). The necrotic core already displays features of "coagulative necrosis" (i.e., a

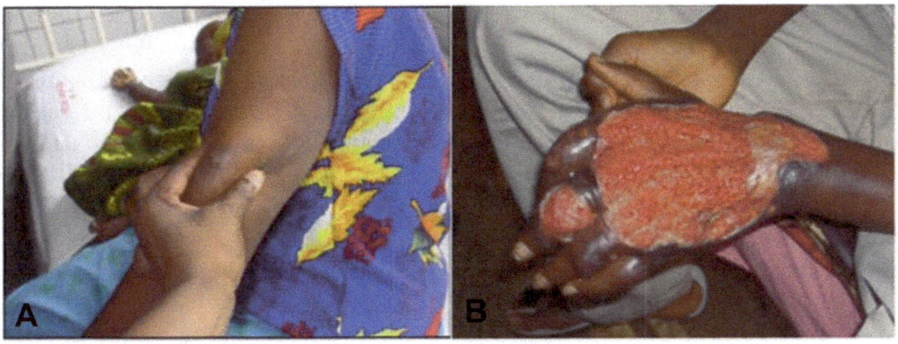

Fig. 5.2 Clinical signs of BU disease. (**a**) Nodule. (**b**) Ulcerative form. Images courtesy of World Health Organization from http://www.who.int/buruli/photos

pale region that is not perfused with blood) in gross pathology. If untreated, the nodule will increase in size with progressive pathology, until one of a number of outcomes emerges.

- Plaque: A larger area of necrotic tissue covered with intact dermis that may be discolored or erythematous.
- Ulceration: The center of the nodule or a region of plaque collapses due to necrosis of the dermis leaving subcutaneous areas exposed. The area of necrosis extends beyond the visible ulcer, which is often described as having "undermined" edges and contains a yellowish woollike "slough."
- Edema: The area of infection is swollen and either may or may not be ulcerated.
- Osteomyelitis: The most severe form of disease where infection has entered the bone, perhaps by lymphohematogenous spread.

The unique pathology of BU is largely due to the production of mycolactone by MU. Usually mycobacterial infections cause a granulomatous disease, the most famous example being tuberculosis that is classically associated with caseous granulomas of the lung. Furthermore, *M. marinum* (or "fish tank" granuloma) causes granulomas of the skin. The difference for MU is the effect that mycolactone has on host cells, since experimental infection with strains that are unable to produce mycolactone causes granulomas (Adusumilli et al. 2005). However, during infection of skin with wild-type pathogenic strains, large clusters of extracellular acid-fast bacilli are observed, and granulomas are absent (George et al. 1999; Silva et al. 2009). Thus, understanding the actions of mycolactone is central to the pathophysiology of this disease.

Under histopathology, several features are commonly seen in BU, notably the presence of coagulative necrosis—a type of tissue necrosis caused by ischemia (lack of oxygen) and characterized by maintenance of cell structures after cell death. This means that cell architecture and connective tissue (i.e., eosin staining) are present in the absence of cell nuclei (i.e., hematoxylin staining) resulting in so-called ghost cells. Large and extensive areas of necrosis are commonly seen in the subcutaneous regions of ulcers and plaques (Schutte and Pluschke 2009; Ruf et al. 2012; Yeboah-Manu et al. 2013), and these contain large clumps of extracellular bacilli that can be detected by Ziehl–Neelsen staining. Necrosis of the dermis is less common but still a common feature. In contrast the skin is psoriatic and the epidermis is hyperplastic due to an over-proliferation of keratinocytes. Even if the epidermis remains closed (such as in a plaque lesion), recent evidence from immunohistochemistry (IHC) has provided further evidence that it is wounded, since these cells variably overexpress cytokeratin, a marker for keratinocyte hyperproliferation (Andreoli et al. 2014). Another classical feature of BU is a relative lack of immune cells infiltrating the infected necrotic tissue (Silva et al. 2009); instead these are restricted to the borders of the infection (Ruf et al. 2011b). Recently, fibrin deposition was shown to be a common feature of BU lesions suggesting that, similarly to tuberculosis, dysregulation coagulation plays a role in pathogenesis (Ogbechi et al. 2015).

These histopathological features are caused by mycolactone, which has been detected at high levels within ulcers (Sarfo et al. 2011, 2014), and can diffuse through tissue some distance from the infecting microbes. Therefore, at least in clinically presented disease, the pathology of BU is more closely associated with the presence of mycolactone than with MU itself. Recent evidence also suggests that mycolactone may have systemic effects (Sarfo et al. 2011).

5.2.5.1 Experimental Models of *M. ulcerans* Infection

There are many challenges in the study of BU, not the least of these being the extended incubation periods and still obscure transmission mode. Therefore a number of experimental models have been developed for laboratory study. The first model developed in Dr. Pam Small's laboratory was a guinea pig model (George et al. 1999). The ulcers that form following injection with MU closely mimic those seen in BU patients, as does injection with purified mycolactone or chemically complemented mutant strains, further evidencing this molecules' essential role in the development of disease. Injection of 100 μg mycolactone in guinea pig skin causes ulceration within 5 days (George et al. 1999).

There is also a mouse footpad model of infection that has been used extensively and has had particular utility in the development of antibiotic treatment regimens (see below). On the other hand, it has several drawbacks for investigation of pathology of BU, since the murine foot is a bony structure with little subcutaneous tissue. Furthermore, high doses of MU are injected leading to footpad swelling, the most common readout of disease. Indeed MU infection has been shown to cause persistent inflammation in mice (Oliveira et al. 2005). This contrasts sharply with the guinea pig model and BU patients in which little over inflammation is observed. Recently, a new model of MU infection in pigs (*Sus scrofa*) was developed by Prof. Gerd Pluschke's group at the Swiss Tropical and Public Health Institute in Basel (Bolz et al. 2014, 2016). This has several advantages over both the mouse footpad and guinea pig models, primarily that pig skin is much more closely matched to human skin. For instance, both the epidermal and dermal thickness of pig and human skin are similar, and pig skin has a more pronounced layer of subcutaneous fat, thought to play a crucial role in the pathogenesis of BU, than does rodent skin. Indeed pigs are widely used in dermatology research and for models of wound healing for these reasons.

5.2.5.2 Mycolactone

Mycolactone was first purified from the acetone-soluble lipids of MU in 1999 (George et al. 1999), although a virulence factors' existence was postulated as long ago as 1965 (Connor and Lunn 1965). The most virulent strains of MU make a 3:2 ratio of mycolactone A/B. The mycolactones made by other MPM are very similar to A/B, differing only by the length and composition of the long polyketide side chain with those differences including the location of hydroxyl groups and double bonds within it. Since these molecules have differing specific activities (Chany et al. 2013, 2014; Scherr et al. 2013), it is thought that biological function is largely conferred by the long polyketide side chain and not the short polyketide chain—although all mycolactones can invoke the same responses once a suitably high dose is present.

Production of mycolactone from the pMUM plasmid involves mycolactone synthase (*mls*) and associated genes (Stinear et al. 2004). These genes are necessary, but probably not sufficient, for mycolactone production (Porter et al. 2009). They have a strong SigA-type promoter that is constitutively expressed during planktonic monoculture under laboratory conditions (Tobias et al. 2009). However, it is not known how this relates to mycolactone expression in "wild" bacteria and how this changes during early infection. In any event, what is clear from the evidence is that at some point (as yet not exactly defined) after infection with MU, sufficient mycolactone is present in the skin to start affecting the phenotype of cells present in the skin milieu. It also seems that mycolactone has the potential for systemic effects, since experimental injection of mice with a high dose of radiolabeled mycolactone has revealed that myolactone can gain access to all bodily compartments with the exception of the brain (Hong et al. 2008), and it can sometimes be detected in the serum of BU patients (Sarfo et al. 2011).

Mycolactone rapidly enters cells, and fluorescently labeled mycolactone has been detected in cell cytosols within 2 min of exposure (Snyder and Small 2003; Chany et al. 2011). In contrast, mycolactone cannot access nuclei (Snyder and Small 2003), although the reason for this is not clear. Indeed, even the cytosolic location of mycolactone may prove to be inaccurate. All such localization experiments to date have labeled mycolactone on the long polyketide side chain, which profoundly alters the characteristics of the molecule. Therefore it remains to be seen whether this is the true subcellular localization of the molecule.

In fact, recent work from one of our laboratories showed that mycolactone targets biological functions at the surface of the endoplasmic reticulum (Hall et al. 2014; McKenna et al. 2016) and is therefore most likely to be located there in its natural state. Detailed molecular investigations in the Simmonds' group showed that it inhibits the Sec61 complex or so-called protein-conducting channel (Dudek et al. 2014). This fundamental cellular process acts to transport (Guenin-Mace et al. 2013) secreted and membrane proteins, as well as many proteins required for its normal function, into the ER. Much (if not all) of the pathology of BU is caused by defective expression of such proteins (described in detail below), strongly suggesting that this mechanism of action is dominant in disease etiology.

This mechanism inhibits the production of a range of proteins that undergo co-translational translocation, meaning that the proteins are synthesized directly into the ER. Co-translational translocation is the reason for the appearance of rough ER because the ribosomes are attached to Sec61 on the ER surface. So far, the identified events occurring during mycolactone-dependent inhibition of Sec61 are as follows: Mycolactone does not affect either transcription of target proteins (Simmonds et al. 2009) or the initiation, rate, or location of protein translation (Hall et al. 2014). However, mycolactone prevents newly synthesized proteins from entering the ER, and so they remain located in the cytosol. Here, the proteins are recognized as being in the wrong cellular compartment and are degraded by the 26S proteasome so rapidly that it becomes hard to detect their synthesis has taken place at all (Hall et al. 2014). Interestingly, if proteasomal activity is inhibited by drugs, it is possible for the affected proteins to be detected once more, although they cannot be redirected into the ER.

By studying glycosylated and secreted proteins as indirect readouts of ER-transiting proteins, we were able show that many such proteins can no longer be synthesized in the presence of inhibitory doses of mycolactone (Hall et al. 2014). One exception to this is a small subset of proteins, less than 100 amino acids in size, that undergo posttranslational translocation due to their small size. These seem to be somewhat resistant to mycolactone inhibiton of translocation, hinting at complexity in mycolactone's interaction with this pathway (McKenna et al. 2016). Nevertheless, approximately one-half of all known mammalian proteins have a signal sequence that would direct them to Sec61-dependent translocation. The cellular consequences of this pathogenic activity are twofold. First, induced ER-transiting proteins cannot be made in the presence of mycolactone. Second, constitutively expressed proteins slowly deplete from cells at the turnover rate, since they cannot be replaced. It is very interesting to consider the known pathological effects in the context of this newly defined and transformative biological activity of mycolactone.

5.2.5.3 Immunosuppression by Mycolactone

In the early stages of experimental infection, before either mycolactone production is induced or while microbial numbers are low, MU can be phagocytosed by macrophages as expected (Coutanceau et al. 2005; Ruf et al. 2012). It is widely believed that this state persists until mycolactone causes either apoptosis or necrosis of the infected macrophage and the bacilli become released into the intercellular space. Infection of tissue with a microbe such as MU would normally be expected to have associated inflammation both due to an innate immune response to the microbe and due to the inflammation caused by the necrosis of cells and tissue. Yet, this inflammation is not seen in clinical presentation of BU and neutrophils cannot access the microbes (Torrado et al. 2007b; Andreoli et al. 2014). Indeed, inflammation at a gross level is rarely reported in BU patients even in the presence of high bacterial loads (Chany et al. 2013). It is now generally accepted that mycolactone is the immunosuppressive factor, being able to suppress both innate and adaptive immune responses. This is central to the indolent nature of the disease, so here we will revisit data from patients and experimental work in light of the ability of mycolactone to inhibit protein translocation.

Mycolactone suppresses the production of cytokines and chemokines (Pahlevan et al. 1999; Torrado et al. 2007a; Phillips et al. 2009; Simmonds et al. 2009) by monocytes and macrophages such as TNF and other cytokines, chemokines, and inflammatory mediators such as IL-6, IL-8, IL-10, and COX-2 (Pahlevan et al. 1999; Torrado et al. 2007a; Simmonds et al. 2009; Hall et al. 2014). Since these mediators are intimately involved in inflammatory infiltrates in tissue, this inhibition probably explains the lack of an inflammatory infiltrate in BU. By extrapolation, the peripheral infiltrate probably indicates the limit of biologically active doses of mycolactone (Silva et al. 2009). Notably all of these cytokines must transit the ER to be produced. Mycolactone also blocks the production of cytokines by CD4+ T cells, which would normally be induced following mitogen or CD3/CD28-depedent activation. Affected cytokines include IL-2, IL-4, IL-10, IL-17, and IFNγ (Phillips et al. 2009). The fact that mycolactone similarly targets anti-inflammatory IL-10 reflects the nonselective nature of its inhibition mechanism. Thus, mycolactone prevents the production of both pro- and anti-inflammatory cytokines, but the net result is no inflammation.

Chemokine production by dendritic cells (DCs), monocytes, macrophages, and T cells is also affected by mycolactone, including IL-8, IP-10, MCP-1, MCP-2, MIP-1α, MIP-1β, and RANTES (Coutanceau et al. 2007; Phillips et al. 2009; Hall et al. 2014). This lack of chemokines in the areas of infection explains the lack of inflammatory infiltrate in the tissue, since neutrophils and other cells require a gradient of chemokines for their expected migration patterns. A discrepancy exists between the chemokine profile of DCs in published work (Coutanceau et al. 2007) and our own observations (Drexler and Simmonds unpublished). These researchers only investigated concentrations up to 25 ng/ml and showed a clear inhibition of MIP-1α, MIP-1β, IP-10, RANTES, and MCP-1 but little effect on IL-8 (Coutanceau et al. 2007). Our results clearly shows that IL-8 production is strongly inhibited by mycolactone. It may be that [similar to macrophages (Hall et al. 2014)] DCs are less susceptible to mycolactone and therefore require a higher suppressive dose than 25 ng/ml; indeed many of the partially reduced cytokines may well have been more strongly suppressed had a higher dose been used. An alternative explanation may be that chemokines are presumptive users of posttranslatioinal translocation—it remains to be established directly if chemokines are subject to Sec61 blockade.

A notable exception to the general inhibition of cytokine and chemokine production is IL-1β; unlike TNF, IL6, IL-8, and IL-10, mycolactone cannot fully prevent IL-1β production from primary human monocytes in response to various different PAMPs (Lopez-Castejon and Brough 2011). This may now be explained by the mechanism of mycolactone action because, unlike other cytokines, IL-1β does not transit the ER during its production. Instead, pro-IL-1β is synthesized as a pro-protein in the cytosol and is released from cells following activation of the NALP-3 inflammasome.

Alongside this, mycolactone also prevents the upregulation of costimulatory molecules (including CD80 and CD40) and other markers (CD83) that otherwise normally would be expected to occur during DC maturation. All of these proteins would have to transit the ER and would therefore potentially be subject to Sec61-dependent activities. The expression of these proteins is associated with a change in the function of DCs, allowing DCs to participate in the functional activation of antigen-specific T cells. If such T cells encounter DCs presenting antigens in the absence of costimulation, a state known as "anergy" is induced. Anergic T cells are unresponsive and immunologically inactive. A further contribution to a lack of T cell responsiveness is an inhibition of lymphocyte homing to peripheral lymph nodes due to mycolactone, in these cases caused by a downregulation of the adhesion receptors L-selectin and LFA-1 on the surface of T cells (Guenin-Mace et al. 2011). Like costimulatory molecules, adhesion receptors such as these also transit the ER in a Sec61-depedendent manner meaning that this inhibitory mechanism may also explain inhibition of lymphocyte homing.

Taken together, it is clear that the inhibition of Sec61-dependent translocation underlies the lack of inflammation and the finding of anergic T cells in BU patients whose T cells no longer respond to mycobacterial antigens after exposure (Gooding et al. 2003). It would also explain the reduced Th1/Th2/Th17 responses seen in whole blood restimulations (Phillips et al. 2009).

5.2.5.4 Cytotoxicity Due to Mycolactone

A very obvious effect of MU on the tissue is the presence of extensive subcutaneous necrosis. Like immunosuppression, there is strong evidence that this is also dependent on mycolactone since mycolactone-negative strains of MU cause granulomas rather than tissue necrosis. However, exactly how this happens is actually far less clear. Injection of purified mycolactone into guinea pig skin was reported to cause increase in TUNEL staining for apoptotic cells (George et al. 2000). More recently, this was confirmed in BU patients by the same method (Bozzo et al. 2010), as well as immunostaining for caspase 3, caspase 8, and Bax (Zavattaro et al. 2012). These latter proteins are all involved in the expression of apoptotic pathways, although it is not clear how the expression of total cellular caspases relates to apoptosis; indeed cleaved caspase 3 (the active form that expresses apoptotic function) is rarely detected in BU lesions (Andreoli et al. 2014). Despite this, all seem to agree that apoptosis cannot be detected in regions of necrosis—presumably because the cells are already destroyed, meaning the markers are not present.

Much work has therefore been carried out using cells cultured in vitro—and much of this on transformed cell lines. Caution is required when directly superimposing in vitro findings on the pathogenesis of the ulcers themselves, for several reasons. First, skin is a complex tissue consisting on multiple layers of different types of cells. This is in stark contrast to tissue culture, where cells are typically grown as monocultures. Second, oxygen and nutrients are supplied to the tissue by the pervasive microvasculature that consists of small vessels and capillaries formed of endothelial cells instead of the monolayers utilized in tissue culture. Here, cells receive oxygen and nutrients from media which is in direct contact with the monolayer. Third, transformed cell lines differ substantially to cells in tissue; their transformed (or cancerous) nature means that they proliferate rapidly and are immortal, not the typical phenotype of healthy skin fibroblasts, apart from the keratinocytes at the basal layer of the epidermis. Indeed "cytotoxicity," by definition, means the killing of cancerous cells, not healthy tissue. Therefore most researchers in the field refer to an in vitro "cytopathic" effect instead. Despite these limitations, experiments testing the effect of mycolactone on the morphology and phenotype of these cells have been highly informative.

The first and foremost cytopathic effect of mycolactone seen following cellular exposure is a change in cellular morphology such that the affected cells take on a rounded appearance and appear to be less strongly attached to the tissue culture (TC) plate after around 24 h (Hockmeyer et al. 1978; George et al. 2000). Between 24 and 48 h of exposure, the cells may detach from the TC plate or be less likely to readhere. However, detached cells have been reported to be viable (George et al. 2000), something that we have also found (Ogbechi et al. 2015). Mycolactone was shown to cause cell cycle arrest in L929 fibroblasts at the G_0/G_1 stage (George et al. 1999), while this specific finding has not been reported for other cell lines, mycolactone treated cells are known to display growth inhibition (George et al. 1999; Hall et al. 2014).

Therefore, despite having the profound tissue effects described, mycolactone is in fact not very fast at killing cells (George et al. 2000; Adusumilli et al. 2005) and little in the way of either apoptosis or cell death is seen for the first 3 days of exposure at low doses (George et al. 2000; Bozzo et al. 2010). Indeed, there is a delay between injection of mycolactone into guinea pig skin and ulcer formation of 5 days (George et al. 1999), although apoptosis can be detected in 2 days (George et al. 2000). It is interesting to note that different effects of mycolactone have been described for each of the cell types studied. While of course it is possible that this relates to lab-to-lab differences as well as perhaps different preparations of mycolactone (synthetic vs natural, obtained from different strains and by different purification methods), at least some of the variation is likely to be due to differences between the cells themselves. For instance, using the same preparation of synthetic mycolactone, 125 ng/ml is required for maximal expression of Sec61 inhibitory activity in RAW264.7 cells (Hall et al. 2014) but only 4 ng/ml in primary human dermal microvascular endothelial cells (Ogbechi et al. 2015). We speculate that such differences may relate to the abundance of either ER or more specifically Sec61 within cells.

However, in the presence of high doses of mycolactone, apoptosis and cell death are rapid (Bozzo et al. 2010). Therefore, different mechansims may underly the tissue necrosis in different areas of clinical lesions depending on the mycolactone concentration. It seems clear that cells exposed to mycolactone in vitro stop dividing and then die. However, little is known about the precise molecular mechanism underlying death. One intriguing possibility is that it involves a process known as anoikis, i.e., death by detachment. What is known of anoikis is from situations where it does not happen. For instance, cancer metastasis is understood to involve an absence of natural anoikis, meaning that the cells can survive after attachment (Paoli et al. 2013). Yet, it seems to sit nicely with the observed cytopathic effects of mycolacton and the documented loss of adhesion molecules such as E-cadherin (Guenin-Mace et al. 2013). Myolactone treatment of HeLa cells causes the cells to rapidly form filopodia. Filopidia are cytoplasmic protrusions containing clusters of actin bundles, commonly associated with wound healing and cell–cell interactions, which in turn may relate to an increase in actin polymerization via the GTPase domain of the Wiskott-Aldrich syndrome protein (WASP) family of proteins (Guenin-Mace et al. 2013). However, inhibition of WASP by wiskostatin does not appear to restore any other activity effected by mycolactone (Hall et al. 2014).

5.2.5.5 Treatment

Prior to the last decade, and still required for larger lesions (Ruf et al. 2011a; Nienhuis et al. 2012; Yeboah-Manu et al. 2013), surgical debridement and skin grafting were the mainstay of BU treatment. Such approaches require skilled personnel as well as supplies and facilities to carry out surgery and adequately care for the wounds postoperatively since the process exposes uninfected tissue including fascia and muscle resulting in a painful wound. In the absence of such treatment, affected limbs can undergo “contractures” that result in serious loss of functionality leading to lifelong disability. Such poor outcomes can be prevented by intense physiotherapy for postoperative patients in combination with appropriate analgesia to allow the patients to mobilize affected joints through their full range of movement. Even where surgical approaches were pursued with the highest possible standards, recurrences were still relatively common and have been reported to vary from 16 to 28% (WHO 2008).

Thus surgical approaches present serious challenges in the poor, rural communities where BU is most common, especially in cases where ulceration is extensive or involves the trunk or face. The introduction of antibiotic therapy in 2004 was therefore a major advance in the treatment of BU. It should be noted that not all ulcers are currently treated with antibiotics; surgery is still frequently required for larger lesions, those that do not respond to antibiotics or patients with osteomyelitis. In addition the antibiotic regimens (long duration, route of application, and potentially toxic) are not always appropriate for all patients. Nevertheless, the majority of cases of BU patients should now receive the WHO-recommended treatment (Converse et al. 2011; WHO 2012; Simpson et al. 2013): a combination of specific antibiotics for 8 weeks as first-line treatment for all forms of active disease, wound care, interventions only to minimize or prevent disability, and surgery to remove necrotic tissue, cover large skin defects, and correct deformities.

5.2.5.6 Antibiotic Treatment Strategies

The biocidal activity of antibiotics against MU in experimental infections has been known since the 1970s (Havel and Pattyn 1975), but it wasn't until this century that the possibilities began to be explored fully, following a recommendation of the WHO Advisory Group on BU to examine the possible benefits of antibiotic treatment in human subjects.

The murine footpad model has been used extensively in the development of antibiotics for BU. In 2000 Dega et al. reported successful treatment of established MU infections in mice with 8-week treatment with either 10 mg/kg rifampicin, or 10 mg/kg rifabutin, or 100 mg/kg amikacin (Dega et al. 2000), whereas other antibiotics were unsuccessful. A number of reports have also shown promise for antibiotic treatment combinations that included rifampicin supplemented with either amikacin or streptomycin and the triple combination of rifampicin supplemented with both clarithromycin and sparfloxacin, suggesting those combinations were not associated with subsequent disease relapse (Bentoucha et al. 2001; Dega et al. 2002; Marsollier et al. 2003).

In 2008, the WHO published provisional guidance on the use of antibiotics to treat BU following a series of investigations showing that 8-week combination therapy with rifampicin and streptomycin was successful (WHO 2008). The first, groundbreaking study showed that such treatment resulted in sterilization of small early lesions (Etuaful et al. 2005). These were soon followed by further studies demonstrating that antibiotic treatment could result in complete healing without the need for surgery in as many as 95% of cases (Chauty et al. 2007; Sarfo et al. 2010a). Recurrence rates are much lower than for surgery (<3%), although posttreatment healing time could be long (up to a year) for larger ulcers (Chauty et al. 2007; Sarfo et al. 2010a), and antibiotics can be used to support surgical intervention.

The treatment regime has continued to evolve. Recently, successful outcomes have been reported for different antibiotic strategies, either replacing or reducing streptomycin in the regime with other orally available antibiotics, again following successful trials in mice (Ji et al. 2007, 2008; Almeida et al. 2011). This is important for two reasons associated with the fact that streptomycin must be injected and thus has both cost and logistical implications and can also be painful (Simpson et al. 2013). Furthermore, the standard 8-week course exposes the patient to a cumulative dose of 56 g streptomycin, which is known to be intrinsically toxic (Klis et al. 2014b). Antibiotic (rifampicin with streptomycin) treatment of BU is associated with long-term ototoxicity (Klis et al. 2014b). Such newer approaches have usually retained rifampicin, as it is the most bactericidal (Ji et al. 2006, 2007), and have largely focused on supplementing the rifampicin with either orally available clarithromycin (Nienhuis et al. 2010; Phillips et al. 2014) or fluoroquinolones (O'Brien et al. 2012; Friedman et al. 2013; Cowan et al. 2015).

In practice, BU treatment in 2015 most commonly involves combination therapy with two or more antibiotics. However, antibiotic regimen choice may be determined by country of residence, wound size at diagnosis, response to antibiotics, as well as patient factors such as allergy and pregnancy. For instance, streptomycin use is contraindicated in pregnancy and so clarythromycin is preferentially prescribed (WHO 2012). As it begins to be adopted more widely, researchers have been urged to monitor rates of compliance with the oral antibiotic treatments, in comparison with those for intramuscular streptomycin injections. Rates of completion have been reported to be as low as ~50% in Ghana (Klis et al. 2014a), many of whom defaulted due to the cost of traveling to the clinic each day for an injection. In a recent study, the economic burden of BU estimated that the household economic cost of BU could be 45% of the household annual income, mostly as a result of transportation costs for nonhospitalized patients (Amoakoh and Aikins 2013). On the other hand, it will be harder to know the rate of treatment completion if patients are given a full course of antibiotics, since they will not be required to attend their health center regularly (Klis et al. 2014a).

There is also much interest in the development of "all-oral" regimens, which so far have only been investigated in small pilot groups and observational studies. Although fluoroquinolones are now commonly used in the treatment of MU infection in Australia, no data at all on their utility in an African setting is available. In Australia the outcomes of such approaches are excellent (up to 100% success rate) (O'Brien et al. 2012) when there is active management of disease and either limited or extensive surgical intervention, as required. However, full adoption of these protocols awaits a randomized control trial to be published. Also, the availability of expensive drugs (moxifloxacin costs AUD$671 for an 8-week course) and surgical facilities means that treatments suitable in Australia, where surgical support is commonplace, may not be applicable in Africa. It remains to be seen whether the proposed algorithm for management of patients with BU in Australia (Simpson et al. 2013) will be applicable in Africa.

Alternative treatments for BU are available, including thermotherapy that heats the skin in order to kill the bacteria. This was proposed as early as the 1950s and was formally shown to be effective in 1974 (Meyers et al. 1974b). However, the equipment required to effect the treatment was cumbersome, and it was not until 2009 that phase change material (PCM, sodium acetate trihydrate, also used for commercial pocket heat pads) was shown to be an easily applied alternative (Junghanss et al. 2009). Thermotherapy was reported to be well tolerated by the patients, and even those with extensive ulcers had complete healing without recurrence when combined with skin grafting (Junghanss et al. 2009).

5.2.5.7 Mechanisms of Resolution and Healing

A chronic ulcer is, by definition, an open area of either skin or mucous membrane that will not heal. As with other types of chronic ulcers, BU ulcers will not heal unless the underlying cause is removed (Trott 1992). It has not been proven systematically whether poor healing is due to MU infection per se or the continued presence of mycolactone in the ulcer. One of the remarkable things about BU is the fact that, even in the face of extensive infections and ulceration, spontaneous healing can sometimes occur (Dobos et al. 2000; van der Werf et al. 2005). Such rare events are associated with a delayed-type hypersensitivity response, suggesting that the body's own immune system has overcome the underlying immunosuppression by mycolactone. Indeed there is some evidence that mycolactone concentrations might decrease during antibiotic treatment and prior to sterilization (Sarfo et al. 2014). On the other hand, some studies have been able to detect mycolactone in ulcer exudates of patients who have completed a sterilizing course of antibiotic therapy (Sarfo et al. 2011), as well as after healing in experimental infections (Sarfo et al. 2013). These results would argue that healing is independent of mycolactone or MU, however unlikely this may seem. The state of the art in this area still awaits a method by which the quantity, and location, of mycolactone in healing ulcers can be assessed.

Whatever the underlying cause, the local immunosuppression seen in the necrotic areas of infected skin is dramatically reversed during antibiotic treatment. Massive infiltration of mononuclear cells and the formation of organized lymphoid structures are seen (Schutte et al. 2007; Schutte and Pluschke 2009). These granulomatous structures (resembling infections with non-MPM and mycolactone-deficient MU mutants) are associated with healing and are due to an antigen-specific reaction of T cells (Schutte et al. 2007).

An important, and relatively common, response to treatment has emerged in recent years. These so-called paradoxical reactions or immune reconstitution inflammatory syndrome reactions have been reported in 9–23% of cases (Nienhuis et al. 2012; O'Brien et al. 2012, 2013; Phillips et al. 2014). They can be severe, requiring treatment with steroids (Friedman et al. 2012; O'Brien et al. 2013) but should not be confused with treatment failure. Paradoxical reactions can manifest themselves in different ways, including the worsening of existing lesion after an initial improvement following the start of antibiotic therapy or the appearance of one or more new lesions. New lesions can be some distance from the original and sometimes even on a different limb. A paradoxical reaction, in contrast with a primary BU lesion, is much more inflammatory in nature—both by gross appearance (including the presence of swelling, pus, and pain) and by histopathology where an "intense inflammatory reaction" is always observed (O'Brien et al. 2013; Phillips et al. 2014). It is thought that paradoxical lesions that appear far from the initial lesion are sites of subclinical MU infection. The hypothesis is that biocidal antibiotics rapidly kill the mycobacteria, leading to a concomitant reduction in mycolactone-dependent immunosuppression and the release of immunostimulators from the dead cells. Thus, normal neutrophil and macrophage function is rapidly reinstated and MU is "unmasked" leading to prompt and dangerously profound immune responses (Trevillyan and Johnson 2013). This hypothesis is supported by the finding that paradoxical lesions can be culture negative but could be PCR and ZN stain positive (Ruf et al. 2011a; Nienhuis et al. 2012; O'Brien et al. 2013) and the detected acid-fast bacilli take on a "beaded" appearance (Ruf et al. 2011a). In contrast to these findings, thermotherapy for BU has not been reported to be associated with paradoxical reactions, although patient numbers are low (Junghanss et al. 2009). The localized method of killing may contribute to this, and healing ulcers show that the undermined edges of the ulcer collapse rapidly (in the first few days of treatment). Histopathological examination revealed a complete lack of inflammation or lymphoid tissue; therefore the process of healing associated with thermotherapy seems quite distinct from that of antibiotics. Most of the ulcers had started re-epithelization within a week, although skin grafting may still be required for larger lesions where re-epithelization does not spontaneously take place.

5.2.6 Epidemiology

Human behavior, specific contact factors, agricultural practices, and environmental conditions have been studied in many BU endemic countries in an effort to determine which activities or factors lead to human infection (Jacobsen and Padgett 2010). Many of these studies were conducted at regional or country-wide scales, while others were at a more focal scale. Jacobson and Padgett systematically reviewed risk factors associated with MU infection worldwide and found both poor wound care, plus either living or working near water that was either slow moving or stagnant, to be increased risk factors, while the usage of protective clothing was associated with decreased risk (Jacobsen and Padgett 2010).

Several case-control studies have also been conducted at the local level. A recent study involving 266 patients and matched controls was conducted in the Eastern Region of Ghana and found that the presence of wetlands, insect bites in water, use of adhesive bandages, washing in the Densu River, and walls built with mud were significant risks for contracting BU (Kenu et al. 2014). The study found faming with long sleeve clothes and rubbing an insect bite with alcohol to be protective; however, the authors acknowledged the possibility of recall bias as most patients had BU for at least 2 years.

A retrospective study was conducted in Australia examining data from 1998 to 2012 within an endemic area to determine risk factors associated with edematous lesions, a less common but more severe form of the disease (O'Brien et al. 2014). The study reported that edematous lesions were frequently misdiagnosed and treated as bacterial cellulitis and that these lesions strongly associated with certain parts of the body. Notably, the dorsum of the hand was affected 85 times more often and ankles and elbows being affected 8 times more often than other parts of the body (O'Brien et al. 2014). The reasons for these particular body location disease associations were not clear, though the authors speculated that either lower skin temperatures or reduced local immune function might have been involved (O'Brien et al. 2014).

Few BU studies show consistent results, presumably due to differences in study design, study area, sample size, definition of patients versus controls, and sampling period. Typically, the conclusions from these studies differ with respect to whether a particular activity or environment poses either an increased or decreased risk or no risk at all. For instance, a case-control study was performed in the endemic region within the Nyong Valley in Central Cameroon in 2006 that found bed nets to be protective independent of socioeconomic status (Pouillot et al. 2007). A similar study was conducted in 2009 in Bankim, Cameroon, showing the same results. Case-control studies conducted in Australia showed mosquito bites were associated with increased risk and mosquito repellent was found to be protective (Quek et al. 2007a, b). However, a study conducted in Benin in 2007 did not show an association with bed net use and infection. Insect bites were also not been independently identified as a risk factor for BU in Ghana (Raghunathan et al. 2005; Merritt et al. 2010).

There have been some consistent findings from studies at local scales in West Africa showing increased risk among activities such as washing clothes or wading in stagnant, marshy areas (Aiga et al. 2004; Raghunathan et al. 2005; Debacker et al. 2006). Despite this, fishermen do not appear to be at risk (Marston et al. 1995; Pouillot et al. 2007). These conflicting results make it necessary for additional, systematic studies to resolve these discrepancies.

5.2.7 Ecology and Transmission

5.2.7.1 Overview

Research on the ecology, distribution, and transmission of MU in the environment had received little attention until the mid-2000s. Until that time nearly all epidemiological studies had associated disease outbreaks with villages in close proximity to human-disturbed freshwater habitats, including marshes, aquatic impoundments, wetlands, and slow-moving riverine environments (Merritt et al. 2005, 2010) (Fig. 5.3). This ecological association with aquatic habitats is generally reported and described as part of the distribution and determinant of the disease but has rarely been quantified. Recent research has suggested that MU DNA can be detected in the detritus of water bodies for over 2 years (Bratschi et al. 2014), suggesting that either the mycobacterium is reproducing in this environment or is being repeatedly introduced there. However, in that same study, it was determined that MU DNA was still detected even when all BU cases had been treated and there were no new cases reported, supporting the hypothesis that MU is reproducing in aquatic habitats.

Fig. 5.3 (**a**) A severe ulceration known as BU disease (*Mycobacterium ulcerans* infection). (**b**) Landscape modifications adjacent to water ways and ponds, wetlands, and low-lying areas are suspected high-risk habitat for BU disease. (**c**) A slow-moving river in Benin where there are many cases of BU and documented high *M. ulcerans* populations in the water sources around this village. (**d**) Biting water bugs (Hemiptera) have been suggested as vectors of *M. ulcerans*; however, studies show contradicting data regarding this hypothesis

Despite such findings, the exact reservoir (or reservoirs) of the causative microorganism has not been definitively characterized, and evidence inferred from the MU genome contends that the pathogen should require a dark and stable niche (Stinear et al. 2007; Demangel et al. 2009); whether this niche is either within a vertebrate host (Durnez et al. 2010; Fyfe et al. 2010), an amoeba (Eddyani et al. 2008; Gryseels et al. 2012; Amissah et al. 2014), or part of the aquatic microbial communities on plants, rocks, and possibly invertebrates that dominate such habitats (Benbow et al. 2008, 2013; Williamson et al. 2008; McIntosh et al. 2014) is still uncertain. A study by Williamson et al. (2012b) revealed a positive relationship with quantitative estimates of MU DNA in the environment using qPCR and BU prevalance, suggesting that while MU may be relatively widespread and found in nonendemic areas, transmission and subsequent symptom presentation in humans likely require increased environmental MU population densities. The hypothesis of there being an ecological threshold population abundance of MU in the environment associated with BU in the human population is supported by evidence from seroepidemiological studies showing that a large number of MU exposures in humans resulted in non-symptomatic infections. Additionally, Röltgen et al. (2014) reported that the infective dose is critical for disease presentation, and Williamson et al. (2014) determined that inoculation into the epidermis through either a deep puncture wound or laceration was necessary for infection, while exposure associated with shallow skin abrasions did not result in symptoms.

Thus, while much is known about the distribution of MU in the environment, much less is known about where the pathogen replicates, the number of cells necessary for progressive infection, and how the mode of inoculation plays a role in transmission, subsequent infection, and symptom presentation. Additionally, the exact role of potential reservoirs and vectors is unclear and may depend on the geograhic location and evolutionary history of *M. ulcerans* serovars: see Röltgen and Pluschke (2015) for a discussion of serovars and potential vectors.

5.2.7.2 Landscape and Climate Associations

The trend in the literature is for BU cases to be associated with human activity that includes disturbed water bodies and altered water quality (WHO 2000b; Merritt et al. 2005). In an early review by Merritt et al. (2005), BU was hypothesized by many authors to be ecologically associated with (1) flooding of lakes and rivers with heavy rainfall; (2) the damming of streams and rivers to create lakes, ponds, and wetlands; (3) changing the landscape in ways that modify wetlands and other low-lying habitats; (4) deforestation, mining, and increased agriculture leading to increased flooding and runoff; (5) construction of agricultural irrigation systems; and (6) population expansion, resettlement, and migration closer to water bodies (Merritt et al. 2010).

Increased human modification of the landscape (e.g., mining) transforms forests that may in turn change water quality through both point and nonpoint sources of pollution, and these transformations have been considered important to BU case emergences (Merritt et al. 2005, 2010). For instance, there have been reported significant spatial relationships of BU with arsenic-enriched surface waters, with soil, and with distance to gold mining that produces tailings enriched in arsenic (Duker et al. 2004, 2006). The authors of these studies suggested that increased BU risk is related to immunosuppression from the consumption of arsenic-enriched drinking water and food crops; however, a direct link to landscape arsenic enrichment and BU has not been demonstrated.

In two studies of BU cases in Benin, West Africa, disease clusters were identified using landscape variables and were found to be associated with the degree of agriculture surrounding communities at low elevations with complex topography (Wagner et al. 2008a, b); however, these associations were only significant at certain spatial scales. These studies concluded that there are specific land uses that lead to local and regional risk for BU, but that the degree of risk depended on the scale of inquiry of land use around a specific village (e.g., 20 km radius rather than 1 km or 100 km).

Although there have been historic reports of a seasonal distribution in BU cases related to rainfall-influenced patterns of village water body usage (Revill and Barker 1972), it has been difficult to understand changes in foci because the incubation period has been estimated to vary from 2 weeks to 3 months (Meyers et al. 1974c), and the environmental time between an ecological event and disease reporting and confirmation was unknown.

The hypothesis of an environmental lag was tested by van Ravensway et al. (2012) in a study that incoporated landscape variables such as land use and topography with climate data to model BU outbreaks in Australia for about a 30-year period, evaluating time lags between climate events and when BU cases were reported and confirmed. In that study they reported that climate conditions during a prior time period of 1.5 years in advance of reported cases were indicative of BU incidence, wherein particularly the sequential occurrence of warmer and wetter than normal conditions at 18–19 months followed by 5 months of dry conditions prior to the reporting period constituted long-term risk conditions for outbreaks. They concluded that understanding coupled landscape-climate factors and associated environmental lag periods may improve the ability of scientists to identify either reservoirs or vectors that respond to changing temperature and rainfall but are not immediately associated with outbreaks because of an unknown "environmental incubation" period. More dynamic and complex modeling studies such as these hold great promise in unraveling the transmission routes of MU from the environment to susceptible human populations. There have been several other landscape investigations (Brou et al. 2008; Carolan et al. 2014b) that support this argument and the accompanying idea that understanding transmission will require the simultaneous examination of abiotic factors such as rainfall and flooding, also with the variation and extent of MU distribution in the environment and biotic interactions that affect the population biology of this pathogen.

5.2.7.3 Environmental Distribution of *M. ulcerans*

Early observations suggested that BU was associated with the regional distribution of different plant species found in rivers and swamps of Uganda, Africa (Barker 1972; Barker et al. 1972). However, only recently has testing detected MU on aquatic plants of several species and in the water column of the water body where the plants were located (McIntosh et al. 2014). There have been numerous studies documenting that MU DNA can be detected in a wide range of habitats and substrates, including detritus, biofilms, fish, insects, snails, other invertebrates, and vertebrates, as reviewed elsewhere (Röltgen and Pluschke 2015). Thus, the continued documentation of MU in many different habitats and substrate types of aquatic, low-lying areas prone to flooding suggests that this mycobacteria is widespread and that a multiple route of transmission hypothesis should be considered in more detail.

In Australia, while MU DNA has been found in possums and other mammals, including fecal samples from these animals, it has also been detected in other aquatic organisms or substrates (Fyfe et al. 2010) suggesting a link between aquatic and terrestrial habitats that might be important to transmission. However, the true densities of the number of MU cells among different substrate types are difficult to estimate and compare. For instance, it is difficult to compare the number of MU genome forming units in one liter of filtered water from a wetland to two grams of possum feces. This difficulty in making quantitative comparisons of MU population densities among different substrates in the environment limits the ability to detect the strongest signals in nature and make strong inference into which substrates or habitats are relatively more significant as "hot spots" of either potential transmission, replication, or virulence. Furthermore, it should be noted that molecular studies target MU DNA rather than viable bacilli. This latter fact makes it difficult to clearly identify specific niches for MU replication. Experimental studies that rigorously evaluate a means to differentiate environmental concentrations of MU are needed in order to better understand the potential vectors associated with transmission.

5.2.7.4 Potential Vectors of *M. ulcerans*

In an excellent review of potential reservoirs and vectors of MU, Röltgen and Pluschke (2015) discuss several proposed transmission hypotheses that support a multiple transmission pathway hypothesis. They also cover an impressive synthesis of integrating what is known about the MU genome and what is understood about its evolution with making predictions about the ecology of the pathogen that potentially are important for transmission, whether transmission is through vectors or through direct contact with environmental sources. Here we provide a summary of the key findings over the last 15 years that have provided the history and background for the current multiple transmission pathway hypothesis.

In 1999, Portaels et al. first suspected that aquatic bugs (Hemiptera) could be reservoirs of *M. ulcerans* as part of a multiple trophic level transmission pathway that leads to humans being bitten by these bugs. This hypothesis was supported in a series of laboratory studies by Marsollier et al. (2002a, b, 2004) who demonstrated experimentally that MU could survive and multiply within the salivary glands of biting aquatic bugs (Naucoridae: *Naucoris cimicoides*) and that aquatic snails could acquire MU after feeding on inoculated aquatic plant biofilms (Marsollier et al. 2004). However, despite these impressive and provocative laboratory findings, some of the first field studies, and even those conducted more recently, have not supported the idea that Hemiptera are potential vectors but do suggest that Hemiptera are probably important to maintain MU in the environment (2002a, b, 2004).

For instance, to evaluate the relationship of MU and the entire aquatic invertebrate communities, Benbow et al. (2008) did not find any clear associations of biting Hemiptera with MU that were greater than MU associations with other aquatic invertebrates like snails, midges, and clams. Indeed, in that study and a more recent large-scale survey of over 90 sites in Ghana (Benbow et al. 2013), many non-hemipteran invertebrates had similar MU positivity rates to the biting Hemiptera, suggesting that MU is not specific to one or two invertebrate taxa and is likely maintained within the environment food webs (see below) (Roche et al. 2013). This suggestion has been supported in other studies as well throughout Africa (Vandelannoote et al. 2010; Williamson et al. 2012a; Roche et al. 2013; Carolan et al. 2014b). There have been several studies that have supported Hemiptera as possible vectors, but they have all suffered from the same limitation in that they only evaluated associations of MU with Hemiptera (Marion et al. 2010; Ebong et al. 2012; Carolan et al. 2014a) while ignoring all of the other invertebrate taxa, and therefore they were unable to rule out that MU was also and perhaps predominately associated with other aquatic invertebrate taxa. In addition, biting water bugs do not actively search for humans. Neither do the water bugs require a protein source such as a blood meal to mature their eggs, nor is there any evolutionary history either suggesting or supporting the idea that these semiaquatic Hemiptera serve as vectors for pathogen transmission nor their having coevolving a host and parasite relationship with MU. As Merritt et al. (2010) discussed in a systematic review of BU ecology, there are series of criteria that are necessary to scientifically identify a new vector association for a new pathogen, and to date these criteria have not been satisfied for any of the invertebrate vectors hypothesized for BU.

In Australia one of the earliest hypotheses of transmission was that aerosols arising from contaminated water disseminated MU to hosts via either the respiratory tract or through contamination of skin lesions and minor abrasions (Hayman 1991). This hypothesis was related to a series of BU cases geographically associated with the installation of an irrigation system that drew water from a human-made aquatic impoundment which had tested positive for MU (Stinear et al. 2000). However, this transmission mechanism was never definitively proven and has been mostly abandoned as a hypothesis in Southeast Australia.

Most recently, MU has been reported in a variety of vertebrate animals and mosquitoes in Southeast Australia (Johnson et al. 2007; Fyfe et al. 2010; Lavender et al. 2011; Carson et al. 2014), with mammals being suspected as amplifying hosts in MU ecology (Fyfe et al. 2010). Additional studies from this region have documented correlations of BU with MU positive mosquitoes (Lavender et al. 2011) and with other vectorborne notifiable diseases (Johnson and Lavender 2009), supporting the possibility that mosquitoes and mammals are important components of MU ecology in that area of the world.

However, even though Wallace et al. (2010) found that MU could be detected in all larval life stages of four mosquito species, the pathogen could not be transferred to either the pupal or adult stages. These findings suggest that adult mosquitoes would require becoming contaminated from the environment rather than becoming inoculated with MU transovarially. Environmental inoculation would likely be from aquatic habitats where *M. ulcerans* has been detected from the water column and biofilms of ponds where mosquitoes are found (Fyfe et al. 2010). In the same study by Wallace et al. (2010), the authors also documented that MU was readily fed upon by the larval mosquitoes and that the mycobacteria could be transferred within the food web to hemipteran predators, supporting the hypothesis that MU is in aquatic invertebrate food webs and can be transferred among trophic levels. Yet, the question of where this pathogen replicates within the food web remains an exciting area of future research, and new results related to replication within the food web would allow researchers to test and identify multiple transmission pathways.

5.2.7.5 Multiple Transmission Pathways of *M. ulcerans*

Certain bugs (Naucoridae, Belostomatidae) are aggressive predators of other aquatic invertebrates and fish but also will accidentally bite humans and other animals visiting freshwater habitats (Merritt and Cummins 1996). Further, while mosquitoes have never been definitively shown to directly transmit a bacterial pathogen, they could act as mechanical vectors much like that possiblity presumed for biting water bugs. In this way, the water bugs and mosquitoes either may be mechanically inoculating MU into the skin or depositing bacterial cells onto the skin surface near sites of trauma including minor abrasions. This would support those studies which showed that either an injection, puncture, or laceration is necessary for MU transmission in the guinea pig model (Williamson et al. 2014). However, a fresh and deep puncture or laceration that subsequently is exposed to MU in the environment may also provide a plausable route of transmission, with scratching or rubbing by the host physically pushing an infective dose of cells into the wound. Alternatively, it is possible that persons with recent skin lesions and lacerations may be inoculated either directly from the water or by contacting biofilms present on aquatic plants and wetted soil. To date, most evidence suggests multiple MU transmission routes.

5.2.7.6 Control and Management

In the absence of a clear path for primary prevention, education and treatment are necessary to prevent progression from early lesions to ulcers. One of the most significant activities impacting awareness is active case detection. A combination of active case findings followed by BU awareness campaign efforts has contributed to the dissemination of information regarding BU and has led to a significant increase in the discovery of MU infections at the stage of early lesions and simple ulcerative forms and a decrease in their discovery as category III ulcers. This has also increased the proportion of healing with reduced complications, as help-seeking behavior is influenced by the perceived effectiveness of treatment (Porten et al. 2009; Ackumey et al. 2011a; Agbenorku et al. 2011; Phanzu et al. 2011). These works advocate the cooperation between national control programs, municipal executives, health staff, teachers, school children, community leaders, and community health volunteers to overcome difficulties associated with health-care costs, accommodations, and adequate infrastructure for surgery as well as postsurgical aftercare.

Compliance with Ethical Standards
Funding: This chapter was funded in part by Grant Number R01TW007550 from the Fogarty International Center through the NIH/NSF Ecology of Infectious Diseases Program and grant number R03AI062719. The content is solely the responsibility of the authors and does not necessarily represent the official views of the Fogarty International Center or the National Institutes of Health.

Conflict of Interest: M. Eric Benbow declares that he/she has no conflict of interest. Rachel Simmonds declares that he/she has no conflict of interest. Richard W. Merritt declares that he/she has no conflict of interest. Heather R. Jordan declares that he/she has no conflict of interest.

Ethical approval: This chapter does not contain any studies with human participants or animals performed by any of the authors.

References

Abass KM, van der Werf TS, Phillips RO et al (2015) Buruli ulcer control in a highly endemic district in Ghana: role of community-based surveillance volunteers. Am J Trop Med Hyg 92:115–117

Ablordey A, Amissah DA, Aboagye IF et al (2012) Detection of *Mycobacterium ulcerans* by the loop mediated isothermal amplification method. PLoS Negl Trop Dis 6:e1590

Ablordey AS, Vandelannoote K, Frimpong IA et al (2015) Whole genome comparisons suggest random distribution of *Mycobacterium ulcerans* genotypes in a Buruli ulcer endemic region of Ghana. PLoS Negl Trop Dis 9:e0003681

Ackumey MM, Gyapong M, Pappoe M et al (2011a) Help-seeking for pre-ulcer and ulcer conditions of *Mycobacterium ulcerans* disease (Buruli ulcer) in Ghana. Am J Trop Med Hyg 85:1106–1113

Ackumey MM, Kwakye-Maclean C, Ampadu EO et al (2011b) Health services for Buruli ulcer control: lessons from a field study in Ghana. PLoS Negl Trop Dis 5:e1187

Ackumey MM, Gyapong M, Pappoe M et al (2012a) Illness meanings and experiences for pre-ulcer and ulcer conditions of Buruli ulcer in the Ga-West and Ga-South Municipalities of Ghana. BMC Public Health 12:264

Ackumey MM, Gyapong M, Pappoe M et al (2012b) Socio-cultural determinants of timely and delayed treatment of Buruli ulcer: implications for disease control. Infect Dis Poverty 1:6

Adusumilli S, Mve-Obiang A, Sparer T et al (2005) *Mycobacterium ulcerans* toxic macrolide, mycolactone modulates the host immune response and cellular location of *M. ulcerans in vitro* and *in vivo*. Cell Microbiol 7:1295–1304

Agbenorku P, Agbenorku M, Amankwa A et al (2011) Factors enhancing the control of Buruli ulcer in the Bomfa communities, Ghana. Trans R Soc Trop Med Hyg 105:459–465

Aiga H, Amano T, Cairncross S et al (2004) Assessing water-related risk factors for Buruli ulcer: a case-control study in Ghana. Am J Trop Med Hyg 71:387–392

Akoachere J-FK, Nsai FS, Ndip RN (2016) A community based study on the mode of transmission, prevention and treatment of Buruli ulcers in Southwest Cameroon: knowledge, attitude and practices. PLoS One 11:e0156463

Al-Anazi KA, Al-Jasser AM, Al-Anazi WK (2014) Infections caused by non-tuberculous mycobacteria in recipients of hematopoietic stem cell transplantation. Front Oncol 4:311

Alexander KA, Laver PN, Michel AL et al (2010) Novel *Mycobacterium tuberculosis* complex pathogen, *M. mungi*. Emerg Infect Dis 16:1296-1299.

Almeida D, Converse PJ, Ahmad Z et al (2011) Activities of rifampin, rifapentine and clarithromycin alone and in combination against mycobacterium ulcerans disease in mice. PLoS Negl Trop Dis 5:e933

Amissah NA, Gryseels S, Tobias NJ et al (2014) Investigating the role of free-living amoebae as a reservoir for *Mycobacterium ulcerans*. PLoS Negl Trop Dis 8:e3148

Amoakoh HB, Aikins M (2013) Household cost of out-patient treatment of Buruli ulcer in Ghana: a case study of Obom in Ga South Municipality. BMC Health Serv Res 13:507

Andreoli A, Ruf MT, Sopoh GE et al (2014) Immunohistochemical monitoring of wound healing in antibiotic treated Buruli ulcer patients. PLoS Negl Trop Dis 8:e2809

Asiedu K, Etuaful S (1998) Socioeconmoic implications of Buruli ulcer in Ghana: a three-year review. Trans R Soc Trop Med Hyg 59:1015–1022

Barker DJP (1972) The distribution of Buruli disease in Uganda. Trans R Soc Trop Med Hyg 66:867–874

Barker DJP, Clancey JK, Rao SK (1972) *Mycobacteria* on vegetation in Uganda. East Afr Med J 49:667–671

Beissner M, Herbinger KH, Bretzel G (2010) Laboratory diagnosis of Buruli ulcer disease. Future Microbiol 5:363–370

Benbow M, Williamson H, Kimbirauskus R et al (2008) Aquatic invertebrates as unlikely vectors of Buruli ulcer disease. Emerg Infect Dis 14:1247–1254

Benbow ME, Kimbirauskas R, McIntosh MD et al (2013) Aquatic macroinvertebrate assemblages of Ghana, West Africa: understanding the ecology of a neglected tropical disease. EcoHealth:1–16

Bentoucha A, Robert J, Dega H et al (2001) Activities of new macrolides and fluoroquinolones against *Mycobacterium ulcerans* infection in mice. Antimicrob Agents Chemother 45:3109–3112

Bolz M, Ruggli N, Ruf MT et al (2014) Experimental infection of the pig with *Mycobacterium ulcerans*: a novel model for studying the pathogenesis of Buruli ulcer disease. PLoS Negl Trop Dis 8:e2968

Bolz M, Ruggli N, Borel N et al (2016) Local cellular immune responses and pathogenesis of Buruli ulcer lesions in the experimental *Mycobacterium ulcerans* pig infection model. PLoS Negl Trop Dis 10:e0004678

Bosio S, Leekha S, Gamb SI et al (2012) *Mycobacterium fortuitum* prosthetic valve endocarditis: a case for the pathogenetic role of biofilms. Cardiovasc Pathol 21:361–364

Bozzo C, Tiberio R, Graziola F et al (2010) A *Mycobacterium ulcerans* toxin, mycolactone, induces apoptosis in primary human keratinocytes and in HaCaT cells. Microbes Infect 12:1258–1263

Bratschi MW, Ruf M-T, Andreoli A et al (2014) *Mycobacterium ulcerans* persistence at a village water source of Buruli ulcer patients. PLoS Negl Trop Dis 8:e2756

Brosch R, Gordon SV, Marmiesse M et al (2002) A new evolutionary scenario for the *Mycobacterium tuberculosis* complex. Proc Natl Acad Sci USA 99:3684–3689

Brou T, Broutin H, Elguero E et al (2008) Landscape diversity related to Buruli ulcer disease in Côte d'Ivoire. PLoS Negl Trop Dis 2:e271

Burback BL, Perry JJ (1993) Biodegradation and biotransformation of groundwater pollutant mixtures by Mycobacterium vaccae. Appl Environ Microbiol 59:1025–1029

Carolan K, Ebong SMÀ, Garchitorena A et al (2014a) Ecological niche modelling of Hemipteran insects in Cameroon; the paradox of a vector-borne transmission for *Mycobacterium ulcerans*, the causative agent of Buruli ulcer. Int J Health Geogr 13:44

Carolan K, Garchitorena A, García-Peña GE et al (2014b) Topography and land cover of watersheds predicts the distribution of the environmental pathogen *Mycobacterium ulcerans* in aquatic insects. PLoS Negl Trop Dis 8:e3298

Carson C, Lavender CJ, Handasyde KA et al (2014) Potential wildlife sentinels for monitoring the endemic spread of human Buruli ulcer in South-East Australia. PLoS Negl Trop Dis 8:e2668

Chany AC, Casarotto V, Schmitt M et al (2011) A diverted total synthesis of mycolactone analogues: an insight into Buruli ulcer toxins. Chemistry 17:14413–14419

Chany AC, Tresse C, Casarotto V et al (2013) History, biology and chemistry of *Mycobacterium ulcerans* infections (Buruli ulcer disease). Nat Prod Rep 30:1527–1567

Chany AC, Veyron-Churlet R, Tresse C et al (2014) Synthetic variants of mycolactone bind and activate Wiskott-Aldrich syndrome proteins. J Med Chem 57:7382–7395

Chauty A, Ardant MF, Adeye A et al (2007) Promising clinical efficacy of streptomycin-rifampin combination for treatment of buruli ulcer (*Mycobacterium ulcerans*) disease. Antimicrob Agents Chemother 51:4029–4035

Clancey JK (1964) Mycobacterial skin ulcers in Uganda: description of a new mycobacterium (*Mycobacterium buruli*). J Pathol Bacteriol 88:175–187

Collins CH, Grange JM, Noble WC et al (1985) *Mycobacterium marinum* infections in man. J Hyg (Lond) 94:135–149

Comas I, Coscolla M, Luo T et al (2013) Out-of-Africa migration and Neolithic coexpansion of *Mycobacterium tuberculosis* with modern humans. Nat Genet 45:1176–1182

Connor DH, Lunn HF (1965) *Mycobacterium ulcerans* infection (with comments on pathogenesis). Int J Lepr 33(Suppl):698–709

Conteh L, Engels T, Molyneux DH (2010) Socioeconomic aspects of neglected tropical diseases. Lancet 375:239–247

Converse PJ, Nuermberger EL, Almeida DV et al (2011) Treating *Mycobacterium ulcerans* disease (Buruli ulcer): from surgery to antibiotics, is the pill mightier than the knife? Future Microbiol 6:1185–1198

Cook KL, Britt JS, Bolster CH (2010) Survival of *Mycobacterium avium* subsp. *paratuberculosis* in biofilms on livestock watering trough materials. Vet Microbiol 141:103–109

Coscolla M, Gagneux S (2014) Consequences of genomic diversity in *Mycobacterium tuberculosis*. Semin Immunol 26:431–444

Coscolla M, Lewin A, Metzger S et al (2013) Novel *Mycobacterium tuberculosis* complex isolate from a wild chimpanzee. Emerg Infect Dis 19:969–976

Coutanceau E, Marsollier L, Brosch R et al (2005) Modulation of the host immune response by a transient intracellular stage of *Mycobacterium ulcerans*: the contribution of endogenous mycolactone toxin. Cell Microbiol 7:1187–1196

Coutanceau E, Decalf J, Martino A et al (2007) Selective suppression of dendritic cell functions by *Mycobacterium ulcerans* toxin mycolactone. J Exp Med 204:1395–1403

Cowan R, Athan E, Friedman ND et al (2015) *Mycobacterium ulcerans* treatment – can antibiotic duration be reduced in selected patients? PLoS Negl Trop Dis 9:e0003503

Dangy J-P, Scherr N, Gersbach P et al (2016) Antibody-mediated neutralization of the exotoxin mycolactone, the main virulence factor produced by mycobacterium ulcerans. PLoS Negl Trop Dis 10:e0004808

Daniel AK, Lee RE, Portaels F et al (2004) Analysis of *Mycobacterium* species for the presence of a macrolide toxin, mycolactone. Infect Immun 72:123–132

de Souza DK, Quaye C, Mosi L et al (2012) A quick and cost effective method for the diagnosis of Mycobacterium ulcerans infection. BMC Infect Dis 12:8

Debacker M, Portaels F, Aguiar J et al (2006) Risk factors for Buruli ulcer, Benin. Emerg Infect Dis 12:1325–1331

Dega H, Robert J, Bonnafous P et al (2000) Activities of several antimicrobials against *Mycobacterium ulcerans* infection in mice. Antimicrob Agents Chemother 44:2367–2372

Dega H, Bentoucha A, Robert J et al (2002) Bactericidal activity of rifampin-amikacin against *Mycobacterium ulcerans* in mice. Antimicrob Agents Chemother 46:3193–3196

Demangel C, Stinear T, Cole S (2009) Buruli ulcer: reductive evolution enhances pathogenicity of *Mycobacterium ulcerans*. Nat Rev Microbiol 7:50–60

Diaz D, Dobeli H, Yeboah-Manu D et al (2006) Use of the immunodominant 18-kiloDalton small heat shock protein as a serological marker for exposure to *Mycobacterium ulcerans*. Clin Vaccine Immunol 13:1314–1321

Dobos KM, Spotts EA, Marston BJ et al (2000) Serologic response to culture filtrate antigens of Mycobacterium ulcerans during Buruli ulcer disease. Emerg Infect Dis 6:158–164

Doig KD, Holt KE, Fyfe JA et al (2012) On the origin of *Mycobacterium ulcerans*, the causative agent of Buruli ulcer. BMC Genomics 13:258

Dudek J, Pfeffer S, Lee PH et al (2014) Protein transport into the human endoplasmic reticulum. J Mol Biol 427(6 Pt A):1159–1175

Duker AA, Carranza EJM, Hale M (2004) Spatial dependency of Buruli ulcer prevalence on arsenic-enriched domains in Amansie West District, Ghana: implications for arsenic mediation in Mycobacterium ulcerans infection. Int J Health Geogr 3:19

Duker AA, Stein A, Hale M (2006) A statistical model for spatial patterns of Buruli ulcer in the Amansie West district, Ghana. Int J Appl Earth Obs Geoinf 8:126–136

Durnez L, Suykerbuyk P, Nicolas V et al (2010) Terrestrial small mammals as reservoirs of *Mycobacterium ulcerans* in Benin. Appl Environ Microbiol 76:4574-4577.

Ebong S, Eyangoh S, Marion E et al (2012) Survey of water bugs in Bankim, a new Buruli ulcer endemic area in Cameroon. J Trop Med 123843:8

Eddyani M, De Jonckheere JF, Durnez L et al (2008) Occurrence of free-living amoebae in communities of low and high endemicity for Buruli ulcer in Southern Benin. Appl Environ Microbiol 74:6547–6553

En J, Goto M, Nakanaga K et al (2008) Mycolactone is responsible for the painlessness of *Mycobacterium ulcerans* infection (buruli ulcer) in a murine study. Infect Immun 76:2002–2007

Etuaful S, Carbonnelle B, Grosset J et al (2005) Efficacy of the combination rifampin-streptomycin in preventing growth of *Mycobacterium ulcerans* in early lesions of Buruli ulcer in humans. Antimicrob Agents Chemother 49:3182–3186

Falkinham III JO (2009) The biology of environmental mycobacteria. Environ Microbiol Rep 1:477–487

Feasey N, Wansbrough-Jones M, Mabey DCW et al (2010) Neglected tropical diseases. Br Med Bull 93:179–200

Feazel LM, Baumgartner LK, Peterson KL et al (2009) Opportunistic pathogens enriched in showerhead biofilms. Proc Natl Acad Sci USA 106:16393–16399

Friedman ND, McDonald AH, Robson ME et al (2012) Corticosteroid use for paradoxical reactions during antibiotic treatment for *Mycobacterium ulcerans*. PLoS Negl Trop Dis 6: e1767

Friedman ND, Athan E, Hughes AJ et al (2013) *Mycobacterium ulcerans* disease: experience with primary oral medical therapy in an Australian cohort. PLoS Negl Trop Dis 7:e2315

Fyfe JAM, Lavender CJ, Handasyde KA et al (2010) A major role for mammals in the ecology of *Mycobacterium ulcerans*. PLoS Negl Trop Dis 4:e791

Gama JB, Ohlmeier S, Martins TG et al (2014) Proteomic analysis of the action of the *Mycobacterium ulcerans* toxin mycolactone: targeting host cells cytoskeleton and collagen. PLoS Negl Trop Dis 8:e3066

George KM, Chatterjee D, Gunawardana G et al (1999) Mycolactone: a polyketide toxin from *Mycobacterium ulcerans* required for virulence. Science 283:854–857

George KM, Pascopella L, Welty DM et al (2000) A *Mycobacterium ulcerans* toxin, mycolactone, causes apoptosis in guinea pig ulcers and tissue culture cells. Infect Immun 68:877–883

Gooding TM, Kemp AS, Robins-Browne RM et al (2003) Acquired T-helper 1 lymphocyte anergy following infection with *Mycobacterium ulcerans*. Clin Infect Dis 36:1076–1077

Gryseels S, Amissah D, Durnez L et al (2012) Amoebae as potential environmental hosts for *Mycobacterium ulcerans* and other mycobacteria, but doubtful actors in Buruli ulcer epidemiology. PLoS Negl Trop Dis 6:e1764

Guenin-Mace L, Carrette F, Asperti-Boursin F et al (2011) Mycolactone impairs T cell homing by suppressing microRNA control of L-selectin expression. Proc Natl Acad Sci USA 108:12833–12838

Guenin-Mace L, Veyron-Churlet R, Thoulouze MI et al (2013) Mycolactone activation of Wiskott-Aldrich syndrome proteins underpins Buruli ulcer formation. J Clin Invest 123:1501–1512

Hall BS, Hill K, McKenna M et al (2014) The pathogenic mechanism of the *Mycobacterium ulcerans* virulence factor, mycolactone, depends on blockade of protein translocation into the ER. PLoS Pathog 10:e1004061

Hall-Stoodley L, Brun OS, Polshyna G et al (2006) *Mycobacterium marinum* biofilm formation reveals cording morphology. FEMS Microbiol Lett 257:43–49

Havel A, Pattyn SR (1975) Activity of rifampicin on Mycobacterium ulcerans. Ann Soc Belg Med Trop 55:105–108

Hayman J (1991) *Mycobacterium ulcerans* infection. Lancet 337:124

Heitkamp MA, Freeman JP, Miller DW et al (1988) Pyrene degradation by a *Mycobacterium sp.*: identification of ring oxidation and ring fission products. Appl Environ Microbiol 54:2556–2565

Hockmeyer WT, Krieg RE, Reich M et al (1978) Further characterization of *Mycobacterium ulcerans* toxin. Infect Immun 21:124–128

Hong H, Coutanceau E, Leclerc M et al (2008) Mycolactone diffuses from *Mycobacterium ulcerans*-infected tissues and targets mononuclear cells in peripheral blood and lymphoid organs. PLoS Negl Trop Dis 2:e325

Horsburgh CR, Meyers WM (1997) Buruli ulcer. In: Horsburgh CR, Nelson AM (eds) Pathology of emerging infections. American Society for Microbiology, Washington, DC, pp 119–126

Hotez PJ, Fenwick A, Savioli L et al (2009) Rescuing the bottom billion through control of neglected tropical diseases. Lancet 373:1570–1575

Huang GKL, Johnson PD (2014) Epidemiology and management of Buruli ulcer. Expert Rev Anti-Infect Ther 12:855–865

Huard RC, Fabre M, de Haas P et al (2006) Novel genetic polymorphisms that further delineate the phylogeny of the *Mycobacterium tuberculosis* complex. J Bacteriol 188:4271–4287

Huminer D, Pitlik SD, Block C et al (1986) Aquarium-borne *Mycobacterium marinum* skin infection. Report of a case and review of the literature. Arch Dermatol 122:698–703

Hurst LC, Amadio PC, Badalamente MA et al (1987) *Mycobacterium marinum* infections of the hand. J Hand Surg [Am] 12:428–435

Islam MS, Richards JP, Ojha AK (2012) Targeting drug tolerance in mycobacteria: a perspective from mycobacterial biofilms. Expert Rev Anti-Infect Ther 10:1055–1066

Jacobs S, George A, Papanicolaou GA et al (2012) Disseminated *Mycobacterium marinum* infection in a hematopoietic stem cell transplant recipient. Transpl Infect Dis 14:410–414

Jacobsen KH, Padgett JJ (2010) Risk factors for *Mycobacterium ulcerans* infection. Int J Infect Dis 14:e677–e681

Janssens P (1972) Skin ulcers caused by acid-fast bacilli. Essays on tropical. Dermatology 2:264–295

Ji B, Lefrancois S, Robert J et al (2006) In vitro and in vivo activities of rifampin, streptomycin, amikacin, moxifloxacin, R207910, linezolid, and PA-824 against *Mycobacterium ulcerans*. Antimicrob Agents Chemother 50:1921–1926

Ji B, Chauffour A, Robert J et al (2007) Orally administered combined regimens for treatment of *Mycobacterium ulcerans* infection in mice. Antimicrob Agents Chemother 51:3737–3739

Ji B, Chauffour A, Robert J et al (2008) Bactericidal and sterilizing activities of several orally administered combined regimens against *Mycobacterium ulcerans* in mice. Antimicrob Agents Chemother 52:1912–1916

Johnson PDR, Lavender CJ (2009) Correlation between Buruli ulcer and vector-borne notifiable diseases, Victoria, Australia. Emerg Infect Dis 15:614–615

Johnson PDR, Veitch MGK, Leslie D et al (1996) The emergence of *Mycobacterium ulcerans* infection near Melbourne. Med J Aust 164:76–78

Johnson PDR, Azuolas J, Lavender CJ et al (2007) *Mycobacterium ulcerans* in mosquitoes captured during outbreak of Buruli ulcer, Southeastern Australia. Emerg Infect Dis 13:1653–1660

Junghanss T, Um Boock A, Vogel M et al (2009) Phase change material for thermotherapy of Buruli ulcer: a prospective observational single centre proof-of-principle trial. PLoS Negl Trop Dis 3:e380

Kamel G, Youssef M, Haidar R et al (2014) Osteomyelitis at two noncontiguous sites caused by *Mycobacterium marinum* in an immunocompetent host case report and literature review. J Med Liban 62:180–182

Kaser M, Rondini S, Naegeli M et al (2007) Evolution of two distinct phylogenetic lineages of the emerging human pathogen *Mycobacterium ulcerans*. BMC Evol Biol 7:177

Kaser M, Hauser J, Pluschke G (2009a) Single nucleotide polymorphisms on the road to strain differentiation in *Mycobacterium ulcerans*. J Clin Microbiol 47:3647–3652

Kaser M, Hauser J, Small P et al (2009b) Large sequence polymorphisms unveil the phylogenetic relationship of environmental and pathogenic mycobacteria related to *Mycobacterium ulcerans*. Appl Environ Microbiol 75:5667–5675

Kenu E, Nyarko KM, Seefeld L et al (2014) Risk factors for buruli ulcer in Ghana-a case control study in the Suhum-Kraboa-Coaltar and Akuapem South Districts of the eastern region. PLoS Negl Trop Dis 8:e3279

Klis S, Kingma R, Tuah W et al (2014a) Compliance with antimicrobial therapy for Buruli ulcer. Antimicrob Agents Chemother 58:6340

Klis S, Stienstra Y, Phillips RO et al (2014b) Long term streptomycin toxicity in the treatment of Buruli Ulcer: follow-up of participants in the BURULICO drug trial. PLoS Negl Trop Dis 8: e2739

Kressel AB, Kidd F (2001) Pseudo-outbreak of *Mycobacterium chelonae* and *Methylobacterium mesophilicum* caused by contamination of an automated endoscopy washer. Infect Control Hosp Epidemiol 22:414–418

Kullavanijaya P, Sirimachan S, Bhuddhavudhikrai P (1993) *Mycobacterium marinum* cutaneous infections acquired from occupations and hobbies. Int J Dermatol 32:504–507

Lakhanpal A, Arfon S, McKeon DJ (2011) So, they thought it was all over. BMJ Case Rep 2011

Lastoria JC, Abreu MA (2014a) Leprosy: a review of laboratory and therapeutic aspects – part 2. An Bras Dermatol 89:389–401

Lastoria JC, Abreu MA (2014b) Leprosy: review of the epidemiological, clinical, and etiopathogenic aspects – part 1. An Bras Dermatol 89:205–218

Lavender CJ, Stinear TP, Johnson PD et al. (2008) Evaluation of VNTR typing for the identification of *Mycobacterium ulcerans* in environmental samples from Victoria, Australia. FEMS Microbiol Lett 287:250-255.

Lavender CJ, Fyfe JAM, Azuolas J et al (2011) Risk of Buruli ulcer and detection of *Mycobacterium ulcerans* in mosquitoes in Southeastern Australia. PLoS Negl Trop Dis 5

Lewis FM, Marsh BJ, von Reyn CF (2003) Fish tank exposure and cutaneous infections due to *Mycobacterium marinum*: tuberculin skin testing, treatment, and prevention. Clin Infect Dis 37:390–397

Linell F, Norden A (1954) *Mycobacterium balnei*, a new acid-fast bacillus occurring in swimming pools and capable of producing skin lesions in humans. Acta Tuberc Scand Suppl 33:1–84

Lopez-Castejon G, Brough D (2011) Understanding the mechanism of IL-1beta secretion. Cytokine Growth Factor Rev 22:189–195

MacCallum P, Tolhurst JC, Buckle G et al (1948) A new mycobacterial infection in man. J Pathol Bacteriol 60:93–122

Marion E, Eyangoh S, Yeramian E et al (2010) Seasonal and regional dynamics of *M. ulcerans* transmission in environmental context: deciphering the role of water bugs as hosts and vectors. PLoS Negl Trop Dis 4:e731

Marsollier L, Legras P, Manceau A-L et al (2002a) Role des punaises d'eau dans la transmission de *M. ulcerans*. BULL ALLF or Bulletin de l'ALLF 10:23–25

Marsollier L, Robert R, Aubry J et al (2002b) Aquatic insects as a vector for *Mycobacterium ulcerans*. Appl Environ Microbiol 68:4623–4628

Marsollier L, Prevot G, Honore N et al (2003) Susceptibility of *Mycobacterium ulcerans* to a combination of amikacin/rifampicin. Int J Antimicrob Agents 22:562–566

Marsollier L, Severin T, Aubry J et al (2004) Aquatic snails, passive hosts of *Mycobacterium ulcerans*. Appl Environ Microbiol 70:6296–6298

Marston BJ, Diallo MO, Horsburgh JCR et al (1995) Emergence of Buruli ulcer disease in the Daloa region of Cote D'ivoire. Am J Trop Med Hyg 52:219–224

Mathers CD, Ezzati M, Lopez AD (2007) Measuring the burden of neglected tropical diseases: the global burden of disease framework. PLoS Negl Trop Dis 1:e114

McGann H, Stragier P, Portaels F et al (2009) Buruli ulcer in United Kingdom tourist returning from Latin America. Emerg Infect Dis 15:1827–1829
McIntosh M, Williamson H, Benbow ME et al (2014) Associations between *Mycobacterium ulcerans* and aquatic plant communities of West Africa: implications for Buruli ulcer disease. EcoHealth 11:184–196
McKenna M, Simmonds RE, High S (2016) Mechanistic insights into the inhibition of Sec61-dependent co-and post-translational translocation by mycolactone. J Cell Sci 129:1404–1415
Merritt RW, Cummins KW (eds) (1996) An introduction to the aquatic insects of North America, 3rd edn. Kendall/Hunt, Dubuque, IA
Merritt RW, Benbow ME, Small PLC (2005) Unraveling an emerging disease associated with disturbed aquatic environments: the case of Buruli ulcer. Front Ecol Environ 3:323–331
Merritt RW, Walker ED, Small PL et al (2010) Ecology and transmission of Buruli ulcer disease: a systematic review. PLoS Negl Trop Dis 4:e911
Meyers W, Connor D, McCullough B et al (1974a) Distribution of *Mycobacterium ulcerans* infections in Zaire, including the report of new foci. Ann Soc Belg Med Trop 54:147–157
Meyers WM, Shelly WM, Connor DH (1974b) Heat treatment of *Mycobacterium ulcerans* infections without surgical excision. Am J Trop Med Hyg 23:924–929
Meyers WM, Shelly WM, Connor DH et al (1974c) Human *Mycobacterium ulcerans* infections developing at sites of trauma to skin. Am J Trop Med Hyg 23:919–923
Moncayo A, Yanine MO. (2007) The neglected diseases and their economic determinants. Encyclopedia of infectious diseases: modern methodologies. Wiley, Hoboken, NJ, pp 603–617.
Mve-Obiang A, Lee RE, Portaels F et al (2003) Heterogeneity of mycolactones produced by clinical isolates of *Mycobacterium ulcerans*: implications for virulence. Infect Immun 71:774–783
Nguyen HH, Fadul N, Ashraf MS et al (2015) Osteomyelitis infection of *Mycobacterium marinum*: a case report and literature review. Case Rep Infect Dis 2015:905920
Nienhuis WA, Stienstra Y, Thompson WA et al (2010) Antimicrobial treatment for early, limited *Mycobacterium ulcerans* infection: a randomised controlled trial. Lancet 375:664–672
Nienhuis WA, Stienstra Y, Abass KM et al (2012) Paradoxical responses after start of antimicrobial treatment in *Mycobacterium ulcerans* infection. Clin Infect Dis 54:519–526
Norden A, Linell F (1951) A new type of pathogenic *Mycobacterium.* Nature 168:826
O'Brien DP, McDonald A, Callan P et al (2012) Successful outcomes with oral fluoroquinolones combined with rifampicin in the treatment of *Mycobacterium ulcerans*: an observational cohort study. PLoS Negl Trop Dis 6:e1473
O'Brien DP, Robson M, Friedman ND et al (2013) Incidence, clinical spectrum, diagnostic features, treatment and predictors of paradoxical reactions during antibiotic treatment of *Mycobacterium ulcerans* infections. BMC Infect Dis 13:416
O'Brien DP, Friedman ND, McDonald A et al (2014) Clinical features and risk factors of oedematous *Mycobacterium ulcerans* lesions in an Australian population: beware cellulitis in an endemic area. PLoS Negl Trop Dis 8:e2612
Ogbechi J, Ruf M-T, Hall BS et al (2015) Mycolactone-dependent depletion of endothelial cell thrombomodulin is strongly associated with fibrin deposition in Buruli ulcer lesions. PLoS Pathog 11:e1005011
Ojha AK, Trivelli X, Guerardel Y et al (2010) Enzymatic hydrolysis of trehalose dimycolate releases free mycolic acids during mycobacterial growth in biofilms. J Biol Chem 285:17380–17389
Oliveira MS, Fraga AG, Torrado E et al (2005) Infection with *Mycobacterium ulcerans* induces persistent inflammatory responses in mice. Infect Immun 73:6299–6310
Pahlevan AA, Wright DJ, Andrews C et al (1999) The inhibitory action of *Mycobacterium ulcerans* soluble factor on monocyte/T cell cytokine production and NF-kappa B function. J Immunol 163:3928–3935
Paoli P, Giannoni E, Chiarugi P (2013) Anoikis molecular pathways and its role in cancer progression. Biochim Biophys Acta 1833:3481–3498

Phanzu DM, Suykerbuyk P, Imposo DB et al (2011) Effect of a control project on clinical profiles and outcomes in Buruli ulcer: a before/after study in Bas-Congo, Democratic Republic of Congo. PLoS Negl Trop Dis 5:e1402

Phillips MS, von Reyn CF (2001) Nosocomial infections due to nontuberculous mycobacteria. Clin Infect Dis 33:1363–1374

Phillips R, Horsfield C, Kuijper S et al (2005) Sensitivity of PCR targeting the IS2404 insertion sequence of *Mycobacterium ulcerans* in an Assay using punch biopsy specimens for diagnosis of Buruli ulcer. J Clin Microbiol 43:3650–3656

Phillips R, Sarfo FS, Guenin-Mace L et al (2009) Immunosuppressive signature of cutaneous *Mycobacterium ulcerans* infection in the peripheral blood of patients with buruli ulcer disease. J Infect Dis 200:1675–1684

Phillips RO, Sarfo FS, Abass MK et al (2014) Clinical and bacteriological efficacy of rifampin-streptomycin combination for two weeks followed by rifampin and clarithromycin for six weeks for treatment of *Mycobacterium ulcerans* disease. Antimicrob Agents Chemother 58:1161–1166

Pidot SJ, Porter JL, Marsollier L et al (2010) Serological evaluation of *Mycobacterium ulcerans* antigens identified by comparative genomics. PLoS Negl Trop Dis 4:e872

Pommelet V, Vincent QB, Ardant MF et al (2014) Findings in patients from Benin with osteomyelitis and polymerase chain reaction-confirmed *Mycobacterium ulcerans* infection. Clin Infect Dis 59:1256–1264

Portaels F, Elsen P, Guimaraes-Peres A et al (1999) Insects in the transmission of *Mycobacterium ulcerans* infection. Lancet 353:986

Porten K, Sailor K, Comte E et al (2009) Prevalence of Buruli ulcer in Akonolinga health district, Cameroon: results of a cross sectional survey. PLoS Negl Trop Dis 3:e466

Porter JL, Tobias NJ, Hong H et al (2009) Transfer, stable maintenance and expression of the mycolactone polyketide megasynthase mls genes in a recombination-impaired *Mycobacterium marinum*. Microbiology 155:1923–1933

Pouillot R, Matias G, Wondje CM et al (2007) Risk factors for Buruli ulcer: a case control study in Cameroon. PLoS Negl Trop Dis 1:e101

Pszolla N, Sarkar MR, Strecker W et al (2003) Buruli ulcer: a systemic disease. Clin Infect Dis 37: e78–e82

Qi W, Kaser M, Roltgen K et al (2009) Genomic diversity and evolution of *Mycobacterium ulcerans* revealed by next-generation sequencing. PLoS Pathog 5:e1000580

Quek TY, Athan E, Henry MJ et al (2007a) Risk factors for *Mycobacterium ulcerans* infection, southeastern Australia. Emerg Infect Dis 13:1661-1666.

Quek TY, Henry MJ, Pasco JA et al (2007b) *Mycobacterium ulcerans* infection: factors influencing diagnostic delay. Med J Aust 187:561–563

Radford AJ (1975) *Mycobacterium ulcerans* in Australia. Aust NZ J Med 5:162–169

Raghunathan PL, Whitney EA, Asamoa K et al (2005) Risk factors for Buruli ulcer disease (*Mycobacterium ulcerans* Infection): results from a case-control study in Ghana. Clin Infect Dis 40:1445–1453

Rallis E, Koumantaki-Mathioudaki E (2007) Treatment of *Mycobacterium marinum* cutaneous infections. Expert Opin Pharmacother 8:2965–2978

Recht J, Kolter R (2001) Glycopeptidolipid acetylation affects sliding motility and biofilm formation in *Mycobacterium smegmatis*. J Bacteriol 183:5718–5724

Reilly AF, McGowan KL (2004) Atypical mycobacterial infections in children with cancer. Pediatr Blood Cancer 43:698–702

Renzaho AM, Woods PV, Ackumey MM et al (2007) Community-based study on knowledge, attitude and practice on the mode of transmission, prevention and treatment of the Buruli ulcer in Ga West District, Ghana. Trop Med Int Health 12:445–458

Revill WDL, Barker DJP (1972) Seasonal distribution of mycobacterial skin ulcers. Br J Prev Soc Med 26:23–27

Roche B, Benbow ME, Merritt RW et al (2013) Identifying Achilles' heel of multi-host pathogens: the concept of keystone "host" species illustrated by *Mycobacterium ulcerans* transmission. Environ Res Lett 8(4):045009

Röltgen K, Pluschke G (2015) Mycobacterium ulcerans disease (Buruli ulcer): potential reservoirs and vectors. Curr Clin Microbiol Rep:1–9

Roltgen K, Qi W, Ruf MT et al (2010) Single nucleotide polymorphism typing of *Mycobacterium ulcerans* reveals focal transmission of buruli ulcer in a highly endemic region of Ghana. PLoS Negl Trop Dis 4:e751

Roltgen K, Stinear TP, Pluschke G. (2012) The genome, evolution and diversity of *Mycobacterium ulcerans*. Infect Genet Evol 12:522–529

Röltgen K, Bratschi MW, Ross A et al (2014) Late onset of the serological response against the 18 kDa small heat shock protein of *Mycobacterium ulcerans* in children. PLoS Negl Trop Dis 8:e2904

Roth A, Fischer M, Hamid ME et al (1998) Differentiation of phylogenetically related slowly growing mycobacteria based on 16S-23S rRNA gene internal transcribed spacer sequences. J Clin Microbiol 36:139–147

Ruf MT, Chauty A, Adeye A et al (2011a) Secondary Buruli ulcer skin lesions emerging several months after completion of chemotherapy: paradoxical reaction or evidence for immune protection? PLoS Negl Trop Dis 5:e1252

Ruf MT, Sopoh GE, Brun LV et al (2011b) Histopathological changes and clinical responses of Buruli ulcer plaque lesions during chemotherapy: a role for surgical removal of necrotic tissue? PLoS Negl Trop Dis 5:e1334

Ruf MT, Schutte D, Chauffour A et al (2012) Chemotherapy-associated changes of histopathological features of *Mycobacterium ulcerans* lesions in a Buruli ulcer mouse model. Antimicrob Agents Chemother 56:687–696

Sakyi SA, Aboagye SY, Darko Otchere I et al (2016) Clinical and laboratory diagnosis of Buruli ulcer disease: a systematic review. Can J Infect Dis Med Microbiol 2016:5310718

Sarfo FS, Phillips R, Asiedu K et al (2010a) Clinical efficacy of combination of rifampin and streptomycin for treatment of *Mycobacterium ulcerans* disease. Antimicrob Agents Chemother 54:3678–3685

Sarfo FS, Phillips RO, Rangers B et al (2010b) Detection of mycolactone A/B in *Mycobacterium ulcerans*-infected human tissue. PLoS Negl Trop Dis 4:e577

Sarfo FS, Le Chevalier F, Aka N et al (2011) Mycolactone diffuses into the peripheral blood of Buruli ulcer patients—implications for diagnosis and disease monitoring. PLoS Negl Trop Dis 5:e1237

Sarfo FS, Converse PJ, Almeida DV et al (2013) Microbiological, histological, immunological, and toxin response to antibiotic treatment in the mouse model of *Mycobacterium ulcerans* disease. PLoS Negl Trop Dis 7:e2101

Sarfo FS, Phillips RO, Zhang J et al (2014) Kinetics of mycolactone in human subcutaneous tissue during antibiotic therapy for *Mycobacterium ulcerans* disease. BMC Infect Dis 14:202

Scherr N, Gersbach P, Dangy JP et al (2013) Structure-activity relationship studies on the macrolide exotoxin mycolactone of *Mycobacterium ulcerans*. PLoS Negl Trop Dis 7:e2143

Schutte D, Pluschke G (2009) Immunosuppression and treatment-associated inflammatory response in patients with *Mycobacterium ulcerans* infection (Buruli ulcer). Expert Opin Biol Ther 9:187–200

Schutte D, Um-Boock A, Mensah-Quainoo E et al (2007) Development of highly organized lymphoid structures in Buruli ulcer lesions after treatment with rifampicin and streptomycin. PLoS Negl Trop Dis 1:e2

Sharma R, Lahiri R, Scollard DM et al (2013) The armadillo: a model for the neuropathy of leprosy and potentially other neurodegenerative diseases. Dis Model Mech 6:19–24

Silva MT, Portaels F, Pedrosa J (2009) Pathogenetic mechanisms of the intracellular parasite *Mycobacterium ulcerans* leading to Buruli ulcer. Lancet Infect Dis 9:699–710

Simmonds RE, Lali FV, Smallie T et al (2009) Mycolactone inhibits monocyte cytokine production by a posttranscriptional mechanism. J Immunol 182:2194–2202

Simpson C, O'Brien DP, McDonald A et al (2013) *Mycobacterium ulcerans* infection: evolution in clinical management. ANZ J Surg 83:523–526

Snyder DS, Small PL (2003) Uptake and cellular actions of mycolactone, a virulence determinant for *Mycobacterium ulcerans*. Microb Pathog 34:91–101
Sopoh GE, Johnson RC, Chauty A et al (2007) Buruli ulcer surveillance, Benin, 2003–2005. Emerg Infect Dis 13:1374–1376
Sopoh G, Johnson R, Anagonou S et al (2011) Buruli ulcer prevalence and altitude, Benin. Emerg Infect Dis 17:153–154
Stahl DA, Urbance JW (1990) The division between fast- and slow-growing species corresponds to natural relationships among the mycobacteria. J Bacteriol 172:116–124
Stelmack PL, Gray MR, Pickard MA (1999) Bacterial adhesion to soil contaminants in the presence of surfactants. Appl Environ Microbiol 65:163–168
Stinear T, Davies JK, Jenkin GA et al (2000) Identification of *Mycobacterium ulcerans* in the environment from regions in Southeast Australia in which it is endemic with sequence Capture-PCR. Appl Environ Microbiol 66:3206–3213
Stinear TP, Mve-Obiang A, Small PL et al (2004) Giant plasmid-encoded polyketide synthases produce the macrolide toxin of *Mycobacterium ulcerans*. Proc Natl Acad Sci USA 101:1345–1349
Stinear TP, Pryor MJ, Porter JL et al (2005) Functional analysis and annotation of the virulence plasmid pMUM001 from *Mycobacterium ulcerans*. Microbiology 151:683–692
Stinear T, Seemann T, Pidot S et al (2007) Reductive evolution and niche adaptation inferred from the genome of *Mycobacterium ulcerans*, the causative agent of Buruli ulcer. Genome Res 17:192–200
Swanson DS (1998) Central venous catheter-related infections due to nontuberculous *Mycobacterium* species. Pediatr Infect Dis J 17:1163–1164
Thomas BS, Bailey TC, Bhatnagar J et al (2014) *Mycobacterium ulcerans* Infection Imported from Australia to Missouri, USA, 2012. Emerg Infect Dis 20:1876–1879
Tobias NJ, Seemann T, Pidot SJ et al (2009) Mycolactone gene expression is controlled by strong SigA-like promoters with utility in studies of *Mycobacterium ulcerans* and buruli ulcer. PLoS Negl Trop Dis 3:e553
Torrado E, Adusumilli S, Fraga AG et al (2007a) Mycolactone-mediated inhibition of tumor necrosis factor production by macrophages infected with *Mycobacterium ulcerans* has implications for the control of infection. Infect Immun 75:3979–3988
Torrado E, Fraga AG, Castro AG et al (2007b) Evidence for an intramacrophage growth phase of *Mycobacterium ulcerans*. Infect Immun 75:977–987
Tran H, Kamino H, Walters RF (2008) *Mycobacterium marinum* infection. Dermatol Online J 14:7
Trevillyan JM, Johnson PD (2013) Steroids control paradoxical worsening of *Mycobacterium ulcerans* infection following initiation of antibiotic therapy. Med J Aust 198:443–444
Trott A (1992) Chronic skin ulcers. Emerg Med Clin North Am 10:823–845
Turankar RP, Lavania M, Singh M et al (2012) Dynamics of *Mycobacterium leprae* transmission in environmental context: deciphering the role of environment as a potential reservoir. Infect Genet Evol 12:121–126.
Uganda Buruli Group (1971) Epidemiology of *Mycobacterium ulcerans* infection (Buruli ulcer) at Kinyara, Uganda, 1971. Trans R Soc Trop Med Hyg 65:763–775
Unknown (2002) The Buruli mysteries: unanswered questins surround a growing epidemic. World Health Organization
Unknown (2005) Buruli ulcer. Centers for Disease Control and Prevention
Unknown (2014) Leprosy. In: Fact sheet Number 101. Centre, World Health Organization Media
Unknown (2015) Tuberculosis (TB). World Health Organization Website
van der Werf TS, Stienstra Y, Johnson RC et al (2005) *Mycobacterium ulcerans* disease. Bull World Health Organ 83:785–791
van Ravensway J, Benbow ME, Tsonis AA et al (2012) Climate and landscape factors associated with Buruli ulcer incidence in Victoria, Australia. PLoS One 7:e51074

van Soolingen D, de Haas PE, Haagsma J et al (1994) Use of various genetic markers in differentiation of Mycobacterium bovis strains from animals and humans and for studying epidemiology of bovine tuberculosis. J Clin Microbiol 32:2425–2433

Vandelannoote K, Durnez L, Amissah D et al (2010) Application of real-time PCR in Ghana, a Buruli ulcer-endemic country, confirms the presence of *Mycobacterium ulcerans* in the environment. FEMS Microbiol Lett 304:191–194

Vergne I, Chua J, Singh SB et al (2004a) Cell biology of Mycobacterium tuberculosis phagosome. Annu Rev Cell Dev Biol 20:367–394

Vergne I, Fratti RA, Hill PJ et al (2004b) *Mycobacterium tuberculosis* phagosome maturation arrest: mycobacterial phosphatidylinositol analog phosphatidylinositol mannoside stimulates early endosomal fusion. Mol Biol Cell 15:751–760

Wagner T, Benbow ME, Brenden TO et al (2008a) Buruli ulcer disease prevalence in Benin, West Africa: associations with land use/cover and the identification of disease clusters. Int J Health Geogr 7:25

Wagner T, Benbow ME, Burns M et al (2008b) A landscape-based model for predicting *Mycobacterium ulcerans* infection (Buruli ulcer disease) presence in Benin, West Africa. EcoHealth 5:69–79

Wallace JR, Gordon MC, Hartsell L et al (2010) Interaction of *Mycobacterium ulcerans* with mosquito species: implications for transmission and trophic relationships. Appl Environ Microbiol 76:6215–6222

Webb BJ, Hauck FR, Houp E et al (2009) Buruli ulcer in West Africa: strategies for early detection and treatment in the antibiotic era. East Afr J Public Health 6:144–147

WHO (2000a) Buruli ulcer: *Mycobacterium infection*. Geneva, Switzerland.

WHO (2000b) Buruli ulcer. *Mycobacterium ulcerans* infection. In: Asiedu K, Scherpbier R, Raviglione M (eds) WHOICDS/CPE/GBUIM. WHO, Geneva

WHO (2004a) Changing history: the world health report 2004. WHO, Geneva

WHO (2004b) Report of the World Health Organization 7th advisory group meeting on Buruli ulcer, 8–11 March 2004, Geneva, Switzerland. World Health Organization, Geneva

WHO (2004c) Resolution WHA571. Surveillance and control of *Mycobacterium ulcerans* disease (Buruli ulcer). In: Fifty-Seventh World Health Assembly, Geneva. World Health Organization, Geneva

WHO (2006) Neglected tropical diseases: hidden successes, emerging opportunities. World Health Organization, Geneva

WHO (2008) Provisional guidance on the role of specific antibiotics in the management of Mycobacterium ulcerans disease (Buruli ulcer). WHO, Geneva

WHO (2012) Treatment of Mycobacterium ulcerans disease (Buruli ulcer); guidance for health workers, Geneva

WHO (2015) Global tuberculosis report. World Health Organization, Geneva

Williamson HR, Benbow ME, Nguyen KD et al (2008) Distribution of *Mycobacterium ulcerans* in buruli ulcer endemic and non-endemic aquatic sites in Ghana. PLoS Negl Trop Dis 2:e205

Williamson HR, Benbow ME, Campbell LP et al (2012a) Detection of *Mycobacterium ulcerans* in the environment predicts prevalence of Buruli ulcer in Benin. PLoS Negl Trop Dis 6:9

Williamson HR, Benbow ME, Campbell LP et al (2012b) Detection of *Mycobacterium ulcerans* in the environment predicts prevalence of Buruli ulcer in Benin. PLoS Negl Trop Dis 6:e1506

Williamson HR, Mosi L, Donnell R et al (2014) *Mycobacterium ulcerans* fails to infect through skin abrasions in a guinea pig infection model: implications for transmission. PLoS Negl Trop Dis 8:e2770

Yamazaki Y, Danelishvili L, Wu M et al (2006a) The ability to form biofilm influences *Mycobacterium avium* invasion and translocation of bronchial epithelial cells. Cell Microbiol 8:806–814

Yamazaki Y, Danelishvili L, Wu M et al (2006b) *Mycobacterium avium* genes associated with the ability to form a biofilm. Appl Environ Microbiol 72:819–825

Yeboah-Manu D, Roltgen K, Opare W et al (2012) Sero-epidemiology as a tool to screen populations for exposure to *Mycobacterium ulcerans*. PLoS Negl Trop Dis 6:e1460
Yeboah-Manu D, Kpeli GS, Ruf MT et al (2013) Secondary bacterial infections of buruli ulcer lesions before and after chemotherapy with streptomycin and rifampicin. PLoS Negl Trop Dis 7:e2191
Zavattaro E, Boccafoschi F, Borgogna C et al (2012) Apoptosis in Buruli ulcer: a clinicopathological study of 45 cases. Histopathology 61:224–236

Chapter 6
Coccidioidomycosis: Increasing Incidence of an "Orphan" Disease in Response to Environmental Changes

Antje Lauer

Abstract The number of cases of reported coccidioidomycosis infection, also known as valley fever, has increased significantly in the southwestern USA since the late 1990s. The causative agent of the disease is a dimorphic, soil-dwelling fungus in the genus *Coccidioides* with two species *C. immitis* and *C. posadasii* that persist in dry soils in the form of arthroconidia that can become airborne when soil is disturbed. Excessive disturbance of native soils in *Coccidioides* endemic areas in California, Arizona, and also Mexico, mainly due to construction and agriculture, has resulted in increased dust emission in these locations. In addition, the prolonged and ongoing drought in the southwestern USA has led to an increased level of PM10 (particulate matter 10 μm or less in diameter) pollution, which resulted in an increase in the amount of airborne arthroconidia of these pathogens which in turn led to the increase in disease incidence. Because coccidioidomycosis is not prevalent in the entire USA, research to elucidate the ecology of the pathogen, medical research to develop a vaccine against the disease, as well as the search for new antifungal drugs with less negative side effects in patients never attracted significant funding in the past. As a result, coccidioidomycosis is often named either an "orphan disease" or a "neglected disease." In particular, a vaccine to protect humans from the pathogen has been elusive to date, and both accurate diagnosis and treatment of the disease have remained difficult. Therefore, prevention of coccidioidomycosis through reasonable reduction of exposure is likely the best way to reduce disease incidence and the associated human, animal, as well as financial losses. Although the causal relationships between environmental factors and disease incidence are not well understood at this time, it is reasonable to assume that arthroconidia of *Coccidioides* spp. will be affected by factors that impact the organism's life cycle as well as human exposure to airborne dust sources. The predicted climate change in the southwestern USA to a drier state in the future and increased soil disturbance due to dramatic population increase in *Coccidioides* endemic areas of California and Arizona will likely lead to further increase of coccidioidomycosis incidence and may even lead to the establishment of the

A. Lauer (✉)
California State University Bakersfield, Department of Biology, Bakersfield, CA, USA
e-mail: alauer@csub.edu

C.J. Hurst (ed.), *Modeling the Transmission and Prevention of Infectious Disease*,
Advances in Environmental Microbiology 4, DOI 10.1007/978-3-319-60616-3_6

pathogen in other areas where disease incidence has been traditionally low or nonexistent.

6.1 Introduction

Coccidioides spp. are considered to be the most virulent of the primary fungal pathogens of humans (Dixon 2001). For example, infection of laboratory monkeys via the pulmonary route with as few as ten arthroconidia of the pathogen has been shown to result in rapidly progressive coccidioidomycosis (Casadevall and Pirofski 2006). The dimorphic, soil-dwelling fungal pathogens *Coccidioides immitis* and *Coccidioides posadasii* are responsible for the disease. These two pathogens are represented by several ecotypes that can be found in semiarid areas usually referred to as the "lower Sonoran life zone," which includes areas of the southwestern USA and large parts of Mexico. These pathogens also prevail in some arid areas of South America and China (Hector and Laniado-Laborin 2005; Wang et al. 2015). Incidence of coccidioidomycosis, also known as valley fever, has increased significantly since the late 1990s. This "orphan" disease that is not well known outside its endemic area is difficult to diagnose because of the variety of symptoms it causes in patients (Saubolle et al. 2007; Nguyen et al. 2013). There is no vaccine available, and treatment is costly, lengthy, and not without side effects (Thompson et al. 2015). The following paragraphs focus on the history of coccidioidomycosis, the distribution of the pathogen, the epidemiology of the disease, treatment options, and the ecology of the pathogen under current and anticipated future climate conditions.

6.2 History of Valley Fever

The first case of valley fever in a human patient was observed and studied in South America by A. Posadas and R. Wernicke (1892). Incidence of the disease in humans and cattle was described shortly after also in California by E. Rixford (1894) and L. T. Giltner (1918) (see Rixford 1931; Rixford and Gilchrist 1931; Rixford et al. 1931). These first attempts to identify the causative agent of valley fever by applying Koch's postulates, and the first epidemiological studies to identify endemic areas of the pathogen, are still frequently cited by the scientific community that is involved in valley fever research. Some groundbreaking discoveries were made in the early years of the twentieth century by a few dedicated scientists after initially misidentifying the fungal pathogen as a member of the protozoa. The following paragraph briefly describes some of the important findings and discoveries of some of these scientists, physicians, and health-care administrators.

In 1891, Alexander Posadas, then a medical intern in Buenos Aires, treated an Argentinean soldier who had experienced a dermatological problem since 1889 that had begun with a lesion on his right cheek. The infection ultimately covered much of his cheek and spread through his body leading to ulcerations on his nose, his extremities, and his trunk by 1893. The disease was diagnosed as a malignant skin disease caused by a protozoan pathogen of the genus *Coccidia*, based on microscopic analyses of diseased skin and tissue samples. The patient eventually died in 1898 after 7 years of suffering from recurrent fever and dissemination of the disease to the entire body (Posadas 1892, 1900). By then, Alexander Posadas had successfully applied Koch's postulates by infecting several different mammalian species in the laboratory with the pathogen to confirm the pathogenic nature of the eukaryotic organism that resembled *Coccidia*, using infectious material from his patient (Posadas 1928).

At about the same time, a second patient who apparently contracted the disease in the San Joaquin Valley of California was identified as showing symptoms comparable to those described for the Argentinian soldier. This individual was a 40-year-old laborer who initially had complained about a sore on his neck, but the disease disseminated quickly, resulting in lesions all over his body and ultimately disfiguring his face. An autopsy after his death in 1895 revealed that the pathogen was established in his lungs and other organs, including lymph nodes, which had resulted in a compromised immune response. Patient tissue specimens including some from the lung revealed lesions that could be compared with progressive tuberculosis when observed microscopically, but a fungal pathogen was not suspected (Baker et al. 1943). Interestingly a fungal culture that was obtained from one of the patient specimens was discarded as contamination. Parasitologists at Johns Hopkins University named the pathogen after both its morphologic and clinical features *Coccidioides* ("resembling Coccidia") *immitis* ("not mild") (Rixford 1931a). At the turn of the twentieth century, the pathogen was finally identified as a fungus by William Ophüls and Herbert C. Moffitt (Ophüls and Moffitt 1900). By infecting laboratory animals with mycelium from a pure fungal culture isolated from patient specimens and subsequently observing the progress of the disease, they documented that *C. immitis* was in fact not related to protozoans at all but was a dimorphic fungus that formed hyphae outside the host on artificial media and spherical protozoan-like bodies, later called "spherules" in diseased tissue. The pathogen multiplies in these "spherules" which render protection from the host's immune system by forming numerous endospores. The endospores then can spread to neighboring tissue and into the blood, as well as lymphatic system when the spherules either burst naturally or are being removed by surgery. In 1905, Ophüls finally published findings in which he referred to the disease as "coccidioidal granuloma" (Ophüls 1905), also referred to as progressive or secondary coccidioidomycosis (Rixford and Gilchrist 1931; Rixford et al. 1931; Carter 1934). Interestingly, two other fungal pathogens that use the lung as their primary attack site and which were first described around the same time in the USA similarly were misidentified as protozoa (Gilchrist 1894; Darling 1906). These fungi which are also dimorphic and soil-dwelling species but are endemic in the southeastern USA are now well described and known as *Blastomyces dermatitidis* and *Histoplasma capsulatum* (Brandt et al. 2006).

In the 1930s, antigen derived from spherules of the pathogen (spherulin) or from the mycelial phase (coccidioidin) was isolated and used in skin tests, similar to the tuberculin skin test, to determine if a patient has formed antibodies against the pathogen. These tests were based on delayed-type hypersensitivity (DTH) reactions to the intradermal inoculation of antigens prepared from the pathogen and were also used for immune-identifying *Coccidioides* spp. cultures (Smith et al. 1956; Cox et al. 1977; Kaufman et al. 1985). The DTH skin tests were used to study the epidemiology of coccidioidomycosis and to assist in the clinical management of patients. However, since the late 1990s, when the patent expired, the commercial source of the reagent became unavailable in the USA but was still in use by physicians in Mexico. Recently a reformed spherulin-based skin test has been approved by the Food and Drug Administration (FDA) and has replaced the previously used ones (Wack et al. 2015).

Once the medical community became alerted about the pathogenic nature of *C. immitis* and how humans and animals can become infected with arthroconidia of the fungus, first steps were made to understand the progress of coccidioidomycosis, including the risk of dissemination from the primary site of infection and how to diagnose the disease correctly. It did not take long until additional cases were documented in the USA and the virulence of the fungal pathogen, the mechanisms behind its pathogenicity, and its life cycle became better understood. By the 1930s the Kern County Department of Public Health (as well as health officials in Arizona, especially in the Tucson and Phoenix area) became alerted about how widespread the disease actually is. In California, coccidioidomycosis became known as either "valley fever," "San Joaquin valley fever," "desert fever," "the bumps," or "desert rheumatism." This disease presents with symptoms that can include an acute cough, chest pain, fever, fatigue, weight loss, and pneumonia. These symptoms may be followed later by erythema which can present as either a nodosum or a multiforme. Multiforme is a condition typically affecting the extremities including hands but, and perhaps more noticeably, can have severe effects upon the lower legs. These erythemas are caused by inflammation of the fat cells under the skin and may be mediated by deposition of immune complexes often resulting in tender hypodermic rounded nodules. Necrosis of skin cells, such as those observed in the two initially described cases of coccidioidomycosis, is comparably rare.

6.2.1 First Epidemiological Studies

In the middle of the 1930s, the Kern County Department of Public Health started to investigate the epidemiology of valley fever by performing skin testing with coccidioidin for all cases that potentially could represent valley fever. Coccidioidal granuloma of the lung is difficult to distinguish from calcified tubercles caused by *Mycobacterium tuberculosis*, a widespread bacterial pathogen which causes the highly contagious disease tuberculosis, now more typically associated with humans who suffer from a compromised immune system and/or humans who are living in poor conditions (disease of the poor). Furthermore, tuberculosis often is associated with living in overcrowded environments, especially prisons and homeless shelters. A laboratory demonstration of the coccidioidal fungus growing in cultivation from patient specimens was often used to confirm that a patient indeed suffered from valley fever and not from tuberculosis, especially when other differentiating symptoms such as erythema nodosum were absent. The development of the tuberculin skin test and the coccidioidin skin test enabled physicians to determine if a patient had been exposed to each pathogen at some time in the past or was suffering from an acute infection accompanied with severe symptoms of the disease. This type of investigation revealed that most coccidioidal patients described a history of dust exposure. Random testing of the general public showed that about 25% of Kern County's inhabitants had been exposed to the *Coccidioides* pathogen at some time in their life and thus had produced antibodies against the fungus. These observations therefore also confirmed that coccidioidomycosis was endemic in this geographical area. In fact, the pathogen had already been isolated from soils near Delano in 1932 where members of a Filipino working crew contracted coccidioidomycosis (Steward and Meyer 1932). One of the first larger-scale epidemiological studies on valley fever was performed shortly after in 1937 by Charles E. Smith, who investigated more than 400 farmworkers from the counties of Tulare and Kern in California who had been infected with the pathogen and where cases were already known to the Health Department (Smith 1940). Several important outcomes derived from this early study: first, it appeared that African American and Filipino Americans had a significantly higher risk of contracting the disseminated form of the disease as compared to white farmworkers; second, immunologic response to an attack, as reflected by skin reactivity to coccidioidin, seemed to protect people from further infection with *C. immitis*; and third, the disease was not contagious; infection arose from the inhalation of arthroconidia especially during summer and fall when the soil was dry and fungal propagules become airborne. At about the same time, coccidioidin skin testing performed by Gifford on 2718 schoolchildren showed that reactivity increased according to length of residence in the area. About 80% of the children had positive results after 10 years of residence, but only about 5% of these 80% had been treated for valley fever, indicating that most infections had been mild (Gifford 1936, see also Thorner 1941). Another study performed by Aronson et al. (1942) on Native Americans of Pima County near Phoenix in Arizona revealed that almost all tested men and women had positive coccidioidin skin test results before the age of 20 years. Several studies by Dickson in the middle of the 1930s (e.g., Dickson 1937) noted that most infections with *Coccidioides* spp. were mild ("benign" form) characterized either by weak symptoms or none at all, sometimes causing a pulmonary infection either with or without erythema nodosum and erythema multiforme. Dickson and Gifford also demonstrated that the life-threatening disseminated form of the disease, which only affects a small percentage of patients in which they suffer from progressive and malignant granulosomas when tissue and lymphatic resistance are overcome and the pathogen spreads via the blood and lymphatic system, was caused by the same pathogen as was the milder infection (Gifford et al. 1937; Dickson 1937).

Most of the early crucial clinical and epidemiological information on coccidioidomycosis derived in fact from events such as the migration of farmers from the "Dust Bowl" area in primarily Oklahoma and Kansas to the southwestern USA, as well as from the movement of military personnel, Japanese internees, and Axis prisoners of war to the San Joaquin Valley and other *Coccidioides* endemic areas during the early and mid-1940s as a consequence of World War II. A large percentage of these displaced persons contracted valley fever, proving again that *Coccidioides* spp. are endemic to some semiarid and arid soils in this area (Smith et al. 1946; Lewis and Mewha 1955). These first observations were confirmed by later epidemiological studies in California and Arizona, but many questions remained with regard to the ecology of the pathogen and the differences in susceptibility among different human ethnicities. The ecology of these pathogens and their association with humans still are not fully understood and are the focus of today's research, more than 120 years after *Coccidioides* spp. were first described (Ampel 2010; Brown et al. 2013; Thompson et al. 2015).

It is now accepted knowledge that coccidioidomycosis is comparably rare in South America where it was first observed. Most of the reported cases from Argentina, Brazil, Bolivia, and Paraguay which were initially suspected to be valley fever are in fact paracoccidioidal granuloma, caused by *Paracoccidioides brasiliensis*, also a dimorphic fungus, but the disease it causes differs in etiology and clinical characteristics (de Almeida 1933; Jordan and Weidman 1936). However, it cannot be denied that *C. posadasii* is established in some semiarid areas in South America (Sifuentes-Osornio et al. 2012).

6.3 Epidemiology Today

6.3.1 Coccidioides *Endemic Areas*

The endemic regions of *Coccidioides* spp. roughly correspond to the "lower Sonoran life zone," as defined by Merriam in the late twentieth century (Daubenmire 1938). Areas that belong to this life zone are characterized by low rainfall, high summer temperatures, and moderate winter temperatures. Regions that fit this description are found in the southern deserts of Arizona (including Maricopa, Pinal, and Pima counties), the Central Valley and southern portions of California (including Kern, Tulare, and San Luis Obispo counties), the southern tip of Nevada, southern Utah, southern New Mexico, western Texas (especially along the Rio Grande), and the northern and Pacific coastal areas of Mexico (Fig. 6.1). Recently, a pocket of *Coccidioides* has been identified in eastern Washington State (Marsden-Haug et al. 2012). Some endemic areas have been identified in Central and South America as well (Sifuentes-Osornio et al. 2012). However, it has to be noted that the endemic

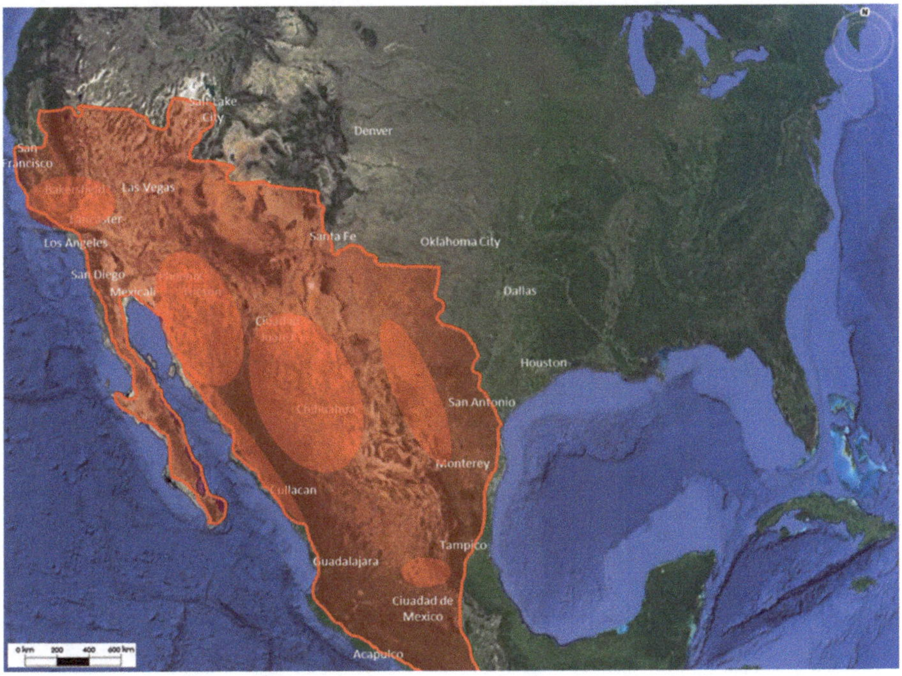

Fig. 6.1 Map of *Coccidioides* spp. endemic areas and suspected endemic areas in the southwestern USA and Mexico based on incidence of coccidioidomycosis and skin testing. Areas of highest incidence (highly endemic) are shaded in *darker red* (map based on Ochoa 1967, Kolivras et al. 2001, and Hector and Laniado-Laborin 2005)

areas of *Coccidioides* are not well defined. Absence of incidence or low incidence in areas outside the "lower Sonoran life zone" does not necessarily mean that the fungal pathogen is not part of the microbial community in the soil; it just has not been investigated. Endemic areas of *Coccidioides* spp., such as Kern County, traditionally known as a hotspot of the pathogen have seen a significant increase in population over the last 15 years (~58%), whereas the general population of California has only increased ~30% (US Census Bureau http://www.census.gov/population/) during that same time period, and the overall population of the USA increased ~24% for comparison. The population of Phoenix, Arizona, an endemic area for *C. posadasii*, has increased ~65% since 1990. In these affected regions, large areas of native soils have been disturbed and pristine land has been converted into agricultural fields. There also has been an establishment of oil rigs, housing complexes, and roads, and more recently large areas in these regions have been disturbed for the construction of photovoltaic stations and wind energy turbines (Carley 2009). This soil disturbance has led to a significant increase in PM10 and PM2.5 (particulate matter of 10 and 2.5 μm or less in diameter) pollution especially during the dry seasons. Small soil particles, such as clay and spores from microorganisms, including arthroconidia from *Coccidioides* spp. can become airborne on windy days and contribute to dust development and air pollution, resulting in low air quality (at least unhealthy for sensitive groups during summer and fall) especially in the Southern San Joaquin Valley. The air quality of the San Joaquin Valley is monitored daily by the San Joaquin Valley Air Pollution Control District and can be assessed at http://www.valleyair.org/Home.htm. For Arizona, information about air quality is available from the Arizona Department of Environmental Quality (ADEQ) and can be assessed at http://www.azdeq.gov/environ/air/.

Until October 2012, *Coccidioides* spp. had been listed as a select agent by both the US Department of Health and Human Services and the US Department of Agriculture and was considered a biosafety level 3 pathogen. The pathogen was delisted, not because it had become less of a concern—in contrary!—but because of financial difficulties for researchers to provide a biosafety 3 laboratory to conduct desperately needed clinical and environmental research.

6.3.2 Incidence of Valley Fever in California and Arizona

A significant increase in valley fever incidence has been documented in *Coccidioides* endemic areas of California, especially in Kern County, but also in Arizona primarily in Pima, Maricopa, and Pinal County (Phoenix and Tucson area) compared to the late 1990s. An alarming increase in the rates of coccidioidomycosis has been documented not only among the general public, construction workers, and farm laborers (Das et al. 2012) but also higher rates of hospitalization for coccidioidomycosis among children (McCarty et al. 2013), and incarcerated populations (Pappagianis 2007; Burwell et al. 2009) have been documented which resulted in increased human suffering and health-care expenses (Sondermeyer et al. 2013).

To characterize long-term national trends of valley fever, the Center for Disease Control and Prevention (CDC) analyzed data from the National Notifiable Diseases Surveillance System (NNDSS) for the period 1998–2011 (Adams et al. 2013). Results of this analysis revealed that the incidence of reported coccidioidomycosis increased substantially during this period, from 5.3 cases per 100,000 population in the endemic area (Arizona, California, Nevada, New Mexico, and Utah) in 1998 to 42.6 cases per 100,000 in 2011. The total number of valley fever cases rose by more than 850% between 1998 and 2011 in the area where valley fever is most common—California, Arizona, Nevada, New Mexico, and Utah—to 22,401 cases. In California, the case counts rose from 719 to 5697 over the 13-year period. In states where the disease is not as common, the number of valley fever cases reported jumped from 6 in 1998 to 240 in 2011. The CDC also pointed out that a 2006 study found only 2 to 13% of patients with signs and symptoms were tested for valley fever, indicating that the disease is likely greatly underreported (Adams et al. 2013; Brown et al. 2013). Figure 6.2 displays the increase in incidence of coccidioidomycosis, comparing Arizona and California, and also compares disease incidence for several counties in California that are known to have high numbers of cases, such as Kern County and the Antelope Valley Area, and includes different Service Planning Areas (SPA) for

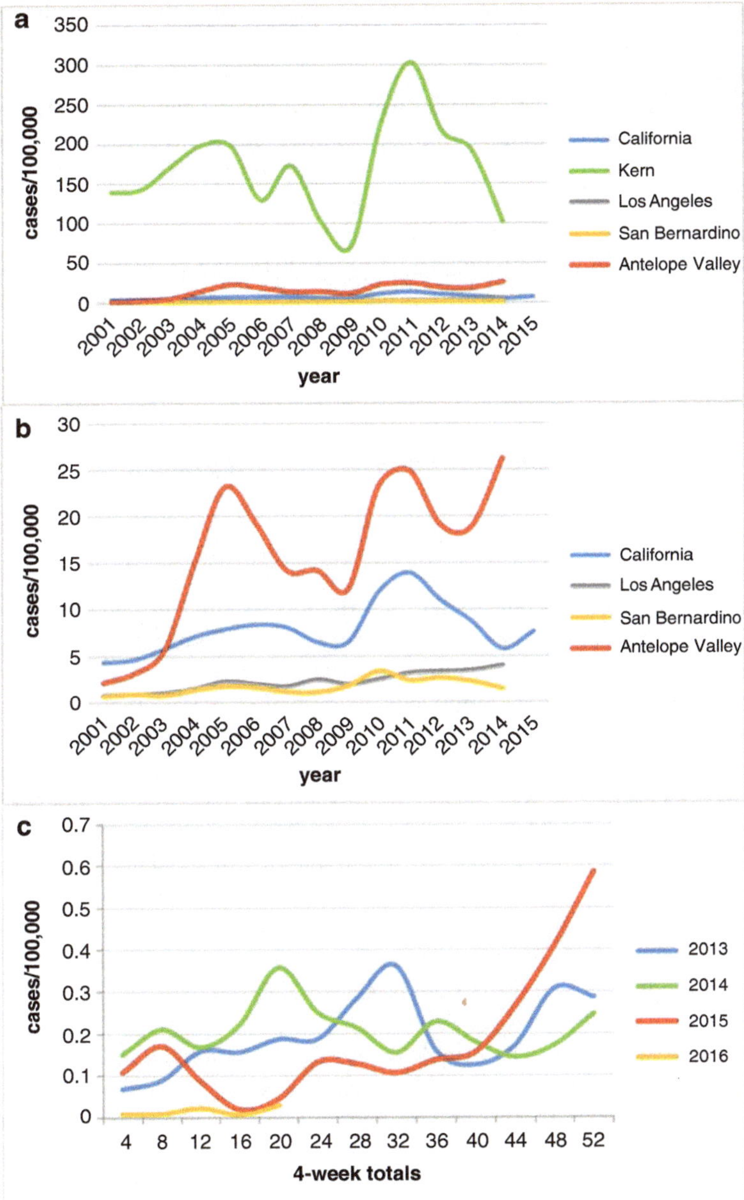

Fig. 6.2 (**a**) Incidence of valley fever in California comparing Kern County, Los Angeles County, San Bernardino County, and the Antelope Valley Area between 2000 and 2014. (**b**) Incidence of valley fever in California without displaying Kern County. (**c**) Incidence of valley fever between 2013 and today (spring 2016) for California (data sources: Center of Disease Control, Morbidity Tables and County of Los Angeles Dept. of Public Health: Annual Morbidity Reports and Special Studies Reports; California Department of Public Health: California Health and Human Services Open Data Portal, Diseases and Conditions)

Los Angeles County. Kern County is obviously the county that always had and still has the highest incidence of coccidioidomycosis. The graphs display the incidence per 100,000 people and not the number of cases; this means that the varying numbers of population in the different counties were taken into account. Within Los Angeles County, the population of the Antelope Valley Area (SPA1) is certainly at a significantly higher risk to contract valley fever, compared to the other SPAs in the same county. Peak infection rates occur during the driest periods of the year. In Arizona, this is the early summer and late fall, whereas in California, it is all throughout the summer. In winter and spring 2016, the increase in precipitation due to El Niño conditions resulted in less fugitive dust development and subsequently a decrease in valley fever cases.

6.3.3 Spectrum of Coccidioidomycosis, Diagnosis, and Therapies

The incubation period of valley fever depends on the number of arthroconidia inhaled and the status of the patient's immune system and can therefore range between 1 and 3 weeks, after which a variety of symptoms develop; among them are extreme fatigue, weight loss, night sweats, and pulmonary symptoms (cough, chest pain, dyspnea, and hemoptysis) accompanied with fever in 50% of the patients and occasionally headaches and joint pain. Signs of the disease unfortunately overlap substantially with the presentation of viral or bacterial respiratory infections. Occasionally, skin manifestations are observed that are not specific for infections with *Coccidioides* spp. and include diffuse rashes, erythema nodosum, and erythema multiforme which are immunologically mediated (Galgiani et al. 2005). Nonwhite race generally is a predictor for dissemination of the disease, and African American patients more often develop disseminated bone disease, while Filipinos are more likely to develop cutaneous or central nervous system disease due to genetic predispositions (Louie et al. 1999; Crum et al. 2004 and references within; Fierer 2007; Viriyakosol et al. 2013).

Coccidioidomycosis is a reportable disease at the national level, and reporting is required in Arizona and California where cases annually number in the thousands. Even though it appears that these numbers are high, it has to be considered that cases are underreported by as high as 90%, because of the lack of technicians testing for coccidioidomycosis and by frequent misdiagnosing of inexperienced physicians (Brown et al. 2013). It has to be noted though that undiagnosed infections are almost certainly not as serious as those that are recognized. Nonetheless, an early diagnosis of the disease reduces health-care costs significantly and might save a patient's life (Hector et al. 2011; Thompson et al. 2011; Huang et al. 2012). The fact that Arizona has approximately twice as many infections as California can be explained by the differences in the population sizes in the highly endemic regions of the two states (Galgiani et al. 2005; Sunenshine et al. 2007). The majority of infected humans might not realize that they had been infected with either *Coccidioides immitis* or *C. posadasii,* because their immune system eliminated the pathogen before it could become established in the body. It is assumed that lifelong immunity results from this immune response (Kirkland and Fierer 1996). Many infected humans realize very mild symptoms, so that they see no need for medical attention. It is estimated that approximately one-third of infected humans who do suffer from valley fever as a clinical illness have symptoms that resemble those of a community-acquired pneumonia including fever, fatigue, and a dry and persistent cough which typically either clears on its own after several weeks of illness or when treated with an antifungal drug (Brown et al. 2013). The successful detection of *Coccidioides* spp. in patient specimens depends on either cultivation success or the sensitivity of DNA-based and serology-based laboratory methods which variously determine the presence of the pathogen in blood, sputum, urine, and spinal fluid. Of course, the success of diagnosis also depends on experience of the physician or laboratory scientist and their recommendation of testing for *Coccidioides* in the first place, because symptoms and signs of valley fever are not uniform among patients (Galgiani et al. 1991; Anstead and Graybill 2006). Whether diagnosed or not, most infections are controlled by induction of immunity, although the associated illness may last for many weeks to many months. There are currently five commercially available oral antifungal drugs with activity for treating coccidioidal infections: ketoconazole, fluconazole, itraconazole, voriconazole, and posaconazole. Treatment with fluconazole or itraconazole for such patients typically involves doses ranging from 200 to 400 mg per day, with treatment durations ranging from several to many months. The cost of therapy is substantial. Drug costs alone range from $2000 to $20,000 per year and patient, depending upon the specific drug and the daily dosage prescribed (Warnock 2006; Hector et al. 2011). If the disease spreads beyond the initial site of the infection, which typically is the lung, then treatment is highly recommended, because complications from such dissemination can lead to chronic illness and in rare cases even death. Amphotericin B is effective only if administered parenterally, and its use is often associated with significant side effects and toxicities (Como 1994). A new antifungal drug, Nikkomycin Z, which has been first described in detail in the 1990s and has less side effects compared to the other antifungals, is finally in clinical trials (Hector et al. 1990; Shubitz et al. 2014). If the disease has disseminated, then treatment usually continues for many months to years. When therapy is discontinued after the apparent successful control of the disease, a relapse of infection occurs in approximately one-third of patients. Therefore, some patients may need lifelong therapy to maintain control. Highly vulnerable to the disseminated form of valley fever are patients with compromised immune systems which include pregnant women and children whose immune system is in the developmental stage. It has been documented that about 5–10% of all *Coccidioides* infections result in pulmonary sequelae, which can mean serious damage to the lung, and 1% or less result in the spread of the infection outside of the lungs. Extrapulmonary infections by these pathogens can lead to destructive lesions in the skin, bones, joints, meninges, and virtually any other organ or tissue in the body to which the infection has spread. The high incidence among incarcerated prisoners is likely the status of their immune system, because drug abuse is common in these facilities. Furthermore, many of these detention centers are located in desolate *Coccidioides* spp. endemic areas in the southwestern USA. Since most coccidioidal infections can only be diagnosed by specific laboratory testing, the lack of clinicians testing for valley fever could easily account for the significant underreporting of illness (McLean 2012). In many cases, patients with persistent respiratory infections often received empiric antibiotics or corticosteroids in ambulatory practices. In one study, 81% of patients with valley fever pneumonia received at least one course, and 31% received multiple courses of antibacterial treatment for their illness (Valdivia et al. 2006). Of concern is here not only the cost of antibiotics but more so the potential increase in antibiotic-resistant bacterial pathogens in the community. Furthermore, it has been observed that the use of anti-inflammatory corticosteroids which are often administered to treat cutaneous or rheumatologic symptoms that may accompany primary coccidioidal infection can have negative impacts on the patient's immune system which can favor spread of the disease beyond the primary site of infection (Shoham and Levitz 2005). Finally, by establishing a diagnosis of coccidioidomycosis early, complications (should they arise) may be more quickly recognized and treated. In summary, the attitude that primary care professionals take regarding early diagnosis of coccidioidal infections is critical to all further discussion about the proper management of this infection in the primary care setting. Recommendations to physicians regarding how to diagnose and treat valley fever are available in more detail at the website of the "Valley Fever Center for Excellence" at the University of Arizona at Tucson (https://www.vfce.arizona.edu/), as well as information for the general public. Development of a vaccine to protect the general public from coccidioidomycosis has not been successful so far, but research in that area is ongoing (Pappagianis and Levine 1975; Cole et al. 2004; Thompson et al. 2015).

6.4 Ecology of *Coccidioides* spp.

6.4.1 Coccidioides *spp. in the Environment*

There are currently two species described in the genus *Coccidioides*, and both of them are able to cause valley fever in humans and animals: *C. immitis* and *C. posadasii*. In the past, all strains were designated as *C. immitis* (the California and the non-California ones), but recent genetic analyses have shown that strains segregate into two distinct groups which represent several ecotypes (Fisher et al. 2002; Whiston and Taylor 2014). Strains now designated *C. immitis* in most cases originate from infections contracted in California, and those designated *C. posadasii* are from infections contracted elsewhere. Hybridization of both species has been confirmed as well, and it is suspected that both species might co-occur in some areas (Neafsey et al. 2010; Johnson et al. 2014). *Coccidioides* spp. are dimorphic fungi that form vegetative hyphae (mycelium) when growing as saprophytes in moist soil, biodegrading dead organic matter. The pathogen is somehow adapted to a variety of environments that could be characterized as the "lower Sonoran life zone" (Merriam 1898) that comprises arid and semiarid areas in the southwestern USA and habitats with similar climate conditions and soils in Mexico and South America (Maddy and Coccozza 1964; Maddy 1965; Kolivras et al. 2001; Fisher et al. 2007; Baptista-Rosas et al. 2007, 2012; Lauer et al. 2012, 2014). *Coccidioides* spp. are not known to be established in any other semiarid or arid environments on other continents. Soil types in *Coccidioides* endemic areas are unique in the Americas; they comprise desert soils and desertlike soils and some of them are alluvial deposits (Fig. 6.3). This figure shows that soils in the Mojave Desert and in endemic areas of Arizona are very similar and characterized by large areas of Mojave soil. The areas of Lancaster in the Antelope Valley located in the western Mojave Desert and Bakersfield in the Southern San Joaquin Valley also share soils characterized as Panoche loam. The map gives a broad overview based on soil parent material. More detailed information can be obtained from the US Department of Agriculture (USDA) websoilsurvey database for all locations. Figure 6.4 shows an example indicating different soil types (separated by orange lines) at a *Coccidioides*-positive site near Lancaster, California. Figure 6.5 shows various locations where *C. immitis* was detected in the western Mojave Desert in recent years.

It is not known how long *Coccidioides* spp. have been established in soils of the Americas. Genomic analyses suggested that *C. immitis* and *C. posadasii* split in two species around 5 million years ago (Sharpton et al. 2009). It has been proposed that *Coccidioides* as a genus has existed much longer—possibly as long as 40–50 million years (Bowman et al. 1992; Kofoupanou et al. 2001; Fisher et al. 2002). Other authors speculate that *Coccidioides* spp. emerged in the early Cenozoic period (Paleocene, ~65 million years ago), corresponding to a point in time where there was rapid diversification of mammalian orders, with the first fossils of *Rodentia* appearing (Rose 2006). Coccidioidal infections of homeotherms in North America have occurred during at least the past 8500 years (early Holocene), suggesting that *Coccidioides* has evolved over a relatively long time span as a pathogen of *Homo sapiens*. Evidence from the fossil record is rare. A pathological, anatomical investigation of early Holocene bison mandibles (*Bison antiquus*) discovered in Nebraska, and dated 6500 A.D., discovered severe, locally extensive, mandibular osteomyelitis with intralesional spherules morphologically consistent with the genus *Coccidioides*. This implies either the fossil home range of *Coccidioides* was larger than it is today or ancient bison migrated between endemic and non-endemic areas during the early Holocene (Morrow 2006). Microscopic investigations of bone thin sections from an ancient Native American skeleton found in northern Arizona found evidence of disseminated coccidioidomycosis. The skeleton dated back to 1000–1400 A.D. and belonged to a 40- to 50-year-old individual of the Sinagua culture which settled in the American southwest. The unexpectedly high preservation of organic material in the skeleton included spherules of the pathogen, endospores, and red and white blood cells. The lesions observed in the bones were so

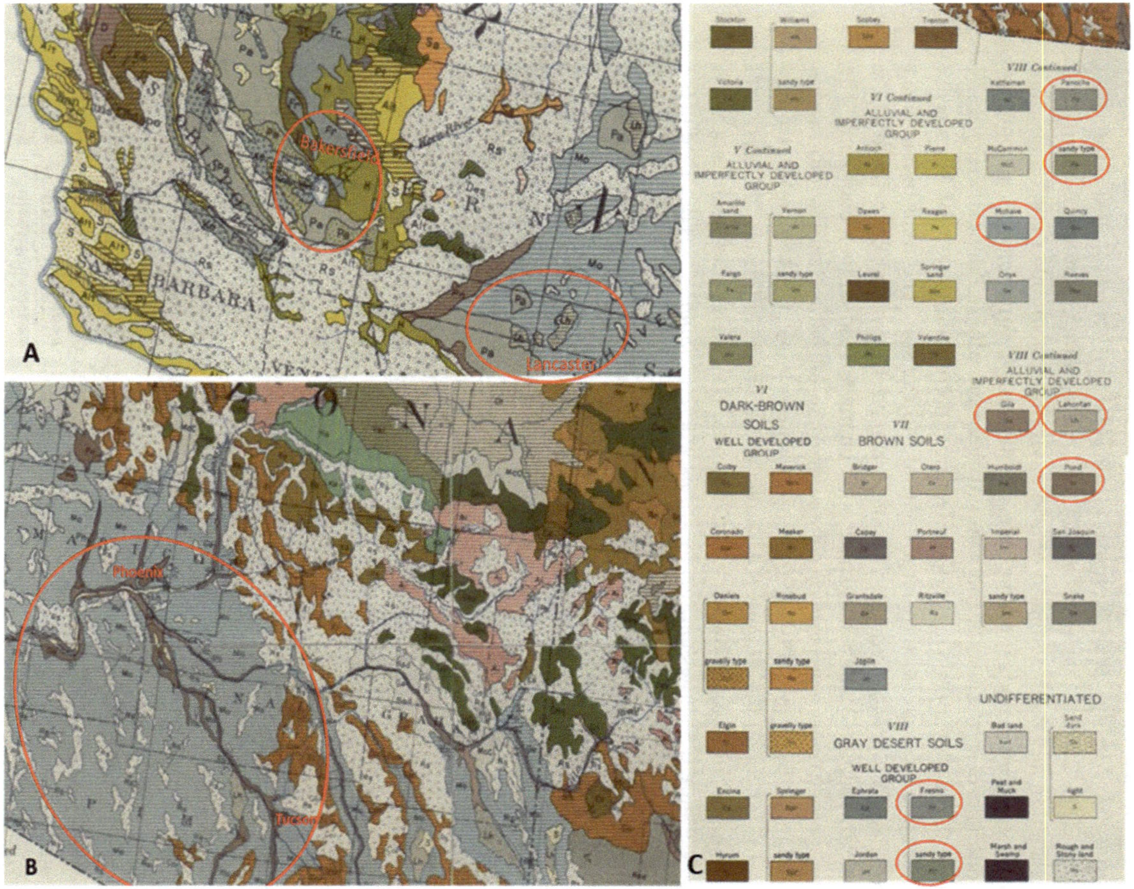

Fig. 6.3 (**a**) Map of soil types in *Coccidioides* endemic areas in California and Arizona. The cities and surrounding areas of Bakersfield, Lancaster, Phoenix, and Tucson which have the highest incidence of valley fever are *encircled in red*. (**b**) Map of soil types in western Arizona with the location of Phoenix and Tucson which have the highest incidence of valley fever is *encircled in red*, as well. The legend indicates soil types that are characteristic for this arid area, such as Mojave soil (Mo), Panoche soil (Pa), and others (*circled in red*) (for more details please see Composite: Soil Map of the United States, scale 1: 2,500,000 US Department of Agriculture, Washington, D.C., available at http://www.davidrumsey.com/)

Antelope Valley Area, California (CA675)			
Map Unit Symbol	Map Unit Name	Acres in AOI	Percent of AOI
HgA	Hesperia loamy fine sand, 0 to 2 percent slopes	18.5	1.8%
HkA	Hesperia fine sandy loam, 0 to 2 percent slopes	177.3	16.8%
HmA	Hesperia fine sandy loam, loamy substratum, 0 to 2 percent slopes	0.2	0.0%
Rm	Rosamond loamy fine sand	14.3	1.4%
Ro	Rosamond fine sandy loam	417.6	39.5%
Rp	Rosamond loam	387.3	36.6%
Rr	Rosamond loam, saline-alkali	4.9	0.5%
Rs	Rosamond loam, sandy loam substratum	20.0	1.9%
Rt	Rosamond silty clay loam	17.1	1.6%
Totals for Area of Interest		1,057.2	100.0%

Fig. 6.4 (**a**) Aerial overview photo of individual sampling spots along Transect F located in the western Mojave Desert (Antelope Valley, California). The Soil Conservation Service Map obtained from the USDA websoilsurvey database indicated two different soil map units (Ro and Rp) which represent different soil types (Rosamond loam and Rosamond sandy loam, respectively). The *red dots* show the location of individual sampling spots. (**b**) Overview of site F showing disturbed soil of a fallow agricultural field that had not been under management for several years. The vegetation consists of mostly invasive grasses and herbs with bare spots throughout. Rodent activity was observed as well

massive and widespread that they suggest a systemic infection rather than a contamination from soil or a postmortem infection (Harrison et al. 1991).

It is not known how long arthroconidia of the pathogen can remain viable in the environment. Revivals from dormant structures such as seeds or spores after short periods of time are well-documented in the scientific literature. Furthermore, cultivation success of microbial species from fossils or ancient soils and sediments that are millions of years old has been published. For example, Sugiyarna and Goto

Fig. 6.5 (**a**) Overview of soil sampling sites located north of the city of Rosamond, California. (**b**) Overview of sampling site Galt. This site also was dominated by different species of saltbushes, herbs, and wildflowers in an area that appeared to have been covered by water in the rainy season, as seen above. (**c**) Overview of sampling site six on the northern end of the Rosamond dry lake bed with saltbushes being the dominant plant species in the area. Creosote bushes were found as well farther away from the dry lake bed at slightly higher elevations. The non-vegetated area in between the saltbushes was colonized by biological soil crusts. (**d**) Sampling site LJ on the western edge of Roger's dry lake bed, showing shallow dunes and saltbushes, as well as scattered Joshua trees which can be seen in the background of the photo

(1969) claim to have cultured various fungal species from the genera of *Penicillium*, *Aspergillus*, *Cladosporium*, *Trichoderma*, *Candida*, and *Rhodotorula* from Pliocene and Miocene soil core samples in Japan (between 5 and 24 million years old, from 600 to 3200 m depth) (see also Kennedy et al. 1994 and references within). Cryptobiologists and astrobiologists have provided evidence that microbes can survive millions of years either in the form of dormant structures or by extremely slowing down their metabolism to survive in either hostile or extreme nutrient-poor environments. Such evidence seems to be abundant but also is skepticism especially when the "revived" microbes resemble ubiquitous microbes that are known to be common contaminants from surrounding soils, water, and air (see Amy and Halderman 1997 and references within; Gilichinsky et al. 2007;

Fig. 6.6 Site near Sharktooth Hill, east of Bakersfield, California, where fossil diggers exposed gray sedimentary soil that contains marine fossils from invertebrates and vertebrates

Onofri et al. 2007). The San Joaquin Valley of California has been a marine environment and later a large freshwater lake in the past. Two-million-year-old sediment layers that include fossils from marine vertebrates and invertebrates have been discovered at a location called Sharktooth Hill in Kern County. This is also one of the first locations where *C. immitis* was isolated from topsoils in the 1950s and 1960s (Egeberg and Ely 1956; Swatek et al. 1967). Deeper soil layers have not been included in such investigation so far (Fig. 6.6).

Since *Coccidioides* spp. are obligate aerobic, oxygen content is a major factor limiting the depth that they can survive in the soil, but other key ecological factors that might be important in supporting the growth of *Coccidioides* spp. in the soil have not been clearly identified. It has been found that *C. immitis* and *C. posadasii* favor somewhat different soils. *C. immitis* has been associated with saline, clay-rich, and alkaline soils in California (Plunkett et al. 1963; Elconin et al. 1964; Lauer et al. 2012), whereas *C. posadasii* was primarily detected in sandier soils (Swatek and Omieczienski 1970; Barker et al. 2012).

The physical and chemical soil parameters that harbor these pathogens were determined in soil samples from Kern County, California (Lauer et al. 2012, 2014). Results from these studies indicate that clay-rich soils (at least 20% clay) with a pH between 7.8 and 8.5, a water supply (15 bar) of 3.5–5 cm, an available water capacity of about 4 cm/cm, and a cation exchange capacity (CEC7) of between

20 and 25 milliequivalents/100 g can be indicative of a suitable habitat for *C. immitis* in the San Joaquin Valley. However, *C. immitis* was also isolated from soils that did not fit into this scheme (Orr 1968). It has to be considered that some of the *C. immitis*-positive sites that were detected in the past might in fact present so-called accumulation sites or dormant sites, where the pathogen has been deposited by the wind but remained inactive and persists in the form of arthroconidia throughout the year because of unfavorable environmental conditions, unable to germinate, in contrast to "active sites" where the fungus has an active metabolism at least at some time during the rainy season (see definition of "active sites" vs. "accumulation sites" in Fisher et al. 2007). A recently completed study which included soils from the western Mojave Desert around Edwards Airforce Base confirmed these observations (Lauer 2015).

The environments that are endemic for *Coccidioides* spp. differ to some degree in regard to soil parameter, climate conditions, and vegetation cover, and it is suspected that different ecotypes of *C. immitis* and *C. posadasii* and their hybrids exist that could also be termed strains of the species. Once additional isolates of *Coccidioides* spp. are investigated genetically in more detail, it might become evident that different populations of *C. immitis* in the Southern San Joaquin Valley and populations in the western Mojave Desert, as well as populations of *C. posadasii* in Arizona and those isolated from Mexico, are indeed equipped with different genes that enable them to thrive successfully in different environments but, once established in a mammalian host, cause the same disease. Ecotypes within several different species of fungi that either can be harmful to humans or are important in bioremediation and biocontrol processes have already been described based on results from detailed phylogenetic studies (e.g., Dettman et al. 2003; Douhan et al. 2008; Colpaert et al. 2011).

Plant species adapted to grow only in specific habitats can be used as indicators for soil physical and chemical parameters. In *Coccidioides* endemic areas of California, Arizona, and Mexico, saltbushes (*Atriplex* spp., *Atripliceae*) and creosote bushes *Larrea tridentata* (*Zygophyllaceae*) are established in non-disturbed areas (Haase 1972; Vasek 1980). *Larrea tridentata* is known for its ability to secure water for itself by inhibiting neighboring plants. It also has been used as medical plant by Native Americans against many different purposes in the past and can serve as an analgesic, antidiarrheal, diuretic, or emetic agent (Moore 1989). It is adapted to well-drained soils and alluvial flats that were once common in the Southern San Joaquin Valley and are still abundant in many areas of the Mojave Desert but can also be found in Mexico and some areas in South America. *Larrea* spp. are able to form colonial clones that can last very long, the oldest one, which was found in the Mojave Desert, being more than 11,000 years old (Vasek 1980; Felger and Moser 1985). The toxin it produces to suppress competing plant growth might be tolerated by *Coccidioides* and opens a niche for this fungal pathogen to persist in an environment that is unfavorable for many other soil microbes. *Coccidioides* spp. are also tolerant to higher concentrations of boron that are natural in many soils of the Mojave Desert (Egeberg and Ely 1956; Kolivras et al. 2001). The natural vegetation in the San Joaquin Valley has been mostly replaced by agricultural fields, oil rigs, houses, etc., and it is sometimes difficult to obtain information about the once established plant associations in a particular area. The best overview of the vegetation of California indicating all plant associations in different climate zones can be found in Sawyer et al. (2008).

Some studies performed in Arizona have associated *C. posadasii* with rodent burrows. An infected rodent that dies at some time because of the infection can shed the pathogen back into the soil, where it can return to its saprophytic stage (Maddy and Crecelius 1967). *Coccidioides* spp. might favor an association with a mammal over its saprophytic stage in the soil, as it had been proposed by Sharpton et al. (2009), because they are apparently lacking essential gene coding for proteins involved in the catabolism of plant-derived material, such as cellulose and pectin, and instead produce enzymes that can degrade keratin. However, more comparative studies including other keratinophilic members of the *Onygenales*, such as the soil-dwelling, nonpathogenic close relative of *Coccidioides* spp. *Uncinocarpus reesii* have to be performed to elucidate alternative strategies those fungi can pursue as soil saprophytes in the absence of keratin.

In regard to seasonal influences, it is understood that rainy periods favor the growth of soil microbes in general, and the addition of moisture to the soil triggers the germination of bacterial spores and conidia of fungi. During the dry season, *Coccidioides* spp. survive as arthroconidia which form when the fungal hyphae stop elongating and cells undergo autolysis, producing spore-like barrel-shaped structures which can easily be disrupted and become airborne when soil is disturbed. These arthroconidia are approximately 3–5 μm in length which is small enough to both remain suspended in the air and be inhaled deep into the lungs, resulting in contamination of lung tissue which might develop into an infection. Once established, arthroconidia can transform into a "spherule" and can enlarge considerably, occasionally to as much as 75 μm in diameter. The spherules provide protection from the host's immune system because they are too large to be engulfed by macrophages. In active infections, mature spherules rupture their outer wall and release the newly formed endospores, each of which can develop into another spherule.

6.4.2 Diversity of Fungi in Coccidioides *Endemic Soils*

Cultivation of bacteria and fungi from the top layers of arid, non-disturbed soils collected in Kern County has indicated the dominance of aerobic and facultative anaerobic bacterial spore formers such as *Bacillus* spp. and *Streptomyces* spp., as well as a few non-fastidious fungal species, among them *Penicillium* and *Aspergillus* spp. When aliquots of diluted soils are plated on artificial media, such as R2A supplemented with 10% soil extracts, a diversity of bacterial and fungal colonies, some of them with antimicrobial properties can be observed (Fig. 6.7).

Few research studies have focused on the diversity of microbes in desert soils of the southwestern USA in the past. Orr (1968) identified members of 34 different

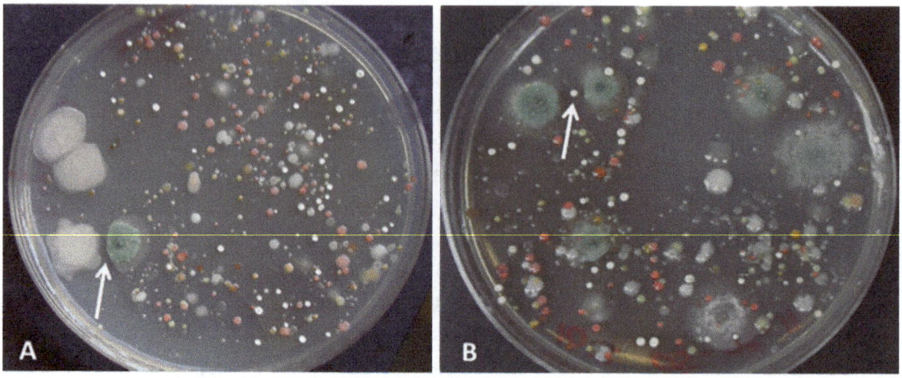

Fig. 6.7 Visible zone of inhibition between bacterial colonies and fungal colonies (samples from a location southwest of Bakersfield, California). (**a**) A large bacterial colony related to *Bacillus* sp. is inhibiting the growth of a *Penicillium* sp. (**b**) A small bacterial colony identified as *Streptomyces* sp. is inhibiting a *Penicillium* colony. The zones of inhibition are indicated by a *white arrow*. The *small white colonies* visible on both plates are *Streptomyces* spp. which are able to produce metabolites that darken the medium

fungal genera from 22 soils collected at six *Coccidioides* endemic areas (Kern, Stanislaus, Merced, and San Diego County) using culture-dependent methods and microscopy. It is of interest to note that three of the six collection sites were associated with Indian camps or middens (burial sites). About 30% of the species identified by Orr (1968) were also isolated in a study from soils of the Sonoran Desert (Ranzoni 1968). Orr also mentioned in his 1967 paper that he had isolated many of these species from soils of the Mojave Desert and the Sonoran Desert himself about a decade earlier.

Polymerase chain reaction (PCR) amplifications using DNA extracts obtained from soil samples of Kern County and the western Mojave Desert 560 in 2014 (English 2010 and unpublished data) revealed the presence of various fungal species known to be adapted to dry conditions and hot climate. Sequencing efforts of these PCR products obtained with primers specific for *Ascomycete* and *Basidiomycete* fungi revealed the presence of members of the *Capnodiales*, especially *Cladosporium* spp., and the *Onygenales*, among them *Coccidioides immitis*, *Uncinocarpus reesii*, and species related to the genera *Aphanoascus*, *Amauroascus*, *Auxarthron*, and *Arthroderma*, and members of the *Eurotiales* related to *Thermomyces* spp.; *Pleosporales*, predominantly *Alternaria* spp.; Pezizales, related to *Ascodesmis* spp.; and *Hypocreales*, especially *Fusarium* spp., among them *Fusarium oxysporum*. These fungi all belong to the *Ascomycetes*. Occasionally members of the *Basidiomycetes* were detected, which are within the order *Filobasidiales* and related to the genus *Cryptococcus*. These results overlap with results obtained by Orr (1968). Frequently encountered were also keratinophilic and dermatophytic species of the genera *Gymnoascus*, *Chrysosporium*, and *Malbranchea*. Members of these same latter three genera have also been isolated from arid soils of other locations, bird feathers, and animal tissues by other researchers (e.g., Emmons 1942, 1954; Rees 1967; Knudtson and Robertstad 1970; Orr and Kuehn 1972; Garg et al. 1985).

Large areas of undisturbed soils in the Mojave Desert also support the growth of biological soil crusts, which typically include lichens and other mutualistic symbioses between fungal species and either photosynthetic bacterial or algal species. These soil crusts generally settle and develop between scattered saltbushes and creosote bushes and can be several hundred years old. Soil crusts often have been observed at *C. immitis*-positive non-disturbed sites in the Mojave Desert (A. Lauer, personal observation). These biological soil crusts are important in preventing soil erosion, and they are also contributing to soil fertility because some of them have a known ability to fix molecular nitrogen (Belnap and Gillette 1997; Billings et al. 2003; Belnap 2003a, b; Thompson et al. 2005). Interestingly, the distribution of biological soil crusts has been assessed via remote sensing in the Mojave Desert (Williams et al. 2012).

6.4.3 Environmental Change and Coccidioidomycosis

The ongoing drought in the southwestern USA has contributed to increased dust development in many areas, and a drier climate is predicted for this area in the near future (Dettman et al. 2003).

Dust clouds generated by storm activity over arid lands can result in soil particulates being transported to altitudes >5 km (Carlson and Prospero 1972; Kellogg and Griffin 2006). Furthermore, dust can be transported several hundreds of km, the finest particles even thousands of km (see review by Péwé 1981). Dust can have effects on climate, ocean and freshwater sedimentation, formation of soil, crop growth, and air pollution. Dust storms develop in naturally semiarid areas that periodically become arid due to seasonal desiccation over the summer months or due to human activity, such as exposing soil surfaces to the elements by either removing larger vegetation, destroying biological soil crusts, or abandoning agricultural fields. Significant disturbance of soil surfaces can also be caused by animals, especially by large colonies of rodents, and desert ants. Dust storms occur several times per year in areas of the southwestern USA, but much more common are dust devils which can be observed daily in disturbed semiarid and arid areas of the Mojave Desert, for example. These dust devils are able to carry clay and silt size particles up to several hundreds of meters into the air and many meters away, occasionally generating huge vortexes of dust (Idso 1975).

Climate change likely affects the ecology of pathogens in the environment including their reproduction success and transport. Weather and climate factors are known to affect dispersion of pathogens either directly or by affecting their vectors and reservoirs. Climate change may therefore have important impacts on incidence of infectious diseases (Boxall et al. 2009). As a spore former, and by being adapted to grow in alkaline desert soils, which are generally poor in organic matter, *Coccidioides* spp. certainly have an advantage over other soil-dwelling fungi that are associated with agricultural fields and which prefer both a pH below 7 and higher content of organic matter. It can be speculated that the ongoing drought in the southwestern USA may result in a shift in the microbial population in soils toward an increase in *Coccidioides* spp. and other drought-tolerant microbes. These effects will affect pathogens and microbial antagonists to these pathogens alike. An ongoing drought will result in increased PM10 pollution from dust that in *Coccidioides* endemic areas will likely contain arthroconidia of the pathogen. The dust predominantly arises from disturbed soils such as eroded agricultural fields which were abandoned because of the scarcity of water in the last few years. Once the topsoil layer is disturbed and the natural vegetation removed, the soil is exposed to the elements and erosion follows. This has been observed in the western Mojave Desert west of the city of Lancaster, California (e.g., see Fig. 6.4). Changes in land use also include the development of urban areas due to population increase. Areas outside the "lower Sonoran life zone" have less dust development and more rain, so that one can conclude that *Coccidioides* arthroconidia, if they were formed, more rarely would become airborne in these geographical areas. One can speculate about the implications of the ongoing drought in the southwestern USA and what that can mean in regard to valley fever in areas that are not known to be endemic, such as Washington State and other states in the northwest.

Ambitious plans in California and Arizona to switch to renewable energy in the near future have put a lot of pressure on the environment. Solar power in California, for example, has been growing rapidly because of high insolation, community support, declining solar costs, and a renewable portfolio standard which requires that 25% of California's electricity come from renewable resources by 2016 and 33% by 2020 (Sherwood 2014). Much of this is expected to come from solar power. As of the end of 2013, California had 490 MW of concentrated solar power and 5183 MW of photovoltaic capacity in operation (Sherwood 2014). The American Solar Energy Industries Association reports that a further 19,200 MW of utility-scale solar projects are under construction or development in the state as of August 2014 (Solar Energy Industries Association [SEIA]) "Major Solar Projects List: Fact Sheet." Some of these photovoltaic plants are situated or are being constructed in highly endemic areas where *Coccidioides* spp. are established. The Antelope Valley Solar Ranch 1 (AVSR1), for example, is a 266-megawatt (MW) photovoltaic power plant near Lancaster within the Antelope Valley, in the western Mojave Desert, Southern California (see Fig. 6.8). The largest solar power installation in the world so far is the 354 MW solar thermal SEGS plant, completed in 1991 whose combined capacity comes from three sites in California. The Ivanpah Solar Electric Generating System (392 MW), located 40 miles (60 km) southwest of Las Vegas, recently overtook the California facility. Although the switch to renewable energy is appreciated to reduce CO_2 emissions from burning fossil fuels, the location for these photovoltaic plants and wind energy turbines should be carefully chosen. Large-scale disturbance of soils that harbor *Coccidioides* spp. can enhance valley fever incidence, because of the risk of arthroconidia becoming airborne when arid soils

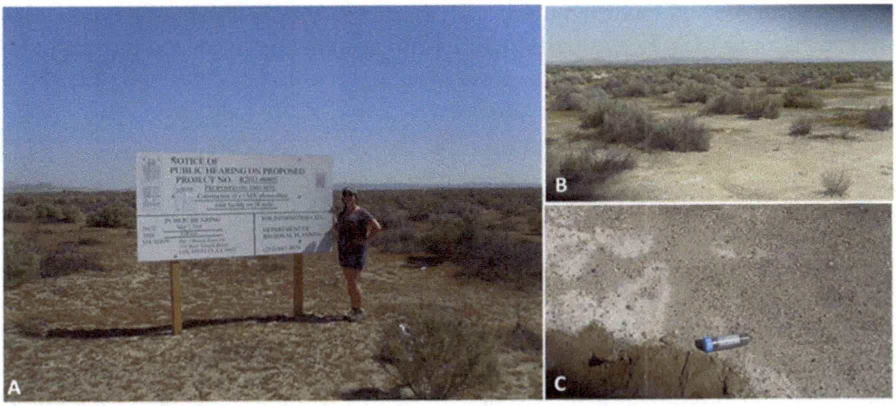

Fig. 6.8 (**a**) Announcement for the development of a photovoltaic system in Lancaster, California, area in the western Mojave Desert, a site that supports the growth of *C. immitis*. (**b**) Overview of the landscape with saltbushes and other drought-resistant annuals. (**c**) Soil sampling site where *C. immitis* was detected. Note the *white patches* of salt crystals on the dry soil surface

are disturbed. Unfortunately, environmental assessments prior to soil disturbance and land development do not include soil testing for potentially airborne pathogens.

6.5 Methods to Detection of *Coccidioides* spp.

6.5.1 Cultivation-Based Methods

A variety of methods are established to detect *Coccidioides* spp. in patient specimen and in environmental samples, such as soil and dust. Methods to detect the pathogen in patient specimen, for example, sputum, blood, spinal fluid, tissue samples, urine samples, etc., conducted before the 1990s were all culture-based and included a variety of laboratory media, typically such as either Sabouraud or brain heart infusion agar usually supplemented with antibiotics to inhibit fast-growing bacteria (Lubarsky and Plunkett 1955; Huppert 1957). Another method relied upon the use of embryonated hen (chicken) eggs (Vogel and Conant 1951). The transformation of arthroconidia into spherules (spherulation) can be induced by incubating at a higher temperature, such as 40 °C, and increased CO_2 concentrations and by adding growth factors such as glutathione (Converse and Besemer 1959; Breslau and Kubota 1964). *Coccidioides* spp. were successfully isolated from soil on media containing ammonium chloride and acetate, as well as growth factors such as acriflavine (Steward and Meyer 1932) on which they form a whitish mycelium. The parasitic stage of the pathogen can also be cultured and investigated on artificial media (e.g., Converse and Besemer 1959). First attempts to isolate *Coccidioides* spp. from soils date back to the beginning of the twentieth century and are still attempted with mixed success today. Obtaining *Coccidioides* spp.

isolates from the soil environment is of important value, because it can prove that arthroconidia are viable and variation in virulence among cultivated pure cultures can be investigated (Huppert et al. 1983).

6.5.2 Culture-Independent Methods

With the start of the modern age of microbiology, PCR-based methods allowed for the detection of *Coccidioides* spp. using diagnostic primer pairs. These methods were initially developed to detect the pathogen in patient specimens and were then discovered, modified, and used to detect *Coccidioides* spp. in DNA extracts from soil and dust. A multiplex PCR was one of the first methods to detect *C. immitis* in soil samples of Kern County and was based on the methods developed by Greene et al. (2000). This PCR method includes two primer pairs, one that amplifies a fragment of the 18S rDNA gene of all fungi (~450 bp) in addition to a shorter fragment (~220 bp) with a *Coccidioides*-specific primer pair. This method was also successfully used on soil samples collected around Bakersfield, California, an area with high incidence of coccidioidomycosis (Lauer et al. 2012). However, the PCR amplicons obtained with the diagnostic PCR in the multiplex PCR are not long enough to distinguish between *C. immitis* and *C. posadasii* (Fig. 6.9).

A variety of primer combinations was developed in the subsequent years to detect *Coccidioides* in soil and dust from different locations in the southwestern USA and Mexico which included diagnostic primer pairs that target the intertranscribed spacer regions (ITS) of the ribosomal gene. Martin and Rygiewicz (2005) developed a nested PCR system to amplify Dikaryomycota efficiently and at the same time reduce unspecific annealing. This nested PCR approach was adapted by Baptista-Rosas et al. (2012) who included a third PCR step with a diagnostic primer pair for *Coccidioides* using a primer pair originally published by Binnicker et al. (2007). The nested PCR approach was also used by Johnson et al. (2014) investigating soils obtained from an archaeological site at Dinosaur Monument in Utah, known as "Swelter Shelter" for the presence of the pathogen, a site of a coccidioidomycosis epidemic in 2001. Johnson et al. developed a new diagnostic primer pair that was long enough to distinguish between the two species *C. immitis* and *C. posadasii* and which was leading to fewer false-positive amplifications as compared to the primer pair published by Binnicker et al. (2007) which was used in the original nested PCR approach published by Baptista-Rosas et al. (2012). All diagnostic primer pairs are vulnerable for unspecific amplification, resulting in occasional false positives. Therefore, sequencing of the PCR amplicons is essential to validate a positive result. Exemplary results of nested PCRs are shown in Figs. 6.10 and 6.11.

Attempts to quantify *Coccidioides* spp. in patient specimens have also been attempted using real-time PCR and different probes. Protocols using real-time PCR have been published by Binnicker et al. (2007, 2011), Castañon-Olivares et al. (2010), and Sheff et al. (2010). Especially beneficial is the hope that real-time PCR assays performed on patient specimens can lead to improved disease surveillance,

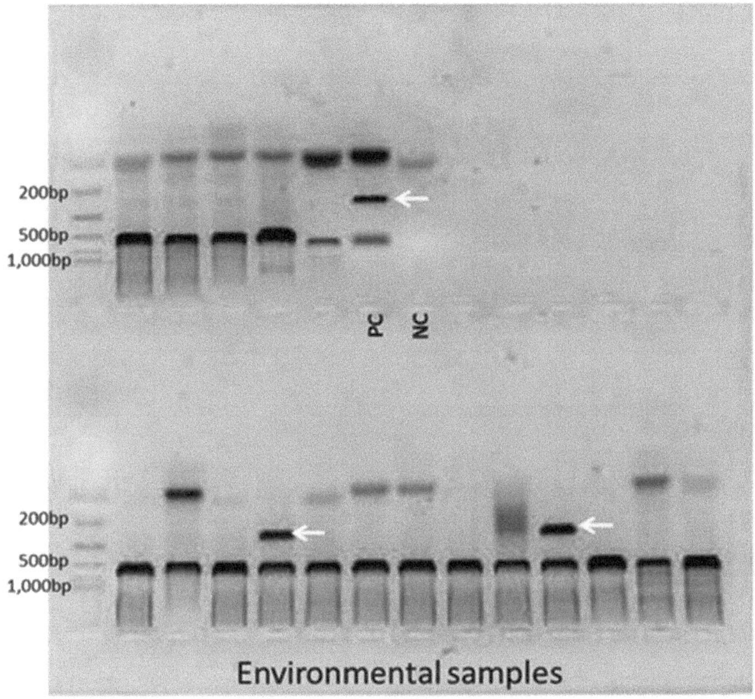

Fig. 6.9 Agarose gel electrophoresis (2%) of PCR products obtained by multiplex PCR with primer pair RDS478/RDS482 and ITSC1A/ITSC2. *White arrows* indicate PCR fragments of correct size (~220 bp) from environmental samples that contained *C. immitis*

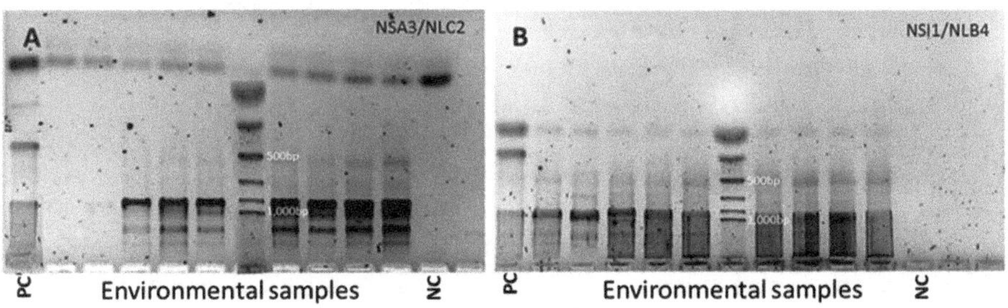

Fig. 6.10 Detecting environmentally present *Coccidioides* by use of PCR analysis technology. (**a**) Example of successful PCR results obtained with primer pair NSA3/NLC2 (~1000 bp). (**b**) Example for a successful nested PCR with primer pair NSI1/NLB4 (~900 bp) using aliquots from the NSA3/NLC2 PCR

increase understanding of the natural history of coccidioidomycosis, and assist in clinical differentiation studies. Real-time PCR has also been adopted to screen environmental samples for the presence of *Coccidioides* spp. These PCR methods are not targeting the ITS regions of the ribosomal gene but instead are based on

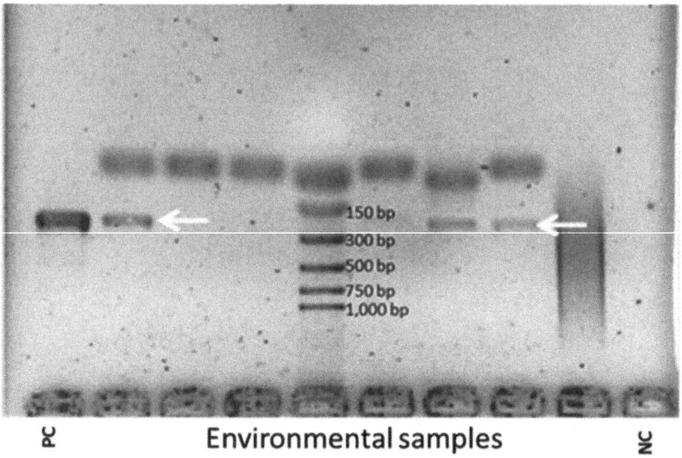

Fig. 6.11 Results of agarose gel electrophoresis (2%) with PCR products obtained with primer pair ITSC1A/ITSC2 with diluted aliquots from amplicons obtained with primer pair NSI1/NLB4 in a nested PCR approach (diagnostic PCR). *White arrows* indicate PCR fragments of correct size (~220 bp) from environmental samples that contained *C. immitis* as confirmed by sequencing of the amplicons

single nucleotide polymorphism (SNP) analyses, which is also able to differentiate *Coccidioides* on the species level (Sheff et al. 2010; Litvintseva et al. 2015).

Screening of hundreds of samples can become labor- and cost-intensive when molecular methods are involved to detect *Coccidioides* spp. A method to screen large amount of samples and at the same time obtaining information about differences in the microbial diversity in the soil samples is denaturing gradient gel electrophoresis (DGGE) which separates PCR products by base composition (Muyzer and Smalla 1998; Kowalchuk et al. 2004). The PCR/DGGE method performed with primers specific for fungi has been used successfully to generate fingerprints of the fungal communities in the soil and can detect *C. immitis* when it is a dominant part of the fungal community. The method is not sensitive enough however, when the pathogen is present in low numbers. The identification of the fungal species that are represented by DGGE fingerprints can be achieved by excising, re-amplification, and sequencing of individual DGGE bands that refer to operational taxonomic units (OTUs) (Heuer and Smalla 1997; Hoshino and Morimoto 2008). Figure 6.12 shows an example of a successful DGGE approach to detect *C. immitis* in soil samples.

Another approach to determine the presence of the pathogen in soils is based upon immunological tests of rodents living in endemic areas. It is hypothesized that small rodents, such as kangaroo rats, pocket mice, ground squirrels, and other animals that dig burrows in soil, cannot avoid being contaminated with arthroconidia of the pathogen. If the pathogen is established in a certain area, then one would expect to find antibodies against it in the blood of rodents living in that area (Catalán-Dibene et al. 2014).

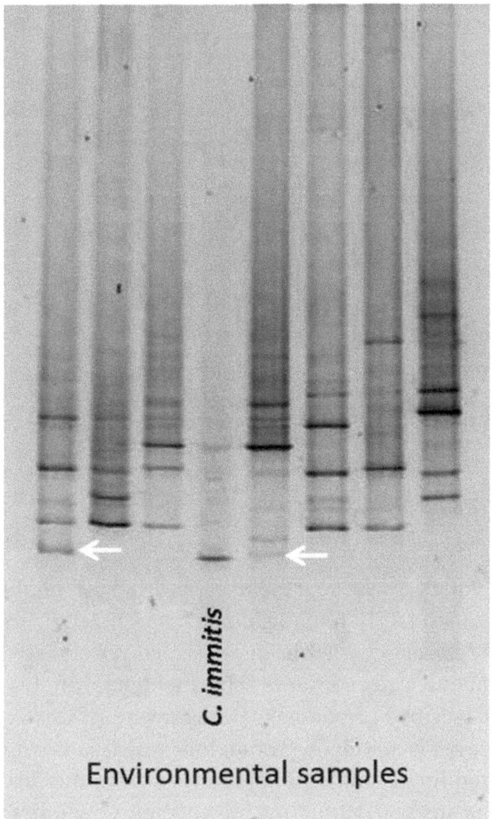

Fig. 6.12 Fingerprint of fungal communities in soil samples collected near Edwards Airforce Base, California, obtained by PCR with primer pair NS1/GC-fung (18S rDNA), followed by DGGE. The individual sites clearly indicate variations in fungal community composition. The *white arrow* points toward bands that appear in the same melting area of the gel as would a PCR fragment from *C. immitis*

Dust samples can be collected using a variety of methods. The most commonly used method includes the spore trap method where air is channeled through filters with different pore size. See Lacey and Venette (1995) for an overview of commonly used dust collection methods. A promising field method is also the Portable In Situ Wind Erosion Laboratory (PI-SWERL®) developed by Etyemezian et al. (2007) which can be easily used in the field mounted on a baby stroller (Fig. 6.13). The PI-SWERL® method can be used to generate windblown dust on test surfaces adjacent to where bulk soil sampling will occur. The PI-SWERL® has been used extensively in prior wind erosion and windblown dust emission studies and provides a portable means to simulate the effect of high wind on soil surfaces. Dust emitted from within the PI-SWERL will be passed through a PM10 size inlet and subsequently onto a filter. The collected dust can then be used for any downstream application, such as DNA extraction and PCR.

Fig. 6.13 Usage of PI-SWERL sampling methodology. (**a**) PI-SWERL equipment in action at a sampling spot near Edwards Airforce Base (EAFB), California. (**b**) Sampling spot where dust had been sampled with the PI-SWERL method at a *C. immitis*-positive site (also located at EAFB)

6.6 Conclusion

Valley fever represents a substantial public health problem, the true burden of which likely remains under-recognized. It has been noted that more cases of valley fever are occurring in areas that previously had low incidence of the disease only about a decade ago. The incidence of this disease is on the rise due to factors described previously. Being aware of this is the first step to action. Clearly, valley fever is an ongoing epidemic which is taking a turn for the worse and which should require sustained interventions from medical and public health communities, such as sustained funding for vaccine development and also research support for ecological studies on *Coccidioides* spp. We need the ability to determine and predict which soil environments support *Coccidioides* growth under current as well as future climate conditions. Clinicians should maintain a high clinical suspicion for valley fever in patients who either live in the endemic regions or who have traveled to those areas.

Compliance with Ethical Standards

Funding: This study was funded by NASA Armstrong Flight Research Center (grant number 42000505751).

Conflict of Interest: Antje Lauer declares that she has no conflict of interest.

Ethical approval: This chapter does not contain any studies with human participants or animals performed by any of the authors.

References

Adams DA, Gallagher KM, Jajosky RA, Kriseman J, Sharp P, Anderson WJ, Aranas AE, Mayes M, Wodajo MS, Onweh DH, Abellera JP (2013) Summary of notifiable diseases-United States, 2011. MMWR Morb Mortal Wkly Rep 60:1–17

Ampel NM (2010) What's behind the increasing rates of coccidioidomycosis in Arizona and California? Curr Infect Dis Rep 12:211–216

Amy PS, Halderman DL (1997) Microbial dormancy and survival in the subsurface. The microbiology of the terrestrial deep subsurface, vol 4. CRC Press, Boca Raton, FL, pp 185–204

Anstead GM, Graybill JR (2006) Coccidioidomycosis. Infect Dis Clin N Am 20:621–643

Aronson JD, Saylor RM, Parr EI (1942) Relationship of coccidioidomycosis to calcified pulmonary nodules. Arch Pathol 34:31–48

Baker E, Mrak M, Smith CE (1943) The morphology, taxonomy and distribution of *Coccidioides immitis* Rixford & Gilchrist 1896. Farlowia 1:199–244

Baptista-Rosas RL, Hinojosa A, Riquelme M (2007) Ecological niche modeling of *Coccidioides* spp. in western North American deserts. Ann NY Acad Sci 1111:35–46

Baptista-Rosas RC, Catalán-Dibene J, Romero-Olivares AL et al (2012) Molecular detection of *Coccidioides* spp. from environmental samples in Baja California: linking valley fever to soil and climate conditions. Fung Ecol 5:177–190

Barker BM, Tabor JA, Shubitz LF et al (2012) Detection and phylogenetic analysis of *Coccidioides posadasii* in Arizona soil samples. Fung Ecol 5:163–176

Belnap J (2003a) The world at your feet: desert biological soil crusts. Front Ecol Environ 1: 181–189

Belnap J (2003b) Biological soil crusts and wind erosion. Biological soil crusts: structure, function, and management. Springer, Berlin, pp 339–347

Belnap J, Gillette DA (1997) Disturbance of biological soil crusts: impacts on potential wind erodibility of sandy desert soils in southeastern Utah. Land Degrad Devel 8:355–366

Billings SA, Schaeffer SM, Evans RD (2003) Nitrogen fixation by biological soil crusts and heterotrophic bacteria in an intact Mojave Desert ecosystem with elevated CO_2 and added soil carbon. Soil Biol Biochem 35:643–649

Binnicker MJ, Buckwalter SP, Eisberner JJ et al (2007) Detection of *Coccidioides* species in clinical specimens by Real-Time PCR. J Clin Microbiol 45:173–178

Binnicker MJ, Popa AS, Catania J et al (2011) Meningeal coccidioidomycosis diagnosed by real-time polymerase chain reaction analysis of cerebrospinal fluid. Mycologia 171:285–289

Bowman BH, Taylor JW, White TJ (1992) Molecular evolution of the fungi: human pathogens. Mol Biol Evol 9:893–904

Boxall ABA, Hardy A, Beulke S et al (2009) Impacts of climate change on indirect human exposure to pathogens and chemicals from agriculture. Environ Health Perspect 117:508–514

Brandt ME, Warnock DW, Murray PR et al (2006) *Histoplasma*, *Blastomyces*, *Coccidioides*, and other dimorphic fungi causing systemic mycoses. In: Manual of clinical microbiology, vol 2, 9th edn. ASM Press, Washington, pp 1857–1873

Breslau AM, Kubota MY (1964) Continuous in vitro cultivation of spherules of *Coccidioides immitis*. J Bacteriol 87:468–472

Brown J, Benedict K, Park BJ et al (2013) Coccidioidomycosis: epidemiology. Clin Epidemiol 5:185–119

Burwell LA, Park BJ, Wannemuehler KA et al (2009) Outcomes among inmates treated for coccidioidomycosis at a correctional institution during a community outbreak, Kern County, California, 2004. Clin Infect Dis 49:e113–e119

Carley S (2009) State renewable energy electricity policies: an empirical evaluation of effectiveness. Energ Policy 37:3071–3081

Carlson TN, Prospero JM (1972) The large-scale movement of Saharan air outbreaks over the northern equatorial Atlantic. J Appl Meteorol 11:283–297

Carter RA (1934) Infectious granulomas of bones and joints, with special reference to coccidioidal granuloma. Radiology 23:1–16

Casadevall A, Pirofski LA (2006) The weapon potential of human pathogenic fungi. Med Mycol 44:689–696

Castañón-Olivares LR, Laniado-Laborín R, Concepcion T, Muñoz-Hernández B, Aroch-Calderón A, Aranda-Uribe IS, Flores-Sánchez MA, del Rocío G-MM, Hernández-Navarez A, Manjarrez-Zavala ME, Miranda-Mauricio S (2010) Clinical comparison of two Mexican coccidioidins. Mycopathologia 169:427–430

Catalán-Dibene J, Johnson SM, Eaton et al (2014) Detection of coccidioidal antibodies in serum of a small rodent community in Baja California, Mexico. Fung Biol 118:330–339

Cole GT, Xue JM, Okeke CN et al (2004) A vaccine against coccidioidomycosis is justified and attainable. Med Mycol 42:189–216

Colpaert JV, Wevers JHL, Krznaric E et al (2011) How metal-tolerant ecotypes of ectomycorrhizal fungi protect plants from heavy metal pollution. Ann For Sci 68:17–24

Como JA (1994) Oral azole drugs as systemic antifungal therapy. N Engl J Med 330:263–272

Converse JL, Besemer AR (1959) Nutrition of the parasitic phase of *Coccidioides immitis* in a chemically defined medium. J Bacteriol 78:231–239

Cox R, Brummer E, Lecara G (1977) *In vitro* lymphocyte responses of coccidioidin skin test-positive and negative-persons to coccidioidin, spherulin and a *Coccidioides* cell wall antigen. Infect Immun 15:751–755

Crum NF, Lederman ER, Stafford CM et al (2004) Coccidioidomycosis: a descriptive survey of a reemerging disease. Clinical characteristics and current controversies. Medicine 83:149–175

Darling ST (1906) A protozoan general infection producing pseudotubercles in the lungs and focal necroses in the liver, spleen, and lymphnodes. J Am Med Assoc 46:283–285

Das R, McNary J, Fitzsimmons K et al (2012) Occupational coccidioidomycosis in California. J Occup Environ Med 54:564–571

Daubenmire RF (1938) Merriam's life zones of North America. Q Rev Biol 13:327–332

De Almeida F (1933) The blastomycoses of Brazil. Ann Fac Med Sa Paulo 9:69

Dettman JR, Jacobson DJ, Taylor JW (2003) A multilocus genealogical approach to phylogenetic species recognition in the model eukaryote *Neurospora*. Evolution 57:2703–2720

Dickson EC (1937) "Valley fever" of the San Joaquin Valley and the fungus *Coccidioides*. Calif West Med 47:151–155

Dixon DM (2001) *Coccidioides immitis* as a select agent of bioterrorism. J Appl Microbiol 91: 602–605

Douhan GW, Smith ME, Huyrn KL et al (2008) Multigene analysis suggests ecological speciation in the fungal pathogen *Claviceps purpurea*. Mol Ecol 17:2276–2286

Egeberg RO, Ely AF (1956) *Coccidioides immitis* in the soil of the Southern San Joaquin Valley. Am J Med Sci 231:151–154

Elconin AF, Egeberg RO, Egeberg MC (1964) Significance of soil salinity on the ecology of *Coccidioides immitis*. J Bacteriol 87:500–503

Emmons CW (1942) Isolation of *Coccidioides immitis* from soil and rodents. Pub Health Rep 27: 109–111

Emmons CW (1954) Isolation of *Myxotrichum* and *Gymnoascus* from the lungs of animals. Mycologia 46:334-338

English LC (2010) Diversity of fungal communities in soil samples of Kern County, CA, with emphasis on detection of *Coccidioides immitis*, the valley fever fungus. Master Thesis, California State University, Bakersfield

Etyemezian V, Nikolich G, Ahonen S et al (2007) The Portable In Situ Wind Erosion Laboratory (PI-SWERL): a new method to measure PM 10 windblown dust properties and potential for emissions. Atmos Environ 41:3789–3796

Felger RS, Moser MB (1985) People of the desert and sea - ethnobotany of the Seri Indians. University of Arizona Press, Tucson, AZ

Fierer J (2007) The role of IL-10 in genetic susceptibility to coccidioidomycosis on mice. Ann NY Acad Sci 111:236–244

Fisher MC, Koenig GL, White TJ et al (2002) Molecular and phenotypic description of *Coccidioides posadasii* sp. nov., previously recognized as the non-California population of *Coccidioides immitis*. Mycologia 94:73–84

Fisher FS, Bultman MW, Johnson SM et al (2007) *Coccidioides* niches and habitat parameters in the southwestern United States - a matter of scale. Ann NY Acad Sci 1111:47–72

Galgiani JN, Grace GM, Lundergan LL (1991) New serologic tests for early detection of coccidioidomycosis. J Infect Dis 163:671–674

Galgiani JN, Ampel NM, Blair JE et al (2005) Coccidioidomycosis. Clin Infect Dis 41:1217–1212

Garg AP, Gandotra S, Mukerji KG et al (1985) Ecology of keratinophilic fungi. Proc Plant Sci 94: 149–163

Gifford MA (1936) Annual report of Kern County Health Department for the fiscal year 1935–1936, pp 22–23

Gifford MA, Buss WC, Douds RJ (1937) Data on *Coccidioides* fungus infection, Kern County, 1901-1936. Annual report of Kern County Department of Public Health 1936–1937, pp 39–54

Gilchrist TC (1894) Protozoan dermatitis. J Cutan Gen Dis 12:496–499

Gilichinsky DA, Wilson GS, Friedmann EI et al (2007) Microbial populations in Antarctic permafrost: biodiversity, state, age, and implication for astrobiology. Astrobiology 7:275–311

Greene DR, Koenig G, Fisher MC et al (2000) Soil isolation and molecular identification of *Coccidioides immitis*. Mycologia 92:406–410

Haase EF (1972) Survey of floodplain vegetation along the lower Gila River in southwestern Arizona. J Arizona Acad Sci 7:75–81

Harrison WR, Merbs CF, Leathers CR (1991) Evidence of coccidioidomycosis in the skeleton of an ancient Arizona Indian. J Infect Dis 164:436–437

Hector RF, Laniado-Laborin R (2005) Coccidioidomycosis—a fungal disease of the Americas. PLoS Med 25;2(1):e2

Hector R, Zimmer BL, Pappagianis D (1990) Evaluation of nikkomycins X and Z in murine models of coccidioidomycosis, histoplasmosis, and blastomycosis. Antimicrob Agents Chemotherapy 34:587–593

Hector RF, Rutherford GW, Tsang CA et al (2011) The public health impact of coccidioidomycosis in Arizona and California. Int J Environ Res Pub Health 8:1150–1173

Heuer H, Smalla K (1997) Application of denaturing gradient gel electrophoresis and temperature gradient gel electrophoresis for studying soil microbial communities. In: Van Elsas JD, Trevors JT, Wellington EMH (eds) Modern soil microbiology. Marcel Dekker, pp 353–373

Hoshino YT, Morimoto S (2008) Comparison of 18S rDNA primers for estimating fungal diversity in agricultural soils using polymerase chain reaction-denaturing gradient gel electrophoresis. Soil Sci Plant Nutr 54:701–710

Huang JY, Bristow B, Shafir S et al. (2012) Coccidioidomycosis-associated deaths, United States, 1990–2008. Center of Disease Control and Prevention (CDC). Emerg Infect Dis. doi:10.3201/eid1811.120752

Huppert M (1957) A technique for safe handling of *Coccidioides immitis* cultures. J Lab Clin Med 50:158–164

Huppert M, Cole GT, Sun SH et al (1983) The propagule as an infectious agent in coccidioidomycosis. Microbiology 10:262

Idso SB (1975) Whirlwinds, density currents, and topographic disturbances: a meteorological melange of intriguing interactions. Weatherwise 28:61–65

Johnson SM, Carlson EL, Fisher FS, Pappagianis D (2014) Demonstration of *Coccidioides immitis* and *Coccidioides posadasii* DNA in soil samples collected from Dinosaur National Monument, Utah. Med Mycol 52:610–617

Jordan JW, Weidman FD (1936) Coccidioidal granuloma. Comparison of the North and South American diseases, etc. Arch Derm Syph 33:31-47, 92

Kaufman L, Standard PC, Huppert M et al (1985) Value of the coccidioidin heat-stable (HS and tube precipitin) antigens in immunodiffusion. J Clin Microbiol 22:515–518

Kellogg CA, Griffin DW (2006) Aerobiology and the global transport of desert dust. Trends Ecol Evol 21:638–644

Kennedy MJ, Reader SL, Swierczynski LM (1994) Preservation records of micro-organisms: evidence of the tenacity of life. Microbiology 140:2513–2529

Kirkland TN, Fierer JF (1996) Coccidioidomycosis: a reemerging infectious disease. Emerg Infect Dis 3:192–199

Knudtson WU, Robertstad GW (1970) The isolation of keratinophilic fungi from soil and wild animals in South Dakota. Mycopathol Mycol Appl 40:309–323

Kofoupanou V, Burt A, Szaro T et al (2001) Gene genealogies, cryptic species, and molecular evolution in the human pathogen *Coccidioides immitis* and relatives (*Ascomycota, Onygenales*). Mol Biol Evol 18:1246–1258

Kolivras KN, Johnson PS, Comrie AC et al (2001) Environmental variability and coccidioidomycosis (valley fever). Aerobiologia 17:31–42

Kowalchuk G, Smit E, Kowalchuk et al (2004) Fungal community analysis using PCR-denaturing gradient gel electrophoresis (DGGE), 2nd edn. Kluwer Academic, pp 771–788

Lacey J, Venette J (1995) Outdoor air sampling techniques. In: Bioaerosols handbook. CRC Lewis, Boca Raton, FL, pp 407–471

Lauer A (2015) Detection of *Coccidioides immitis* in soils around Edwards Airforce Base, California. Final report to the NASA Armstrong Flight Research Center (AFRC), Edwards, CA

Lauer A, Baal JDH, Baal JCH et al (2012) Detection of *Coccidioides immitis* in Kern County, California, by multiplex PCR. Mycologia 104:62–69

Lauer A, Talamantes J, Castañón Olivares LR et al (2014) Combining forces - the use of landsat TM satellite imagery, soil parameter information, and multiplex PCR to detect *Coccidioides immitis* growth sites in Kern County, California. PLOS One. doi:10.1371/journal.pone.0111921

Lewis GG, Mewha J (1955) History of prisoner of war utilization by the United States Army 1776-1945. No. DA-PAM-20-213. Department of the Army, Washington, DC

Litvintseva AP, Marsden-Haug N, Hurst S, Hill H, Gade L, Driebe EM, Ralston C, Roe C, Barker BM, Goldoft M, Keim P (2015) Valley fever: finding new places for an old disease: *Coccidioides immitis* found in Washington State soil associated with recent human infection. Clin Infect Dis 60:e1–e3

Louie L, Ng S, Hajjeh R et al (1999) Influence of host genetics on the severity of coccidioidomycosis. Emerg Infect Dis 5:672–680

Lubarsky R, Plunkett A (1955) In vitro production of the spherule phase of *Coccidioides immitis*. J Bacteriol 70:182–186

Maddy KT (1965) Observations on *Coccidioides immitis* found growing naturally in soil. Arizona Med 22:281–288

Maddy KT, Coccozza J (1964) The probable geographic distribution of *Coccidioides immitis* in Mexico. Bol Oficina Sanit Panam 57:44–54

Maddy KT, Crecelius HG (1967) Establishment of Coccidioides immitis in negative soil following burial of infected animals and animal tissue. In: Proceedings of 2nd coccidioidomycosis symposium, University of Arizona Press, Tucson, Arizona, pp 309–312

Marsden-Haug N, Goldoft M, Ralston C et al (2012) Coccidioidomycosis acquired in Washington State. Clin Infect Dis. doi:10.1093/cid/cis1028

Martin KJ, Rygiewicz PT (2005) Fungal-specific PCR primers developed for analysis of the ITS region of environmental DNA extracts. BMC Microbiol 5:28

McCarty JM, Demetral LC, Dabrowski L et al (2013) Pediatric coccidioidomycosis in central California: a retrospective case series. Clin Infect Dis 56:1579–1585

McLean ML (2012) The epidemiology of coccidioidomycosis – 15 California Counties, 96 2007-2011. Kings County Department of Public Health. http://www.co.kings.ca.us/Health/pdf/2014.01_The%20Epidemiology%20of%20Coccidioidomycosis_%20Collaborative%20County%20Report.pdf

Merriam CH (1898) Life zones and crop zones of the United States. US Government Printing Office

Moore M (1989) Medicinal plants of the desert and Canyon West Santa Fe. Museum of New Mexico Press, New Mexico, pp 27–32

Morrow W (2006) Holocene coccidioidomycosis: valley fever in early Holocene bison (*Bison antiquus*). Mycologia 98:669–677

Muyzer G, Smalla K (1998) Application of denaturing gradient gel electrophoresis (DGGE) and temperature gradient gel electrophoresis (TGGE) in microbial ecology. Antonie van Leeuwenhoek 73:127–141

Neafsey DE, Barker BM, Sharpton TJ et al (2010) Population genomic sequencing of *Coccidioides* fungi reveals recent hybridization and transposon control. Genome Res 20:938–946

Nguyen C, Barker BM, Hoover S, Nix DE, Ampel NM, Frelinger JA, Orbach MJ, Galgiani JN (2013) Recent advances in our understanding of the environmental, epidemiological, immunological, and clinical dimensions of coccidioidomycosis. Clin Microbiol Rev 26:505–525

Ochoa AG (1967) Coccidioidomycosis in Mexico. In: Ajello L (ed) Proceedings of the symposium on coccidioidomycosis. University of Arizona Press, Phoenix, AZ

Onofri S, Selbmann L, de Hoog GS et al (2007) Evolution and adaptation of fungi at boundaries of life. Adv Space Res 40:657–1664

Ophüls W (1905) Further observations on a pathogenic mould formerly described as a protozoon (Coccidioides immitis, Coccidioides pyogenes). J Exp Med 6:443

Ophüls W, Moffitt HC (1900) A new pathogenic mould (formerly described as a protozoan: *Coccidioides immitis pyogenes*): preliminary report. Phila Med J 5:1471–1472

Orr GF (1968) Some fungi isolated with *Coccidioides immitis* from soils of endemic areas in California. Bull Torrey Bot Club 95:424–431

Orr GF, Kuehn HH (1972) Notes on *Gymnoascaceae*. II. Some Gymnoascaceae and keratinophilic fungi from Utah. Mycologia 64:55–72

Pappagianis D (2007) Coccidioidomycosis serology L. coccidioidomycosis in California state correctional institutions. Ann NY Acad Sci 1111:103–111

Pappagianis D, Levine HB (1975) The present status of vaccination against coccidioidomycosis in man. Am J Epidemiol 102:30–41

Péwé TL (1981) Desert dust: an overview. In: Desert dust: origin, characteristics, and effect on man, no 186, pp 1–10

Plunkett OA, Walker L, Huppert M (1963) An unusual isolate of *Coccidioides immitis* form the Los Banos area of California. Sabouradia 3:16–20

Posadas A (1892) Un nuevo caso de micosis fungoides con posrospermias. An Cir Med Argent 15:585–597

Posadas A (1900) Psorospermiose infectante généralisée. Rev Chir Paris 21:277–282

Posadas A (1928) Obras completes. Imprenta de la Universidad, Buenas Aires, pp 278–303

Ranzoni FV (1968) Fungi isolated in culture from soils of the Sonoran Desert. Mycologia 60: 356–371

Rees RG (1967) Keratinophilic fungi from Queensland – isolation from feathers of wild birds. Sabouraudia 6:14–18

Rixford E (1931) Early history of coccidioidal granuloma in California. In: Coccidioidal granuloma. CA State Dept Publ Health Spec Bull 57:5–8

Rixford E, Gilchrist TC (1931) Two cases of protozoan (coccidioidal) infection of the skin and other organs. Johns Hopkins Hosp Rep 10:209–268

Rixford E, Dickson EC, Beck MD (1931) Coccidioidal granuloma. Special Bulletin no. 57. California Department of Public Health, pp 1–43

Rose KD (2006) The beginning of the age of mammals. Johns Hopkins University Press, Baltimore, MD

Saubolle MA, McKellar PP, Sussland D (2007) Epidemiologic, clinical, and diagnostic aspects of coccidioidomycosis. J Clin Microbiol 45:26–30

Sawyer JO, Keeler-Wolf T, Evens JM (2008) A manual of California vegetation, 2nd edn. California Native Plant Society Press, CNPS Publication Committee, Sacramento, CA

Sharpton TJ, Stajich JE, Rounsley SD et al (2009) Comparative genomic analyses of the human fungal pathogens *Coccidioides* and their relatives. Genome Res 19:1722–1731
Sheff KW, York ER, Driebe EM et al (2010) Development of a rapid, cost-effective TaqMan Real-Time PCR assay for identification and differentiation of *Coccidioides immitis* and *Coccidioides posadasii*. Med Mycol 48:466–469
Sherwood L (2014) U.S. solar market trends 2013. Interstate Renewable Energy Council (IREC). http://sherwoodassociates.com/PDF/2011%20ASES%20Paper-Market%20Trends.pdf
Shoham S, Levitz SM (2005) The immune response to fungal infections. Br J Haematol 129: 569–582
Shubitz LF, Hien TT, Perrill RH et al (2014) Modeling Nikkomycin Z dosing and pharmacology in murine pulmonary coccidioidomycosis preparatory to human phase II trials. J Infect Dis. doi:10.1093/infdis/jiu029
Sifuentes-Osornio J, Corzo-León DE, Ponce-de-León LA (2012) Epidemiology of invasive fungal infections in Latin America. Curr Fungal Infect Rep 6:23–34
Smith CE (1940) Epidemiology of acute coccidioidomycosis with erythema nodosum ("San Joaquin" or "Valley Fever"). Am J Public Health Nation's Health 30:600–611
Smith CE, Beard RR, Whiting EG et al (1946) Varieties of coccidioidal infection in relation to the epidemiology and control of the diseases. Am J Pub Health 36:1394–1402
Smith CE, Saito MT, Simons SA (1956) Pattern of 39,500 serologic tests in coccidioidomycosis. J Am Med Assoc 160:546–552
Solar Energy Industries Association [SEIA] (2014) Major solar projects list: fact sheet. http://www.seia.org/research-resources/major-solar-projects-list. Accessed 1 Apr 2017
Sondermeyer G, Lee L, Gilliss D et al. (2013) Coccidioidomycosis-associated hospitalizations, California, USA, 2000–2011. Emerg Infect Dis 19:1590–1597
Steward RA, Meyer KF (1932) Isolation of *Coccidioides immitis* (Stiles) from the soil. Exp Biol Med 29:937–938
Sugiyarna J, Goto S (1969) Mycoflora in core samples from stratigraphic drillings in middle Japan. IV. The yeast genera *Candida* Berkhout, *Trichosporon* Behrend, and *Rhodotorula* Harrison EM. Lodder from core samples. J Fac Sci Univ Tokyo III 10:97–118
Sunenshine RH, Anderson S, Erhart L et al (2007) Public Health surveillance for coccidioidomycosis in Arizona. Ann NY Acad Sci 1111:96–102
Swatek FE, Omieczienski DT (1970) Isolation and identification of *Coccidioides immitis* from natural sources. Mycopathol Mycol 41:155–166
Swatek FE, Omieczienski DT, Plunkett OA (1967) *Coccidioides immitis* in California. In: Proceedings of 2nd Coccidioidomycosis sympsoium. University of Arizona Press, Tucson, AZ, pp 255–264
Thompson DB, Walker LR, Landau FH et al (2005) The influence of elevation, shrub species, and biological soil crust on fertile islands in the Mojave Desert, USA. J Arid Environ 61:609–629
Thompson GR, Lunetta JM, Johnson SM et al (2011) Early treatment with fluconazole may abrogate the development of IgG antibodies in coccidioidomycosis. Clin Infect Dis 53:20–24
Thompson GR III, Stevens DA, Clemons KV, Fierer J, Johnson RH, Sykes J, Rutherford G, Peterson M, Taylor JW, Chaturvedi V (2015) Call for a California coccidioidomycosis consortium to face the top ten challenges posed by a recalcitrant regional disease. Mycopathologia 179:1–9
Thorner J (1941) Coccidioidomycosis – relative value of coccidioidin and tuberculin testing among children of the San Joaquin Valley. California West Med 54:12–15
Valdivia L, Nix D, Wright M et al (2006) Coccidioidomycosis as a common cause of community-acquired pneumonia. Emerg Infect Dis 12:958–962
Vasek FC (1980) Creosote bush: long-lived clones in the Mojave Desert. Am J Bot 67:246–255
Viriyakosol S, del Pilar Jimenez M et al. (2013) Dectin-1 is required for resistance to coccidioidomycosis in mice. mBio 4(1). doi:10.1128/mBio.00597-12
Wack EE, Ampel NM, Sunenshine RH, Galgiani JN (2015) The return of delayed-type hypersensitivity skin testing for Coccidioidomycosis. Clin Infect Dis 15:civ388

Wang XL, Wang S, An CL (2015) Mini-review of published reports on coccidioidomycosis in China. Mycopathologia 180:299–303

Warnock DA (2006) Fungal diseases: an evolving public health challenge. Med Mycol 44: 697–705

Wernicke R (1892) Über einen Protozoenbefund bei mycosis fungoides. Centralbl Bakt 12: 859–861

Whiston E, Taylor JW (2014) Genomics in *Coccidioides*: insights into evolution, ecology, and pathogenesis. Med Mycol 52:149–155

Williams AJ, Buck BJ, Beyene MA (2012) Biological soil crusts in the Mojave Desert, USA: micromorphology and pedogenesis. Soil Sci Soc Am J 76:1685–1695

Chapter 7
Antibiotic-Resistant Environmental Bacteria and Their Role as Reservoirs in Disease

Marilyn C. Roberts

Abstract Antibiotic use has steadily increased since its introduction in the 1940s. This has led to millions of metric tons of antibiotics produced and used worldwide with contamination of both natural and man-made environments, as well as domestic and wild animals, man, and plants. This contamination has influenced the increase in antibiotic resistance over time. It is now recognized that a global "One Health" approach is needed to understand how antibiotic-resistant bacteria and resistance genes spread among and between animals, humans, and the environment. In this chapter, I will discuss specific examples of how environmental bacteria have played roles in the development of specific antibiotic resistance genes as well as their roles as reservoirs for these genes which have impacted the health of humans and animals globally.

7.1 Introduction

The discovery and use of antibiotics has been one of the greatest public health achievements of the twentieth century. The majority of antibiotics currently in use today were primarily produced by living organisms and then modified by man. These molecules either inhibit or kill other microbes by interacting with specific microbial targets. Antibiotics are used for the treatment of animal and human infections, as well as prevention of infections in patients undergoing surgery, chemotherapy, and either organ or tissue transplants. One major nonhuman use of antibiotics like tetracycline in the USA and some other countries is as growth promoters given to animals at subtherapeutic levels in their feed. Back in the 1940s, it was discovered that there were growth promotion effects on chickens from feeding them *Streptomyces aureofaciens* biomass left after fermentation. It was thought that the growth promotion effect was due to low levels of chlortetracycline left in the mixture which lead to the development and wide use of antibiotics especially tetracyclines as antibiotic feed additives. The levels used were at

M.C. Roberts (✉)
Department of Environmental and Occupational Health Sciences, School of Public Health, University of Washington, Seattle, WA 98195, USA
e-mail: marilynr@uw.edu

C.J. Hurst (ed.), *Modeling the Transmission and Prevention of Infectious Disease*, Advances in Environmental Microbiology 4, DOI 10.1007/978-3-319-60616-3_7

subtherapeutic levels often for the life span of the animal. These levels are now known to efficiently select for increased drug-resistant normal flora within the animal as well as drug resistance in the surrounding environmental microbiota. This practice has become increasingly controversial because of the potential that use of low levels of antibiotics is not needed for growth promotion and that the practice contributes to the emergence of antibiotic resistance in both animal and human bacterial populations, which in turn can reduce the range and success of therapeutic options for disease treatment. Because of this concern, the European Union has banned the use of antibiotics for growth promotion (Aaretrup 2012). Volunteer legislation to reduce antibiotic use as growth promoters has been passed in the USA (Food and Drug Administration 2013).

Antibiotics have saved millions of human and animal lives, reduce losses to agricultural and horticultural crops, and contribute to increased food productivity. Antibiotics have extended the lives of people with genetic conditions such as cystic fibrosis and have become indispensible in modern human and animal medicine. There are over ten major classes of antibiotics and hundreds to thousands of derivatives that have been generated over the last 70 years. Antibiotics have non-antibiotic effects which have been used to treat nonbacterial conditions and thus are also used to treat noninfectious conditions. There are a list of other noninfectious diseases and conditions that tetracycline can be used for and a recent review on the non-antibiotic properties of tetracycline which are detailed in the recent review (Garrido-Mesa et al. 2013). A recent randomized double-blind study found that minocycline given to children and adolescents with fragile X syndrome had greater global improvement than placebo treatment over a 3-month treatment period (Leigh et al. 2013). Early studies suggested that tetracyclines may inhibit cancer cell growth. Exploration of tetracyclines to expand activity to cancer targets has been done (Sun et al. 2008) which has led to one randomized phase II trial (Dezube et al. 2006). Other drug classes also have non-antibiotic properties and are used in treatment of noninfectious diseases (Parnham et al. 2014).

Along with all the positives surrounding antibiotic use, there have also been abuses which have led to increases in the level of antibiotic-resistant bacteria (ARB) and antibiotic resistance genes (ARGs) isolated in agricultural as well as aquacultural settings, the food chain, man, and the environment (CDC 2013; Van Boeckel et al. 2015; WHO 2014), though the level of environmental contamination is just now being appreciated (Berendonk et al. 2015). In the early years of antibiotic usages, there were new antibiotics available to replace the older antibiotics as bacteria became resistant. Thus, when one antibiotic did not work, another was available to take its place. However, today there are very few new antibiotics in development to replace the less effective older antibiotics (Davies and Davies 2010). The current lack of new and novel antibiotics coming into the market has led some researchers to anticipate a "post-antibiotic era," where animals, plants, and people will die of common infections that were once easily treatable with antibiotics (Collignon 2013). The factors that contribute to the emergence and dissemination of bacterial resistance are complex, and over the last few years, both CDC (2013) and WHO (2014) have issued reports on the current and potential public health threats due to antibiotic-resistant pathogens in the USA and around the world.

Concerns over the spread of antibiotic resistance have fueled other groups to assess the impacts of ARB/ARGs on human and animal health, agricultural and food production, and agricultural and human waste management (National strategy for combating antibiotic-resistant bacteria 2014; Taylor et al. 2014). One of the primary outcomes of these reports is a call for increased surveillance of ARB and ARGs in agricultural and environmental settings, with a particular interest in identifying transmission routes of ARB and ARGs throughout the world (Collignon 2013; National strategy for combating antibiotic-resistant bacteria 2014). Key to the success of current and future surveillance efforts are strategies to identify which particular types of resistant genes to measure. The known ARGs are not randomly distributed among bacteria, and there is a clear link between bacterial taxonomy and specific types of ARGs (Durso et al. 2012; Roberts et al. 2012). This phenomenon has been particularly well documented with tetracycline resistance (Chopra and Roberts 2001; Roberts 2012; Roberts et al. 2015; Sloan et al. 1994).

The environment dissemination of ARB could be due to horizontal gene transfer. We know less about environmental pathogen transmission routes than we do about transmission within groups of either animals or humans or between animals and humans. Durso et al. (2012) have suggested that the same gene might have different risks for environmental transmission depending on the specific bacterial taxa carrying the gene and as compared to risk of transmission among bacteria from animals to man or between animals and man. It is also critical whether the gene of interest is associated with a mobile element and whether that element has a narrow or broad host range. Thus, it is important to know what specific antibiotic resistance genes are found in what specific bacterial species and genera within the environment just as it is important to know that information for human and animal populations as assessed for regional, national, and international surveillance studies.

Environmental surveillance studies hopefully will allow the identification of major gaps in our understanding of all the forces which work on selection and transmission of bacterial resistance and even reveal how to either slow or stop the march to a "post-antibiotic era" when common infections and minor injuries kill as they did prior to the introduction and widespread use of antibiotics. Hence it is important to know how ARB and ARGs move through the environment as well as populations of people and animals as illustrated by the recent spread of NDM-1 β-lactamase-carrying bacteria (Nordmann et al. 2011). It is clear that a global "One Health" approach is needed where animal and human usage and environmental contamination are all considered together. In this chapter, I will discuss specific examples of how environmental bacteria have played roles in the development of specific antibiotic resistance genes and their roles as reservoirs of these genes which have impacted the health of humans and animals globally.

7.2 Bacterial Resistance: Historic Perspective

The production of antibiotics has steadily increased since its introduction some 70 years ago. This has led to millions of metric tons of antibiotics produced and used worldwide with the contamination of both natural and man-made environments as well as domestic and wild animals, man, and plants which has all influenced the increase in antibiotic resistance over time. The historic evidence for increased antibiotic-resistant bacteria as a product of human activity is illustrated by the study of Hughes and Datta (1983) which found that 24% of the *Enterobacteriaceae*, isolated between 1917 and 1954, carried conjugative plasmids, but only 2% were tetracycline resistant (Tc^r). This study included isolates from the genera *Proteus*, a common environmental bacteria. None of the *Salmonella*, *Shigella*, *Escherichia*, or *Klebsiella* isolates were Tc^r. However, by the mid-1950s, Tc^r- and multidrug-resistant, including ampicillin (Ap^r) and chloramphenicol (Cm^r), *Escherichia coli* and *Shigella* were described and shown to be transferrable. Later, it was found that all three resistance genes were associated with a conjugative plasmid (Watanabe 1963). A similar study shows increased carriage of Tc^r genes over time in enterococci (Atkinson et al. 1997). The *tet*(M) gene coding for tetracycline had been identified in clinical *Enterococcus* spp. isolated between 1954 and 1955, which is approximately the same time as the first Gram-negative *tet* efflux genes were identified. The *Enterococcus* spp. study was not published until 1997, while the Gram-negative studies were published 30 years earlier (Atkinson et al. 1997; Watanabe 1963). Thus, both the *tet* efflux and *tet* ribosomal protection genes have been in bacterial populations for >60 years. A similar study has been done with the pathogenic *Neisseria gonorrhoeae* (Cousin et al. 2003). These and many other studies strongly suggest that antibiotic resistance genes are becoming more common among bacterial populations over time due to antibiotic use by man.

It is clear that use of antibiotics is a selective pressure that has influenced the increase in antibiotic resistance among clinically important bacteria as well as commensal and opportunistic bacteria. However, in some cases, antibiotic resistance was present in bacteria prior to human use of the antibiotic. For example, before the introduction of penicillin therapy, bacteria carrying a penicillinase that was able to inactive penicillin was identified (Davies and Davies 2010). Identification of bacterial resistance either prior to or just after the introduction of either a new or modified antibiotic has since been repeated suggesting that influences besides human use do have a role. It is also clear that antibiotics in the environment can select for resistant bacterial populations and may promote bacterial evolution toward increasing antibiotic resistance for bacteria in the environment. One potential reservoir that has been more recently recognized is the natural putative antibiotic resistance genes present in the environment (Davies and Davies 2010). The actinomycetes and other microbes produce antibiotics and other bioactive small molecules that either may kill or inhibit growth of other microbes. These antibiotic-producing microbes also code for resistance genes which protect themselves from the antibiotics that they produce (Davies and Davies 2010). For example, antibiotic-producing

Streptomyces also carry "genes" which protect the host bacteria from the action of the antibiotic(s) they produce either by reduction or prevention of drug-target interaction with the producing organism (Cundliffe and Demain 2010; D'Costa et al. 2006). Usually these genes are linked to the antibiotic synthesis genes and may resemble classical bacterial resistance genes (Davies and Davies 2010).

7.3 Mechanism of Bacterial Resistance and Its Acquisition

There are a number of different ways that bacteria can become resistant to antibiotics. One way is due to random chromosomal mutations during replication that lead to changes in the gene product which may either alter or eliminate the expression of a protein. Another mechanism is by acquisition of new DNA that is available to a limited number of bacteria which are naturally transformable. Some organisms such as *Streptococcus pneumoniae* and *Neisseria meningitidis* have successfully used transformation to create mosaic genes that have increased resistance to penicillin. Mutations and transformation also may have increased resistance to extended-spectrum cephalosporins and carbapenem antibiotics (Bharat et al. 2015; Philippe et al. 2015). These bacteria have receptors which allow them to take up DNA from either related strains or species and integrate this foreign DNA, which may be parts of genes, complete genes, or even defined elements, into their genome. The integration of new pieces of a gene creates a mosaic gene composed of the host's and foreign bacterial DNA which produces a modified protein which can reduce the antibiotic susceptibility of the host bacteria to an antibiotic. Some species of bacteria are able to acquire foreign DNA by transduction which uses bacteria phage for transmission of the DNA from one host to another host. This occurs during phage replication and is due to packaging host bacterial DNA into the phage protein coat which, when injected into a new host bacterium, allows the foreign DNA to be incorporated into the host's genome (Di Luca et al. 2010). Both of these methods normally transfer DNA between closely related strains or species. Mosaic acquired tetracycline resistance genes have also been identified where two or three different tetracycline ribosomal protection genes have recombined (Stanton and Humphrey 2003; Van Hoek et al. 2008). Recent data suggest that transduction is likely important in transmission of ARGs within the soil microbiome (Perry and Wright 2013). One study of bacteriophage from environmental urban sewage and river water found an abundance of the TEM and CTX-M β-lactamase genes (Colomer-Lluch et al. 2011). More recently, an intragenic recombination leading to a mosaic gene has been identified in an aminoglycoside-resistant *aph(3′)-II*a gene from an environmental *Pseudomonas aeruginosa* plasmid (Woegerbauer et al. 2015).

However, the most common way bacteria acquire antibiotic resistance is by acquisition of new genes associated with mobile elements (plasmids, transposons, and integrons). These mobile elements may carry genes for metal resistance, use of alterative carbon sources, and classical virulence genes as well as a variety of different antibiotic resistance genes. Mobile elements are the main driving force in horizontal gene transfer between strains, species, and genera and are the reason that ARGs can spread quickly through an ecosystem and move across the world. The ARGs are found on wide or narrow-host-range mobile elements and are able to achieve rapid spread through an ecosystem and around the world (Nordmann et al. 2011). Studying these mobile elements has increased our understanding of some of the mechanisms by which bacteria adapt to their changing environment in many cases more quickly than man can respond.

7.4 Mutation

Mutations occur during normal replication and generally result in a low to moderate increase in the level of antibiotic resistance, though exceptions can be found. Mutations may also occur due to exposure to ultraviolet light, chemicals, and insertion of mobile elements, each of which alters the DNA composition of a gene resulting in an alteration of the amino acid composition of the protein. Point mutation changes are due to DNA base pair nucleotide substitution. Point mutations can lead to silent mutations, where, because of code redundancy, the mutation codes for either the same or a closely related amino acid and thus does not alter its function. Other alterations result in missense mutations which are due to substitution of a very different amino acid which results in altering the protein's function. Some changes to the DNA result in nonsense mutation because the change creates a stop codon, and thus, the altered gene results in the production of truncated protein which is normally not functional. If mutations occur in the regulator region, then the protein may either not be produced or produced in higher levels. A good example is AmpC β-lactamases which can be found on both the chromosome and plasmids. The plasmid-mediated AmpC is the second most common mechanism of cephalosporin resistance in the USA (Edquist et al. 2013). Mutations due to either insertions or deletions also occur, and these are often due to either the insertion or deletion of mobile elements such as transposons. These changes may result in either greatly altered or nonfunctional proteins which in some cases may be lethal to the host.

Mutations are passed on to daughter cells most commonly during cell division and only more rarely by transformation and transduction. Mutations can occur in specific structural proteins such as penicillin-binding proteins (PBPs), in RNA molecules such as the 23S rRNA or 16S rRNA and in genes that regulate the expression of other proteins. Mutations in any of these genes may alter susceptibility to either single or multiple classes of antibiotics. However, for most classes of antibiotics, multiple mutations are required to confer clinically relevant levels of resistance. Subtherapeutic levels of antibiotics given in animal feed as growth promoters, and for noninfectious diseases such as skin conditions, end up contaminating the environment and provide selective pressure for the accumulation of mutations in exposed bacteria (Davies and Davies 2010). Resistance due to mutations are passed down to daughter cells during cell replication and generally are not responsible for rapid shifts in antibiotic resistance in bacterial communities because of their limited ability to be spread horizontally among bacterial communities.

7.5 Transformation and Transduction

Bacterial transformation is due to the direct uptake and incorporation of exogenous DNA from its environment. Estimate regarding the number of identified species that are naturally transformable has risen to 82 species, many from the environment (Johnston et al. 2014). Transformation enables individual bacterium to take up and internalize exogenous DNA and then integrate the DNA into the recipient genome by homologous recombination. Normally, foreign DNA from either the same or related species is internalized using specific host cell receptors. The DNA is then integrated into the host creating mosaic proteins which often change the phenotype of the host bacteria (see above). Both Gram-positive and Gram-negative bacteria are naturally transformable. This requires the recipient host bacteria to be viable but does not require the donor bacterium to be alive (Johnston et al. 2014).

The famous Avery et al. (1944) experiments demonstrated that transformation can occur in vivo. Those experiments mixed a nonpathogenic noncapsulated *S. pneumoniae* and a heat-killed virulent encapsulated strain together and injected the mixture into mice. Only encapsulated *S. pneumoniae* are able to kill the mice, and surprisingly some of the mice died. When the bacteria were isolated from these animals, encapsulated *S. pneumoniae* were recovered. The isolates had the same capsule type as was found in the heat-killed strain suggesting that something from this mixture which we now know was DNA was incorporated into the live *S. pneumoniae* allowing them to make capsule, survive, and kill the mice. This illustrates the potential opportunities for evolution due to acquisition of new and modified traits and demonstrated the genetic plasticity that transformation provides for bacteria to survive under changing conditions. It is assumed that transformation occurs not only in living animals but also in the environment.

Transduction is the mechanism of gene transfer whereby DNA from one related bacteria is transferred to another bacteria by transducing phage, which are defective bacterial phage that incorrectly contain packaged bacterial DNA rather than phage DNA. Transducing particles are produced when a phage infects a bacterial cell, and during normal phage replication, either some of the host chromosomal or small plasmid DNA erroneously is packaged into the phage head in substitution for phage DNA. The resulting phage then are released from the infected cell. When a transducing phage reaches the next bacterial host cell, the bacterial DNA carried within the transducing phage enters that new host bacterium and the introduced DNA may then become part of the new host's genome. Generalized transduction allows either chromosomal or plasmid DNA to be transferred to another bacterium. Specialized transduction transfers bacterial genes that were located near the chromosomal insertion site used by a temperate phage. Specialized transduction occurs when the prophage excises imprecisely from the chromosome so that bacterial genes lying adjacent to the prophage are included in the excised DNA, packaged and transferred into a new host cell. Transduction is limited to closely related strains that share common phage. *Staphylococcus aureus* plasmids, carrying a variety of different antibiotic resistance genes, and perhaps some of the *mecA* elements which code for an alternative PBP that confers resistance to all staphylococcal β-lactams are thought to be spread between strains by transduction. More recently, the *mef*(A)/*tet*(O) element has been shown to be transferred by phage among group A streptococci (Di Luca et al. 2010).

Carbadox, a growth-promoting antibiotic added to livestock feed, and fluoroquinolones have been shown to induce phage-mediated gene transfer resulting in generalized transduction of prophage in *Salmonella* Typhimurium DT120 and DT104. Both chromosomal and plasmid DNA carrying antibiotic resistance genes have been transferred (Bearson and Brunelle 2015). The usage of these antibiotics in agriculture could lead to unintended changes to the microbiome by inducing generalized gene transduction. This may also occur in the environment as a result of either veterinary treatment of diseased livestock or use of antibiotics as growth promoter, prevention of disease, or in contaminated feed stocks (Bearson and Brunelle 2015).

7.6 Conjugation

Conjugative gene exchange is the main process of horizontal gene transfer between bacteria strains, species, and genera. It requires transfer of genetic material between live bacterial cells by direct cell-to-cell contact. Antibiotic resistance in most bacteria occurs due to the acquisition of new genes via conjugal transfer which provides new proteins to the host bacterium. Conjugation can occur between either the same or different strains, species, and genera of bacteria, and it is clear that once a gene, on a conjugative element, is in a bacterial community, the gene can be spread to many unrelated ecosystems. This may lead to bacterial phenotypic changes because many ARGs are able to be expressed in large numbers of different host bacteria. Conjugation can also occur between bacteria and eukaryotic cells (Waters 2001). Conjugation is a key element in bacterial evolution and notably can act by facilitating the spread of antibiotic resistance genes within and between different ecosystems and their associated bacteria. Other genes such as virulence, resistance to heavy metals, and pathogenicity genes are also transferrable by conjugation. Conjugation allows for multiple genes, including multiple different antibiotic resistance genes, to be transferred as a single event. Conjugation is the primarily mechanism for the horizontal gene transfer of most ARGs in most genera (Davies and Davies 2010). Conjugation can lead to a rapid dissemination of ARGs within and between bacterial communities and between different ecosystems. In the laboratory, conjugal transfer occurs under a variety of conditions. Transfer frequency can be modified by physical proximity of two different bacterial cells to each other as what occurs either in biofilms or in the laboratory on filters or agar surfaces. The presence of subinhibitory concentrations of antibiotics may enhance the frequency of gene transfer and recombination (Davies and Davies 2010; Facinelli et al. 1993). Conjugal exchange of ARGs allows for rapid genetic response to changing conditions within a community, thought to be the major way by which bacteria acquire new antibiotic resistance genes. However, though very important, conjugation is most likely only one factor influencing the evolution of environmental bacterial communities.

Today acquired antibiotic resistance genes are associated with plasmids which can be conjugative, nonconjugative, or mobilizable. Antibiotic resistance genes may also be associated with transposons, conjugative transposons, and integrons found in the chromosome or associated with plasmids. It is currently assumed that the mobility mechanism and type of element that a specific antibiotic resistance gene is associated with may directly influence both the gene's host range and the gene's ability to spread to new genera and multiple ecosystems. However, that clearly is not always the case. For example, some acquired genes such as the *tet*(H) gene has moved around the world and into different environments but no plasmids or gene transfer has been observed under laboratory conditions (Miranda et al. 2003). Many antibiotic resistance genes may primarily be associated with the environment and only on rare occasions introduced into pathogenic bacteria (Chopra and Roberts 2001; Roberts 1997; see http://faculty.washington.edu/marilynr/). ARGs are now often linked with a variety of additional antibiotic resistance genes, heavy metal resistance genes, virulence factors, and other types of genes. These other genes may keep specific ARGs in bacterial communities even without direct selective pressure due to lack of use of a particular antibiotic in the community, for example, we are still finding Cm^r *S. pneumoniae* in children coming for their well-baby checkups even though these children had never been exposed to antibiotics, let alone chloramphenicol, which has been banned from human use in the USA (Luna and Roberts 1998).

7.7 Mobile Genetic Elements (MGE)

Bacterial mobile genetic elements include plasmids, conjugative transposons, transposons, and integrons. All of these elements have been involved in lateral gene exchange between strains, species, genera, from one ecosystem to another, and between man and animals and the environments and vice versa. These elements are able to carry a variety of different ARGs, as well as genes that confer resistance to disinfectants, heavy metals, genes that produce toxins, virulence factors, and genes for using alternative energy sources. All of the genes carried on any one of these elements may influence the carriage of the entire mobile unit or help maintain those genes in the bacterial population even when that individual gene may not function in its current host (Speer et al. 1991) or may represent an antibiotic class that is no longer in widespread use. These mobile units also allow nonfunctional genes to become stable in the bacterial population without selective pressure. A good example of this is the presence of the *tet*(X) gene which codes for an enzyme that breaks down tetracycline in the presence of oxygen but is nonfunctional in its *Bacteroides* host, the first bacteria to be shown to carry the *tet*(X) gene. The *tet* (X) gene was linked to the *erm*(F) gene in the *Bacteroides* host. The *erm*(F) gene codes for macrolide-lincosamide-streptogramin B resistance which is functional in anaerobes and beneficial in protecting the host in the presence of clindamycin therapy (Speer et al. 1991). Thus, the assumption has been that the *tet*(X) gene was transferred and maintained in an anaerobic host because of association with the beneficial *erm*(F) gene. This may also explain why other antibiotic resistance genes which code for antibiotics that are either no longer or rarely used (such as aminoglycosides and chloramphenicol) are still found in bacterial populations that have never been exposed to those antibiotics (Luna and Roberts 1998). Brief descriptions of the major mobile elements that carry antibiotic resistance gene are provided below.

7.7.1 Plasmids

Plasmids were first described in the 1950s, are mostly circular DNA molecules that range in size, and can replicate independently of the bacterial chromosome (Watanabe 1963). Plasmids can integrate into the chromosome as illustrated by the integration of the staphylococcal plasmid pT181 carrying the *tet*(K) gene within the SSC*mecA* type III 67 kb element (Grimdmann et al. 2006). Plasmids may include transposons and integrons and are able to carry a variety of different antibiotic and heavy metals resistance genes as well as genes that either code for toxins, code for the degradation of various compounds such as alternative carbon sources, or code for virulence factors. All the genes within a mobile element normally move as a unit from one bacterium to another bacterium (Chopra and Roberts 2001). Plasmids are very important for carriage of antibiotic resistance genes in Gram-negative bacteria such as *E. coli* and related genera.

Plasmids normally carry a single or multiple entry exclusion genes which reduces the host cell's ability to acquire other plasmids with the same origin of replication as the resident plasmid. Thus, two plasmids with the same origin of replication are considered incompatible with each other irregardless of whether they are conjugative or mobilizable (Garcillian-Barcia and de la Cruz 2008). Incompatibility of plasmids may limit the spread of some of plasmids between strains, species, and genera, and this partially could explain the differences in distribution found among the different antibiotic resistance genes. Some *S. aureus* isolates have dealt with plasmid incompatibility by integrating their plasmids into their chromosome where the plasmid incompatibility does not function. This strategy allows the host to acquire multiple plasmids with the same origin of replication (Gillespie et al. 1986). In contrast, plasmids with different origins of replication can coexist within a single host bacterium. The plasmids occur in a variety of sizes, from the small 4.45 kb *tet*(K)-positive *S. aureus* pT181 plasmid to large plasmids of $\geq$300 kb.

7.7.2 Transposons

Transposons are discrete pieces of DNA that are able to move from one location in the bacterial genome to another location. Transposons are flanked by terminal inverted or direct repeats. Insertion sequences [*IS* elements] are a family of small elements that range from 768 to 1426 bp in size and flank composite transposons, while unit transposons do not carry insertion sequences. Transposons vary in the genes they carry including the genes they use for insertion and excision into the bacterial genome, and in the antibiotic resistance genes present, and if they carry other genes for heavy metal resistance. As with plasmids, transposons continue to evolve due to their ability to recombine and exchange their genes with other elements. The typical composite transposon carries a gene that codes for an enzyme, such as either a site-specific recombinase or resolvase, which is involved in excision and integration within the host genome, and one or more other genes which may code for traits such as antibiotic resistance which are flanked by IS sequences.

The best characterized of the transposons is Tn*10* which carries the *tet*(B) gene that codes for the most widely distributed Gram-negative tetracycline efflux protein (Lawley et al. 2000). The Tn*10* transposon is most frequently associated with Gram-negative plasmids but can also be found in the chromosome of Gram-negative bacteria such as *H. influenzae* (Marshall et al. 1984). Transposons are found on plasmids, and multiple copies may be distributed throughout the bacterial chromosome. The ability to be either on the chromosome or plasmids allows flexibility as well as stability for these genes. Insertion of a transposon within the genome may lead to mutations, either the loss of gene function or modification of gene expression.

7.7.3 Conjugative Transposons [CTns]

Conjugative transposons are self-transmissible integrating elements that can have broad host ranges, are found in a large number of genera, and carry all the genes required to move from one bacterial cell to another by cell-to-cell contact. Conjugative transposons have fewer restrictions in moving between unrelated bacteria than do plasmids because these transposons lack incompatibility exclusion systems. As a result, multiple copies of the same or related conjugative transposons can be found within a single bacterium (Norgren and Scott 1991). The first conjugative transposons were identified in the 1980s and were members of the Tn*916*-Tn*1545* transposon family which normally carries the *tet*(M) gene coding for a ribosomal protection gene. The conjugative Tn*916*-Tn*1545* transposon family is very promiscuous group which can be found in a large number of different Gram-positive and Gram-negative species. It has also been one of the best studied group of transposons (Rice 2007). The Tn*916*-Tn*1545* transposon family integrates site specifically in some species and relatively nonspecifically in other species (Rice 2007). The Tn*916* transposon is 18 kb and has relatively few restriction sites. Some investigators have suggested that the surrounding flanking regions represent ancient structures evolved for broad-host-range transfer, and the presence of the *tet* (M) gene in the Tn*916* transposon is relatively recent in the evolutionary progression of this element (Rice 2007). Low-dose exposure to tetracycline promotes the conjugal transfer of transposons to neighboring bacterial cells (Facinelli et al. 1993). The Tn*916* can also mobilize co-resident plasmids.

The Tn*916*-Tn*1545* conjugative transposons are adaptable and able to form composite elements by integration of one transposon within another transposon. Both transposons encode for their own transfer, and either the complete composite element may be transferred to another bacterium intact or the embedded transposons can be transferred separately. Composite transposons may have multiple mobile elements, various types of insertion sequences [IS], as well as regions from plasmids and genes from different genera of bacteria. For example, the 65 kb composite Tn*5385* transposon carries resistance genes for penicillin, erythromycin, gentamicin, streptomycin, tetracycline, and mercury and has genetic elements related to those found in three different transposons isolated from enterococci, streptococci, and staphylococci (Rice 2007).

7.7.4 Integrons

Integrons are two-component systems that include an integrase (*intI*) and an *attI* sequence which is the site for integration of cassettes which may contain different antibiotic, heavy metal, and disinfectant resistance genes (Recchia and Hall 1995; Vo et al. 2007). Integrons are found as part of plasmids, chromosomes, and transposons. The large mosaic 86 kb chromosomal antibiotic resistance island which includes both integrons and non-integron regions, carried in *Acinetobacter baumannii* strain AYE genome, is a good example of a multidrug-resistant integron isolated from an environmental bacteria that has become a recent human pathogen (Fournier et al. 2006). This element carries a large number of different genes coding for antibiotic and heavy metal resistance including multiple copies of the same antibiotic resistance genes. The homologous chromosomal region in the antibiotic susceptible *A. baumannii* has the transposases but no antibiotic resistance genes connected to the transposases, suggesting that this regional structure was present prior to acquisition of the integrons and antibiotic resistance genes. It is not clear if these 86 kb elements can be moved as a single unit between strains, but data from Adams et al. (2010) shows that there is variability in the composition of the resistance genes within this island between various *A. baumannii* strains suggesting that specific resistance genes are able to come into and out of the element as they do in smaller integrons.

7.8 The Environment

Most studies on ARGs, over the last 70 years, have focused on clinically important bacteria, and only recently are environmental ARB and ARGs being studied (Berendonk et al. 2015). This is important because it is estimated that there are ~5×10^{30} bacteria in the world with only a small subset adapted to live either in or on humans and animals. It has been estimated that less than 1% of the total number of bacteria in world have been cultured (Kümmerer 2004). The natural world is rich in chemicals made by living organisms as well as those made by human endeavor that in turn influence the natural microbiome evolution (Wright 2007). Knowledge of which type of bacteria carry which resistance genes is important for future environmental studies to correctly select which ARGs are most important in a particular ecosystem especially when using molecular methods of detection (Pei et al. 2006; Smith et al. 2004).

Environmental studies are moving away from culturing bacteria, and determining what ARGs are present in a particular sample is often done with either PCR or qPCR molecular techniques, and sometimes both techniques, without cultivation of the organisms. These molecular assays have been used for direct detection of ARGs in food (Guarddon et al. 2011), animal feeding facilities (Stine et al. 2007), and agricultural soils amended with manure (Peng et al. 2015) and used as indicators for water quality changes (Harnisz et al. 2015). In each of these studies, known ARGs were used without necessarily determining their likely distribution in the particular sample source, and that lack of key information can lead to biased results. For example, if only ARGs present in Gram-negative bacteria are used for screening, then nothing will be known about what is going on in the Gram-positive component of the sample source.

The term antibiotic "resistome" is the collection of all genes that can either directly or indirectly contribute ~~to~~ antibiotic resistance to its' bacterial host (Perry and Wright 2013). Groups have been examining the microbial resistome of natural and clinical environments (D'Costa et al. 2006; Forsberg et al. 2012; Perry et al. 2014; Thaker et al. 2010; Wright 2007). Studies have looked for ARGs in samples linked to human activity such as food production (Donato et al. 2010; Facinelli et al. 1993; Jacobs and Chenia 2007; Kerry et al. 1995; Miranda et al. 2003; Nonaka et al. 2000) and polluted waterways (Adelowo and Fagade 2009). The resistome work suggests that environmental bacteria can be highly resistant to antibiotics carrying both characterized and unknown resistant genes (D'Costa et al. 2006; Thaker et al. 2010; Wright 2007). Metagenomic analyses have identified genes that either are highly related or identical to currently known ARGs in culturable bacteria. Cryptic resistance genes which do not confer resistance and precursor genes that may ultimately be modified and evolve into actual antibiotic resistance genes of clinical importance have also been identified (Wright 2007). One example of a ARG associated with these unculturable bacterial studies is the *tet*(43) gene, which was isolated from metagenomic analysis of soil from an apple orchard which repeatedly had been treated with streptomycin (Donato et al. 2010).

It is unknown what type of bacteria actually carries *tet*(43), and little else has been done to look at the distribution of either this or other genes that have been identified only by molecular methods.

More recently, functional genomic studies have been used to study a variety of microbial environments (Moore et al. 2013). As the name implies, the assay determines if the cloned DNA is able to be expressed and confer resistance allowing the host *E. coli* to grow in the present of antibiotic-supplemented media. This identifies resistance-conferring DNA fragments which can then be sequenced and compared to known ARGs. Antibiotic Resistance Genes Database lists ~20,000 potential resistance genes (Liu and Pop 2009), while the Comprehensive Antibiotic Resistance Database [CARD] also has a large number of accessible resistance genes that can be used to screen sequences (McArthur et al. 2013). A variety of potential new ARGs have been identified using this method (D'Costa et al. 2006; Sommer et al. 2009). One issue with these databases rely on GenBank information which often does not have the correct nomenclature for specific ARGs. Consulting other sources such as http://faculty.washington.edu/marilynr/ should be done to get the correct names for tetracycline and macrolide-lincosamide-streptogramin genes. One disadvantage of these molecular methods is that the knowledge of which type of bacteria carry which resistance genes is not usually determined. Knowing which bacteria, in a particular ecosystem, carry which ARGs is important for future studies. Selecting ARGs that are prevalent or not prevalent in a particular ecosystem will impact the data of future surveillance studies (Pei et al. 2006; Smith et al. 2004). If different genes are selected for surveillance studies, then different results may be obtained and depending on which ARGs are selected may bias the data collected and the conclusions drawn.

7.9 Examples of Host Bacterial Genes Becoming Widespread Resistance Genes

Many bacteria including environmental bacteria encode for β-lactamases which hydrolyze and inactivate β-lactam antibiotics. They are the most internationally distributed of all resistance genes (Davies and Davies 2010). One example is the *ampC* gene which was originally an inducible chromosomal cephalosporinase found in a variety of *Enterobacteriaceae* including opportunistic pathogens belonging to normal intestinal floral of humans and animals, bacterial species which normally lived in either natural soil or water, and both pathogenic and nonpathogenic bacteria (van Hoek et al. 2015; Rubin and Pitout 2014). These β-lactamase genes confer β-lactam resistance and do so with greater activity against cephalosporins than penicillins. It has been proposed that the a*mpC* genes primarily originated in environmental bacteria. The first AmpC-positive clinical strains were from *E. coli* isolated in the 1940s just as the first antibiotics were being developed and used. In a host background which has porin deficiencies, the *ampC* gene when expressed confers carbapenem resistance due to increased production of the AmpC β-lactamase. Increased production of the AmpC β-lactamase is usually due to mutations in genes that upregulate the expression of the enzyme. Now the chromosomal AmpC β-lactamases are found associated with plasmids, a concurrence which was first identified in the 1980s. These plasmids are mobile, often tend to be large, and carry multiple different antibiotic resistance genes. Plasmid-mediated AmpC β-lactamases have greatly expanded the host range of this group of enzymes including epidemic human pathogens such as *E. coli* ST131 which has been isolated from fresh vegetables, food-producing animals, fish farms, pets, and water environments (Capkin et al. 2015; Rubin and Pitout 2014; van Hoek et al. 2015).

Many resistance genes are associated with soil antibiotic producers such as *Streptomyces*. Some of these natural resistance genes have the same mode of action as those found in clinical-resistant bacteria (Davies and Davies 2010). In the past, it was assumed that most environmental bacteria were not well adapted to live in humans or animals. However, this idea is changing as progress in medical science allows severely immunocompromised patients to live in the community. In addition, people who have implanted foreign objects permanently present in their bodies and various types of occupational exposure have provided new opportunities for environmental bacteria to cause disease (Rowlinson et al. 2006; Trujillo and Goodfellow 2003). Other factors have also made the distinction between environmental and non-environmental bacteria more difficult because the mixing of the two has become increasingly common as more "environmental" bacteria are found to cause disease especially in humans and animals. In addition, human contamination of the environment is widespread which has resulted in very few ecosystems left around the world that have not been touched by the activities of human civilization—whether it is in the polar regions or the Amazon jungle (Pallecchi et al. 2012; Rahman et al. 2008). As a result, there is continual mixing of environmental and non-environmental bacteria which provides multiple opportunities for horizontal genetic exchange of ARGs between man, animal, and environmental bacteria.

Both antibiotics, ARB and ARGs, are moved by water and wind (Allen et al. 2010), wastewater treatment discharges (Berglund et al. 2015), biosolids, and manure applications (Agersø et al. 2002) and isolated from recreational beaches (Roberts et al. 2009; Soge et al. 2009a, b). In addition, antibiotics, ARB and ARGs, are moved by transportation of goods and people around the world (Dobbs et al. 2013; MacFadden et al. 2015; Nurjadi et al. 2015). One result of this has been the spread of specific strains worldwide such as *Clostridium difficile* NAP1/027/BI (Gould and Limbago 2010). Originally *C. difficile* was classified as causing nosocomial diseases associated with the hospital settings. In contrast, today *C. difficile* is considered a foodborne and community pathogen. Similarly, 25 years ago, *Acinetobacter baumannii* was a rarely identified human pathogen with *Acinetobacter* spp. isolates primarily found in the environment where they were well adapted to grow at a range of different temperatures and pH values, could use a variety of carbon and energy sources, and persist in both moist and dry places for extended time periods. However, today multidrug-resistant *A. baumannii* is considered an opportunistic pathogen that has become a major concern for military trauma patients because they are very difficult infections to treat (Abbo et al. 2005). *Salmonella typhimurium* (now classified as a serovar of *Salmonella enterica* subsp. *enterica*) DT104 is another example of a global epidemic multidrug-resistant strain that has been identified in animals and people and most likely contaminates the environment which may represent an important unrecognized reservoir (Mather et al. 2013).

Soil bacteria have been screened for their ability to degrade or inactivate antibiotics. Strains were randomly isolated from 11 diverse rural and urban soils and then tested for their ability to utilize 18 different antibiotics as sole source of carbon and nitrogen. Many of the bacteria were *Burkholderia* spp. and *Pseudomonas* spp. The bacteria that could grow on antibiotics was resistant to multiple antibiotics at clinically relevant concentrations suggesting unappreciated reservoir of antibiotic resistance genes in these soils (Dantas et al. 2008).

Antibiotics are used for both human and agricultural activities for prevention and treatment of infections, food additives and growth promoters in food production, for insect pest control by reducing or eliminating the insect symbionts, cloning of plants, and as biocides in toiletries as well as household cleaning products, sterility of cell cultures by removing bacterial contaminates such as mycoplasma that persist is cell culture lines without proper antibiotic treatment, and in research including industry research. All of these activities contaminate both built and natural environments either directly or indirectly and provide selective pressure on the resident environmental bacteria to become antibiotic resistant (Davies and Davies 2010). In some cases, transfer of specific antibiotic resistance genes is increased with exposure to low levels of antibiotics (Facinelli et al. 1993). Antibiotics can be found in domestic animal manure which may be transferred when this manure is applied to fields or stored in lagoons. Antibiotics are sprayed onto crops which then contaminates the surrounding soil, sediment, and groundwater. Antibiotics may be incorporated into the food given to farm animals and fish which will, in turn, contaminate the surrounding area. Antibiotics from human therapeutic use, especially from hospital effluents, are a continual source of pollution and are considered part of the "emerging contaminants" in municipal waste with concentrations of tetracycline varying from ng/L to μg/L (Verlicchi et al. 2010). At these levels, antibiotics may select for Tc^r environmental bacteria which, once present, may persist for extended time periods in the environment and become a reservoir for resistance to tetracycline as well as other antibiotic resistance genes. Antibiotic-resistant bacteria and residues have been identified in tap water, urban water supplies, milk, meat, vegetables, and both processed and unprocessed foods (Hao et al. 2015).

Human activity may directly influence the development of ARB in built environments. For example, a couple of studies have found antibiotic resistance Gram-negative *E. coli* and Gram-positive *S. aureus* from air isolated in homes vs samples outside of the home, the later having higher bacterial levels. However, there was some variability both in the design of the studies and their results (Allen et al. 2010). Potentially, ARB may contaminate the environment either directly, as occurs when manure is applied to enrich agricultural fields, or indirectly due to sewage contamination of receiving waters. The first description of the *tet*(M) gene in

Bacillus spp. and of Tc^r *Bacillus cereus* strains carrying the *tet*(M) gene, on a functional Tn*916* element, was found in animal manure and in the fields where the manure was spread. These results suggest that presence of the *tet*(M)-carrying *B. cereus* in the fields was a direct result of manure application to the soil. Whether or not the *tet*(M)-carrying *B. cereus* would be able to act as a donor and transfer the *tet*(M) gene to either related *B. anthracis* or *B. thuringiensis* is unknown; however, some toxin-encoding plasmids are shared between these three species (Agersø et al. 2002).

An example of how human wastes can increase ARB was illustrated over 30 years ago in a 1980s study which observed three groups of wild baboons in Kenya. Two of the groups lived in their natural habitat with either limited or no human contact, and these groups had low levels of antibiotic-resistant Gram-negative enteric bacteria. The third group lived close to a tourist lodge that provided opportunities for daily contact with unprocessed human refuse, and from these animals, high levels of antibiotic-resistant Gram-negative enteric bacteria were identified with >90% Tc^r. This study suggested that contact with human refuse greatly increased the carriage of Tc^r bacteria in these wild animals (Rolland et al. 1985). Unfortunately, the surrounding environmental bacteria were not sampled in any of these studies, but one could speculate that the level of environmental antibiotic-resistant bacteria was most likely higher around the human refuse site than in the areas where the two other baboon groups lived in a more natural setting. Other studies have found antibiotic-resistant *E. coli* from arctic and subarctic seals (Glad et al. 2010), wild boars (Poeta et al. 2009), and wild rabbits (Silva et al. 2010). More recently, bacteria carrying extended-spectrum beta-lactamase (ESBLs) have been isolated from water birds in remote locations (Ardiles-Villegas et al. 2011). Birds and wild animals can also be found feeding either in or around wastewater treatment ponds, waste landfill sites, and septic tank discharges. Birds have the potential for long-distance dissemination of ARB and ARGs from agricultural settings to remote environments. Such spreading which may explain why ARB and ARGs can be found in environments with little anthropogenic activity such as the remote arctic (Allen et al. 2010).

In many studies, it has been assumed that ARG flowed from humans and/or animals to the environment. But in other cases, the use of antibiotics for food production has created antibiotic-resistant bacteria in the animals and farm environment that has spread to man. One classic example of animal to human spread is the use of avoparcin in farm animals in the EU and much of the world. Avoparcin is closely related to human drug vancomycin and was extensively used in most parts of the world other than the USA and Canada (Nilsson 2012). Vancomycin-resistant enterococci [VRE] develop on these farms contaminating the farm ecosystem including animal, environmental, and human microbiomes. The VRE strains were passed on to the farm workers and families living on the farm. In other cases, the plasmids carrying the *vanA*/*vanB* genes were transmitted from animal to human enterococci (Roberts et al. 2009). In contrast, VRE development in hospital settings in North America has occurred because vancomycin was commonly used in hospitalized individuals but not in the general community population. More recently, VRE strains have spread to the environment in the USA where they now can be isolated in a variety of setting from recreational beaches to birds to farms (Gordoncillo et al. 2012; Oravcova et al. 2014; Roberts et al. 2009).

Tetracyclines have been used extensively in aquaculture, and Tc^r bacteria have been characterized from numerous sources, including fish pathogens and environmental bacteria associated with finfish aquaculture from around the world (Akinbowale et al. 2007; DePaola and Roberts 1995; Furushita et al. 2003; Jacobs and Chenia 2007; Nawaz et al. 2008). It is possible to find Tc^r bacteria in fish feed, the sediment under the fish pens, as well as the water entering and leaving the freshwater ponds (Kerry et al. 1995; Miranda and Zemelman 2002). Some of the greatest diversity in Tc^r genes carried has been identified in the aquaculture environment. In one of our studies, we found that 40% of the Tc^r bacteria isolated from Chilean salmon fish farms carried unidentified Tc^r genes suggesting diversity in the types of *tet* resistance genes present in this ecosystem, which is higher than routinely found in collections from either man or food animals (Miranda et al. 2003). We also identified new bacterial genera carrying previously characterized *tet* genes. Many of these *tet* genes were not readily transferred under laboratory conditions, thereby raising the question of how some of the genes were being transferred to bacteria across the world and from very different environments (Miranda et al. 2003). The diversity of type and number of Tc^r bacteria found in the aquaculture setting suggests that this may be one environment where there is rapid evolution of Tc^r bacteria.

7.10 Genes Found in Environmental Bacteria

A lot of work has been done with looking at tetracycline resistance genes in nature. Bacteria carrying Tc^r are widely distributed throughout the world. They have been isolated from deep subsurface trenches, in wastewater, surface water, and groundwater, sediments and soils, and places which are relatively untouched by human civilization such as penguins in Antarctica and seals from the Arctic (Donato et al. 2010; Glad et al. 2010; Kümmerer 2004; Rahman et al. 2008). Seventeen (39%) of the 43 known *tet* genes including 12 (44%) of the efflux, 3 (25%) of the ribosomal protection, and 2 (66%) of the enzymatic *tet* genes are uniquely ascribed to environmental bacteria. Whether this represents either a true separation between *tet* genes with some truly "unique" to environmental bacteria or is because these 17 genes previously have not been used in surveillance studies of either animal or human bacteria is unclear. As of 2016, there are now 59 *tet* genes, but many of the new genes have not been identified in specific bacteria (http://faculty.washington.edu/marilynr/).

Five different resistance genes from *Streptomyces*, designated *otr*(A), *otr*(B), *otr* (C), *tcr3*, and *tet*, have been identified in the chromosome of antibiotic-producing strains. Today the *otr*(A) and *otr*(B) are now found in *Bacillus* and *Mycobacterium*, the later which can cause animal and human disease. It is possible that over time, other environmental "*tet* genes" will move into bacteria of clinical importance associated with animals and man. For example, *Clostridium* spp. are found in the environment but are also associated with the intestinal tract of humans and animals. The *tetA*(P) and *tetB*(P) genes appear to be unique to *Clostridium* spp. Other environmental genes included are the *tet*(V) gene that has been found in *Mycobacterium smegmatis*, which primarily may be an environmental bacteria (De Rossi et al. 1998), the *tet*(30) gene in *Agrobacterium* (Luo and Farrand 1999), the *tet*(33) which has been found in environmental *Arthrobacter* and *Corynebacterium* spp. (Agersø and Sandvang 2005; Tauch et al. 2002), the *tet* (35) gene in *Vibrio* and *Stenotrophomonas* spp. (Teo et al. 2002) which can cause human disease, the *tet*(41) gene in *Serratia* spp. (Thompson et al. 2007) which rarely causes human disease, and the *tet*(42) gene found in *Bacillus*, *Microbacterium*, *Micrococcus*, *Paenibacillus*, and *Pseudomonas* spp. (Brown et al. 2008) isolated in a deep-sea trench. The *tet*(34) gene was first described in *Vibrio* spp. and more recently identified in *Pseudomonas* and *Serratia*. The *tet*(43) gene was identified from metagenomic analysis of soil from an apple orchard, and it is unknown what species or genus carries this gene (http://faculty.washington.edu/marilynr/). Since this was first written, a number of other *tet* genes have been identified and are now up to *tet*(60); many of the new genes have no known bacterial host having been isolated and characterized from functional metagenomic studies as was *tet*(43) (http://faculty.washington.edu/marilynr/).

Among the 92 genes that confer resistance to one or more macrolide-lincosamide-streptogramin (MLS) antibiotics, there are a number of resistance genes that are exclusively identified in the *Streptomyces* spp. including rRNA methylase genes [*erm*(H), *erm*(I), *erm*(N), *erm*(O), *erm*(S), *erm*(U), *erm*(Z), *erm* (30), *erm*(31), and *erm*(32)], ATP-binding transporters [*car*(A), *ole*(C), *srm*(B), and *tlr*(C)], and a major facilitator [*lmr*(A)] gene. Other rRNA methylases are found innately in various environmental *Mycobacterium* spp., [*erm*(37) to *erm*(41)], while environmental bacteria carry a variety of the known MLS and tetracycline resistance genes (http://faculty.washington.edu/marilynr/). Other than genes associated with *Streptomyces* spp. and *Mycobacterium* spp., there are relatively few genes exclusively associated with environmental bacteria. Why the difference in distribution between *tet* and MLS genes is not clear.

β-lactamases are enzymes that provide resistance to β-lactam antibiotics like penicillins, cephamycins, and carbapenems (ertapenem). These β-lactamase enzymes have random mutations which encode for enzymes with modification in their spectrum of resistance to varying classes of this group of antibiotics. There are hundreds of these modified β-lactamase genes. β-lactamase genes are ancient and have been identified in remote and isolated environments suggesting that β-lactamases occur in nature (Allen et al. 2009; Barlow and Hall 2002). Another class of β-lactamases, the CTX-M genes, which are able to hydrolyze expanded-spectrum cephalosporins originated in environmental *Kluyvera* spp. Bacteria with CTX-M genes were first identified in 1989. Today, these genes can be found across the world (Davies and Davies 2010). The *qnr* genes are from waterborne *Aeromonas*, *Shewanella*, and *Vibrio* spp. (Perry and Wright 2013). Data from a

30,000-year-old permafrost showed that genes conferring resistance to a variety of different classes of antibiotics (β-lactams, tetracycline, and glycopeptides) existed in the environment before antibiotics were used by man, suggesting that ARG linkage between the environment and either human or animal medicine does exist.

7.11 Conclusion

The environmental microbiome remains largely unexplored. However, a few studies suggest the wide distribution of ARB and ARGs in the world. For example, antibiotic-resistant marine bacteria have been isolated 522 km offshore and at depths of 8200 m (Aminov 2011). There was a correlation between degree of pollution in the environment and the level of resistance, suggesting that over time even the more "pristine" environments will become increasingly contaminated with ARGs and ARB. This will ultimately increase resistance in opportunistic and pathogenic bacterial species of humans and animal importance. Increased pressure for antibiotic resistance in environmental microorganisms is likely to continue given the fact that human activities will likely continue to pollute the environment not only locally but around the world. Natural forces such as wind and movement of water will continue to contaminate areas of relatively uninhabited environments. Since the dawn of antibiotic era, the amount of antibiotics produced worldwide and used and ultimately released into the environment has steadily increased. This in turn has provided an increasing selection and maintenance pressure on ARB and ARGs in all environments (animal, human, and environmental).

The overlap between these three disciplines of animal, human, and environmental microbiology has led to the "One Health" concept which is a worldwide strategy for expanding interdisciplinary collaborations and communications in all aspects of health care for humans, animals, and the environment. The aim is for inclusive collaborations dedicated to improving the lives of all species—human and animal—through the integration of human medicine, veterinary medicine, and environmental science. This concept recognizes that using compartmentalized (silo) mentality to approach the three disciplines individually is not adequate since the distinction of environmental from non-environmental bacteria has become increasingly difficult. In addition, the introduction of a new ARG in any humans, animals and agriculture, or the environment ecosystem often leads to cross transmission and dissemination of ARGs and ARB within and between ecosystems (Davies and Davies 2010).

The data summarized in this chapter indicates that the environment is an important reservoir for ARGs and ARB. There is a large diversity of resistance genes in the environment of which many have yet to be identified or characterized. Horizontal gene transfer within the microbial world knows few boundaries, and our ability to experimentally mimic what occurs in nature has significant limitations. The role that the natural environment plays in the evolution, maintenance, and transmission of ARB and ARGs is just now being examined. However, many researchers agree that the human anthropogenic changes are impacting the natural ecosystems which will ultimately impact human and animal health.

Compliance with Ethical Standards
Conflict of Interest: Marilyn C. Roberts declares that she has no conflict of interest.

Ethical approval: This chapter does not contain any studies with human participants or animals performed by any of the authors.

References

Aaretrup F (2012) Get pigs off antibiotics. Nature 486:465–466

Abbo A, Navon-Venezia S, Hamemer-Muntz O et al (2005) Multidrug-resistant *Acinetobacter baumannii*. Emerg Infect Dis 11:22–29

Adams MD, Chan ER, Neil D et al (2010) Genome wide analysis of divergence of antibiotic resistance determinants in closely related isolates of *Acinetobacter baumannii*. Antimicrob Agents Chemother 54:3569–3577

Adelowo OO, Fagade OE (2009) The tetracycline resistance genes tet39 is present in both gram-negative and gram-positive bacteria from a polluted river, Southwestern Nigeria. Lett Appl Microbiol 48:167–172

Agersø Y, Sandvang D (2005) Class 1 integrons and tetracycline resistance genes in *Alcaligenes, Arthrobacter and Pseudomonas* spp. isolated from pigsties and manured soil. Appl Environ Microbiol 71:7941–7947

Agersø Y, Jensen LB, Givskov M et al (2002) The identification of a tetracycline resistance gene *tet*(M), on a Tn*916*-like transposon, in the *Bacillus cereus* group. FEMS Microbiol Lett 214: 251–256

Akinbowale OL, Peng H, Barton MD (2007) Diversity of tetracycline resistance genes in bacteria from aquaculture sources in Australia. J Appl Microbiol 103:2016–2025

Allen HK, Moe, LA, Rodbumrer J et al (2009) Functional metagenomics reveals diverse beta-lactamases in a remote Alaskan soil. ISME J 3:243-251

Allen HK, Donato J, Wang HH et al (2010) Call of the wild: antibiotic resistance gene in natural environments. Nat Rev Microbiol 8:251–259

Aminov RI (2011). Horizontal gene exchange in environmental microbiota. Front Microbiol 2:158 doi:10.3389/fmicb.2011.00158

Ardiles-Villegas KM, Gonzalez-Acuna D, Waldenstrom J et al (2011) Antibiotic resistance patterns in fecal bacteria isolated from Christmas shearwater (*Puffinus nativitatis*) and masked booby (*Sula dactylatra*) at remote Easter Island. Avian Dis 55:486–489

Atkinson BA, Abu-Al-Jaibat A, LeBlanc DJ (1997) Antibiotic resistance among enterococci isolated from clinical specimens between 1953 and 1954. Antimicrob Agents Chemother 41: 1598–1600

Avery OT, MacLeod CM, McCarty M (1944) Studies on the chemical nature of the substance inducing transformation of pneumococcal types. J Exp Med 79:137–158

Barlow M, Hall BR (2002) Phylogenetic analysis shows that the OXA beta-lactamase gens have been on plasmids for millions of hears. J Mol Evol 55:314–321

Bearson BL, Brunelle BW (2015) Fluoroquinolone induction of phage-mediated gene transfer in multidrug-resistant *Salmonella*. Int J Antimicrob Agents 46:201–204

Berendonk TU, Manaia CM, Merlin C et al (2015) Tackling antibiotic resistance: the environmental framework. Nat Rev Microbiol 13:310–317

Berglund B, Fick J, Lindgren PE (2015) Urban wastewater effluent increases antibiotic resistance genes concentrations in a receiving northern European river. Environ Toxicol Chem 34: 192–196

Bharat A, Demczuk W, Martin I et al (2015) Effect of variants of penicillin-binding protein 2 on cephalosporin and carbapenem susceptibilities in *Neisseria gonorrhoeae*. Antimicrob Agents Chemother 59:5003–5006

Brown MG, Mitchell EH, Balkwill DL (2008) Tet 42, a novel tetracycline resitance determinant isolated from deep terrestrial subsurface bacteria. Antimicrob Agents Chemother 52: 4518–4521

Capkin E, Terzi E, Altinok I (2015) Occurrence of antibiotic resistance genes in culturable bacteria isolated from Turkish trout farms and their local aquatic environment. Dis Aquat Organ 114: 127–137

Center for Disease Control (2013) Antibiotic resistant threats in the United States. http://www.cdc.gov/drugresistance/threat-report-2013/pdf/ar-threats-2013-508.pdf. Accessed 12 July 2016

Chopra I, Roberts MC (2001) Tetracycline antibiotics: mode of action, applications, molecular biology and epidemiology of bacterial resistance. Microbiol Mol Biol Rev 65: 232–260

Collignon P (2013) The importance of a one health approach to preventing the development and spread of antibiotic resistance. Curr Top Microbiol Immunol 366:19–36

Colomer-Lluch M, Imamovic L, Jofre J et al (2011) Bacteriophages carrying antibiotic resistance genes in fecal waste from cattle, pigs, and poultry. Antimicrob Agents Chemother 55: 4908–4911

Cousin SL Jr, Whittington WL, Roberts MR (2003) Acquired macrolide resistance genes in pathogenic *Neisseria* spp. isolated between 1940 and 1987. Antimicrob Agents Chemother 47: 3877–3880

Cundliffe E, Demain AL (2010) Avoidance of suicide in antibiotic-producing microbes. J Ind Microbiol Biotechnol 37:643–672

D'Costa V, McGrann KM, Hughes DW et al (2006) Sampling the antibiotic resistome. Science 311:374–377

Dantas G, Sommer MOS, Oluwasegun RD et al (2008) Bacterial subsisting on antibiotics. Science 320:100–103

Davies J, Davies D (2010) Origins and evolution of antibiotic resistance. Microbiol Mol Biol Rev 74:417–433

DePaola A, Roberts MC (1995) Class D and E tetracycline resistance determinants in Gram-negative catfish pond bacteria. Mol Cell Probes 9:311–313

Dezube BJ, Krown SE, Lee JY et al (2006) Randomized phase II trial of matrix metalloproteinase inhibitor COL-3 in AIDS-related Kaposi's sarcoma: an AIDS Malignancy Consortium Study. J Clin Oncol 24:1389–1394

Di Luca MC, D'Ercole S, Petrelli D et al (2010) Lysogenic transfer of *mef*(A) and *tet*(O) genes carried by Phim46.1 among group A streptococci. Antimicrob Agents Chemother 54: 4464–4466

Dobbs FC, Goodrich AL, Tomson FS III et al (2013) Pandemic serotypes of *Vibrio cholerae* isolated from ships' ballast tanks and coastal waters: assessment of antibiotic resistance and virulence genes (*tcpA* and *ctxA*). Microbiol Ecol 65:969–974

Donato JJ, Moe LA, Converse BJ et al (2010) Metagenomic analysis of apple orchard soil reveals antibiotic resistance genes encoding predicted bifunctional proteins. Appl Environ Microbiol 76:4396–4401

Durso LM, Miller DN, Wienhold BJ (2012) Distribution and quantification of antibiotic resistant genes and bacteria across agricultural and non-agricultural metagenomes. PLoS One. doi:10.1371/journal.pone.0048325. Accessed 9 June 2016

Edquist P, Ringman M, Liljequist BO et al (2013) Phenotypic detection of plasmid-acquired AmpC in *Escherichia coli*-evaluation of screening criteria and performance of tow commercial methods for the phenotypic confirmation of AmpC production. Eur J Clin Microbiol Infect Dis 32:1205–1210

Facinelli B, Roberts MC, Giovanetti E et al (1993) Genetic basis of tetracycline resistance in food borne isolates of *Listeria innocua*. Appl Environ Microbiol 59:614–616

Food and Drug Administration (2013) 21 CFR Parts 514 and 558 [Docket No. FDA–2010–N–0155]. RIN 0910–AG95. Veterinary Feed Directive. https://s3.amazonaws.com/public-inspection.federalregister.gov/2013-29696.pdf. Accessed 10 July 2016

Forsberg KJ, Reyes A, Wang B et al (2012) The shared antibiotic resistome of soil and human pathogens. Science 337:1107–1111

Fournier P-E, Vallenet D, Barber V et al. (2006) Comparative genomics of multidrug resistance in *Acinetobacter baumannii*. PLoS Genet. http://www.ncbi.nlm.nih.gov/pmc/articles/PMC1326220/pdf/pgen.0020007.pdf. Accessed 10 July 2016

Furushita M, Shiba T, Maeda T et al (2003) Similarity of tetracycline resistance genes isolated from fish farm bacteria to those from clinical isolates. Appl Environ Microbiol 69:5336–5342

Garcillian-Barcia MP, de la Cruz F (2008) Why is entry exclusion an essential feature of conjugative plasmids? Plasmid 60:1–18

Garrido-Mesa N, Zarzuelo A, Galvez J (2013) Minocycline: far beyond an antibiotic. Br J Pharmacol 169: 337–352

Gillespie MT, May JW, Skurray RA (1986) Detection of an integrated tetracycline resistance plasmid in the chromosome of methicillin-resistant *Staphylococcus aureus*. J Gen Microbiol 132:1723–1728

Glad T, Kristiansen VF, Nielsen KM et al (2010) Ecological characterization of the colonic microbiota in arctic and sub-arctic seals. Microb Ecol 60:320–330

Gordoncillo MJN, Donabedian S, Bartlett PC et al (2012) Isolation and molecular characterization of vancomycin-resistance *Enterococcus faecium* from swine in Michigan, USA. Zoonoses Pub Health 60:319–326

Gould LH, Limbago B (2010) *Clostridium difficile* in food and domestic animals: a new foodborne pathogen? Clin Infect Dis 51(5):577–582

Grimdmann H, Aires-de-Sousa M, Boyce J et al (2006) Emergence and resurgence of methicillin-resistant *Staphylococcus aureus* as a public-health threat. Lancet 368:874–885

Guarddon M, Miranda JM, Rodriquez JA et al (2011) Real-time polymerase chain reaction for the quantitative detection of *tetA* and *tetB* bacterial tetracycline resistance genes in food. Int J Food Microbiol 146:284–289

Hao R, Zhao R, Qiu S et al (2015) Antibiotics crisis in China. Science 348:110–1101

Harnisz M, Korzeniewska E, Ciesielski S et al (2015) *tet* genes as indicators of changes in the water environment: relationships between culture-dependent and culture-independent approaches. Sci Total Environ 505:704–711

Hughes VM, Datta N (1983) Conjugative plasmids in bacteria of the "pre-antibiotic" era. Nature 301:725–726

Jacobs L, Chenia HY (2007) Characterization of integrons and tetracycline resistance determinants in *Aeromonas* spp. isolated from South African aquaculture systems. Int J Food Microbiol 114:295–306

Johnston C, Martin B, Fichant G et al (2014) Bacterial transformation: distribution, shared mechanism and divergent control. Nat Rev Microbiol 12:181–196

Kerry J, Hiney M, Coyne R et al (1995) Fish feed as a source of oxytetracycline-resistant bacteria in the sediments under fish farms. Aquaculture 240:89–100

Kümmerer K (2004) Resistance in the environment. J Antimicrob Chemother 54:311–320

Lawley TD, Burland V, Taylor DE (2000) Analysis of the complete nucleotide sequence of the tetracycline-resistance transposon Tn10. Plasmid 43:235–239

Leigh MJS, Nguyen DV, My Y et al (2013) A randomized double-blind, placebo-controlled trial of minocycline in children and adolescents with Fragile X syndrome. Devel Behav Pediatr 34: 147–155

Liu B, Pop M (2009) ARDB-antibiotic resistance genes database. Nucleic Acids Res 37(Database issue):D443–D447. doi:10.1093/nar/gkn656

Luna VA, Roberts MC (1998) The presence of the *tetO* gene in a variety of tetracycline resistant *Streptococcus pneumoniae* serotypes from Washington State. J Antimicrob Chemother 42: 613–619

Luo Z-Q, Farrand SK (1999) Cloning and characterization of a tetracycline resistance determinant present in *Agrobacterium tumefaciens* C58. J Bacteriol 181:618–626

MacFadden DR, Bogoch II, Brownstein JS et al (2015) A passage from India: association between air traffic and reported cases of New Delhi metallo-beta-lactase 1 from 2007 to 2012. Travel Med Infect Dis 13:295–299

Marshall B, Roberts M, Smith A et al (1984) Homogeneity of transferable tetracycline-resistance determinants in *Haemophilus* species. J Infect Dis 149:1028–1029

Mather AE, Reid SWJ, Maskell DJ et al (2013) Distinguishable epidemics of multidrug-resistant *Salmonella typhimurium* DT104 in different hosts. Science 341:154–1517

McArthur AG, Waglechner N, Nizam F et al (2013) The comprehensive antibiotic resistance database. Antimicrob Agents Chemother 57:3348–3335

Miranda CD, Zemelman R (2002) Bacterial resistance to oxytetracycline in Chilean salmon farms. Aquaculture 212:31–47

Miranda CD, Kehrenberg C, Ulep C et al (2003) Diversity of tetracycline resistance genes in bacteria from Chilean salmon farms. Antimicrob Agents Chemother 47:883–888

Moore AM, Patel S, Forsberg KJ et al (2013) Pediatric fecal microbiota harbor diverse and novel antibiotic resistance genes. PLoS One 8:e78822. http://www.plosone.org/article/fetchObject.action?uri=info:doi/10.1371/journal.pone.0078822&representation=PDF. Accessed 10 July 2016

National strategy for combating antibiotic-resistant bacteria. 2014. https://www.whitehouse.gov/sites/default/files/docs/carb_national_strategy.pdf. Accessed 10 June 2016

Nawaz M, Khan AA, Khan S et al (2008) Isolation and characterization of tetracycline-resistant *Citrobacter* spp. from catfish. Food Microbiol 25:85–91

Nilsson O (2012) Vancomycin resistant enterococci in farm animals-occurrence and importance. Infect Ecol Epidemiol 2:16959. doi:10.3402/lee.v2i0.16959. Accessed 10 June 2016

Nonaka L, Ti I, Suzuki S (2000) The occurrence of the oxytetracycline resistant bacteria in the fish intestine and seawater environment. Microbiol Environ 15:223–228

Nordmann P, Naas T, Poirel L (2011) Global spread of carbapenemase-producing *Enterobacteriaceae*. Emerg Infect Dis 17:1791–1798

Norgren M, Scott JR (1991) The presence of conjugative transposon Tn*916* in the recipient strain does not impede transfer of a second copy of the element. J Bacteriol 173:319–324

Nurjadi D, Friedrich-Janicke B, Schafer J et al (2015) Skin and soft tissue infections in intercontinental travelers and the import of multi-resistant *Staphylococcus aureus* to Europe. Clin Microbiol Infect 21:567.el–567.e10

Oravcova V, Zurek L, Townsend A et al (2014) American crows as carriers of vancomycin-resistant enterococci with *vanA* gene. Environ Microbiol 16:939–949

Pallecchi L, Bartonloni A, Riccobono E et al (2012) Quinolone resistance in absence of selective pressure: the experience of a very remote community in the Amazon forest. PLoS Neglect Trop Dis 6:e1790. doi:10.1371/journal.pntd.0001790. Accessed 10 July 2016

Parnham MJ, Haber VE, Giamarellos-Bourboulis EJ et al (2014) Azithromycin: mechanism of action and their relevance for clinical applications. Pharmacol Ther 143:225-245

Pei R, Kim SC, Carlson KH et al (2006) Effect of river landscape on the sediment concentrations of antibiotics and corresponding antibiotic resistance genes (ARG). Water Res 40:2427–2435

Peng S, Wang Y, Zhou B et al (2015) Long-term application of fresh and composted manure increase tetracycline resistance in the arable soil of eastern China. Sci Total Environ 506–507: 279–286

Perry JA, Wright GE (2013) The antibiotic resistance "mobiolime": searching for the link between environment and clinic. Front Microbiol 4:138. doi:10.3389/fmicb.2013.00138

Perry JA, Westman EL, Wright GE (2014) The antibiotic resistome: what's new? Curr Opin Microbiol. 21:45–50

Philippe J, Gallet B, Morlot C et al (2015) Mechanism of β-lactam action in *Streptococcus pneumoniae*: the piperacillin paradox. Antimicrob Agents Chemother 59:609–621

Poeta P, Radhouani H, Pinto L et al (2009) Wild boars as reservoirs of extended-spectrum beta-lactamase (ESBL) producing *Escherichia coli* of different phylogenetic groups. J Basic Microbiol 49:584–588

Rahman MH, Sakamoto KQ, Nonaka L et al (2008) Occurrence and diversity of tetracycline *tet* (M) in enteric bacteria of Antarctic Adelie penguins. J Antimicrob Chemother 62:627–628

Recchia GD, Hall RM (1995) Gene cassettes: a new class of mobile element. Microbiology 141: 3015–3027

Rice LB (2007) Conjugative transposons. In Bonomo RA, Tolmasky M (eds) Enzyme-mediated resistance to antibiotics: mechanisms, dissemination, and prospects for inhibition. ASM, Washington DC, pp 271–284

Roberts MC (1997) Genetic mobility and distribution of tetracycline resistance determinants. In: Antibiotic resistance: origins, evolution, selection and spread. CIBA Foundation symposium 207. Wiley, Chichester, pp 206–218

Roberts MC (2012) Mechanisms of bacterial antibiotic resistance and lessons learned from environmental tetracycline resistant bacteria. In: Keen P, Montforts M (eds) Antibiotic resistance in the environment. Wiley, New York, pp 93–121

Roberts MC, Soge OO, Giardino MA et al (2009) Vancomycin resistant *Enterococcus* spp. in marine environments from the west coast of the USA. J Appl Microbiol 107:300–307

Roberts MC, Schwarz S, Aarts H (2012) Acquired antibiotic resistance genes: an overview. Front Microbiol 3:384; 1–17. doi:10.3389/fmicb.2012.00384.; http://www.frontiersin.org/Antimicrobials,_Resistance_and_Chemotherapy/10.3389/fmicb.2012.00384/full. Accessed 12 June 2016

Roberts MC, No D, Kuchmiy E et al (2015) The tetracycline resistant *tet*(39) gene identified in three new genera of bacteria isolated in 1999 from Chilean salmon farms. J Antimicrob Chemother 70:619–620

Rolland RM, Hausfater G, Marshall B et al (1985) Antibiotic-resistant bacteria in wild primates: increased prevalence in baboons feeding on human refuse. Appl Environ Microbiol 49: 791–794

Rowlinson M-C, Bruckner DA, Hinnebusch C et al (2006) Clearance of *Cellulosimicrobium cellulans* bacteremia in a child without central venous catheter removal. J Clin Microbiol 44: 2605–2654

Rubin JE, Pitout JDD (2014) Extended-spectrum β-lactamases, carbapenemase and AmpC production Enterobacteriaceae in companion animals. Vet Microbiol 170:10–18

Silva N, Igrejas G, Figueiredo N et al (2010) Molecular characterization of antimicrobial resistance in enterococci and *Escherichia coli* isolates from European wild rabbit (*Oryctolagus cuniculus*). Sci Total Environ 408:4871–4876

Sloan J, McMurry LM, Lyras D et al (1994) The *Clostridium perfringens* Tet P determinant comprise two overlapping genes: *tetA*(P), which mediates active tetracycline efflux, and *tetB* (P), which is related to the ribosomal protection family of tetracycline-resistance determinants. Mol Microbiol 11:403–415

Smith MS, Yang RK, Knapp CW et al (2004) Quantification of tetracycline resistance genes in feedlot lagoons by real-time PCR. Appl Environ Microbiol 70:7372–7377

Soge OO, Meschke JS, No DB et al (2009a) Characterization of methicillin-resistant *Staphylococcus aureus* (MRSA) and methicillin-resistant coagulase-negative *Staphylococcus* spp. (MRCoNS) isolated from West Coast public marine beaches. J Antimicrob Chemother 64:1148–1155

Soge OO, Tivoli L, Meschke JS et al (2009b) A conjugative macrolide resistance gene, *mef*(A), in environmental *Clostridium perfringens* carrying multiple macrolide and/or tetracycline resistance genes. J Appl Microbiol 106:34–40

Sommer MO, Dantas G, Church GM (2009) Functional characterization of the antibiotic resistance reservoir in the human microflora. Science 28:1128–1131

Speer BS, Bedzyk L, Salyers AA (1991) Evidence that a novel tetracycline resistance gene found on two *Bacteroides* transposons encodes an NADP-requiring oxidoreductase. J Bacteriol 173: 176–183

Stanton TB, Humphrey SB (2003) Isolation of tetracycline-resistant *Megasphaera elsdenii* strains with novel mosaic gene combinations of *tet*(O) and *tet*(W) from swine. Appl Environ Microbiol 69:3874–3882

Stine OC, Johnson JA, Keefer-Norris A et al (2007) Widespread distribution of tetracycline resistance genes in a confined animal feeding facility. Int J Antimicrob Agents 29:348–352

Sun Q, Wei X, Feng J et al (2008) Involvement of insulin-like growth factor-insulin recptor signal pathway in the transgenic mouse model of medulloblastoma. Cancer Sci 99:234–240

Tauch A, Gotker S, Puhler A et al (2002) The 27.8-kb R-plasmid pTET3 from *Corynebacterium glutamicum* encodes the aminoglycoside adenyltransferase gene cassette *aadA9* and the regulated tetracycline efflux system Tet 33 flanked by active copies of the widespread insertion sequence IS*6100*. Plasmid 48:117–129

Taylor, J, Hafner M, Yerushalmi E et al (2014) Estimating the economic costs of antimicrobial resistance: model and results. RAND. http://www.rand.org/content/dam/rand/pubs/research_reports/RR900/RR911/RAND_ RR911.pdf. Accessed 9 July 2015

Teo JWP, Tan TMC, Poh CL (2002) Genetic determinants of tetracycline resistance in *Vibrio harveyi*. Antimicrob Agents Chemother 46:1038–1045

Thaker M, Spanogiannopoulos P, Wright GD (2010) The tetracycline resistome. Cell Mol Life Sci 67:419–431

Thompson SA, Maani EV, Lindell AH et al (2007) Novel tetracycline resistance determinant isolated from an environmental strain of *Serratia marcescens*. Appl Environ Microbiol 73: 2199–2206

Trujillo ME, Goodfellow M (2003) Numerical phenetic classification of clinically significant aerobic sporoactinomycetes and related organisms. Antoinie Van Leeuwenhoek 84:39–68

Van Boeckel TP, Brower C, Gilbert M et al (2015) Global trends in antimicrobial use in food animals. Proc Natl Acad Sci USA 112:5649–5654

Van Hoek AHAM, Mayrhofer S, Doing KJ et al (2008) Mosaic tetracycline resistance gene and their flanking regions in *Bifidobacterium thermophilum* and *Lactobacillus johnsonii*. Antimicrob Agents Chemother 52:248–252

Van Hoek AHAM, Veenman C, van Overbeek WM, Lynch G, de Roda Husma AM, Blaak H (2015) Prevalence and characterization of ESBL-and AmpC-producing Enterobacteriaceae on retail vegetables. Int J Food Microbiol 204:1–8

Verlicchi P, Galletti A, Masotti L (2010) Management of hospital wastewaters: the case of the effluent of a large hospital situated in a small town. Water Sci Technol 61:2507–2519

Vo ATT, van Duijkeren E, Fluit AC et al (2007) A novel *Salmonella* genomic island 1 and rare integron types in *Salmonella typhimurium* isolates from horses in The Netherlands. J Antimicrob Chemother 59:594–599

Watanabe T (1963) Infective heredity of multiple drug resistance in bacteria. Bacteriol Rev 27: 87–115

Waters VL (2001) Conjugation between bacterial and mammalian cells. Nat Genet 29:375–376

Woegerbauer M, Kuffner M, Domingues S et al (2015) Involvement of *aph(3′)-IIa* in the formation of mosaic aminoglycoside resistance genes in natural environments. Front Microbiol 6:442. http://journal.frontiersin.org/article/10.3389/fmicb.2015.00442/abstract. Accessed 9 June 2016

Wright GD (2007) The antibiotic resistome: the nexus of chemical and genetic diversity. Nat Rev Microbiol 5:175–186

WHO (2014) WHO antibiotic resistance global report of surveillance 2014. http://apps.who.int/iris/bitstream/10665/112642/1/9789241564748_eng.pdf?ua=1. Accessed 8 June 2016

Chapter 8
Population Dynamics, Invasion, and Biological Control of Environmentally Growing Opportunistic Pathogens

Veijo Kaitala, Lasse Ruokolainen, Robert D. Holt, Jason K. Blackburn, Ilona Merikanto, Jani Anttila, and Jouni Laakso

Abstract In this chapter, we analyze environmentally growing opportunistic diseases, which are a growing threat to human health, food production, and wildlife. The traditional treatment methods for opportunist diseases often fail because environmentally growing opportunist pathogens can utilize outside-host environmental resources for growth and can reside indefinitely in the environmental reservoirs. Moreover, environmental growth may promote high virulence because the trade-off between virulence and transmission can be weak or absent. Moreover, as multicellular organisms are a relatively rich resource compared to the surrounding environment, transition from free-living organism into an opportunist pathogen is a potential pathway through which novel, obligate pathogens emerge. We show that the environmental growth strategy can profoundly change the epidemiological dynamics because the pathogen faces environmental variability and outside-host food web interactions such as competition, predation, and parasitism. The contrasting evolutionary challenges between the outside-host environment and host immune responses likely play an important role in the epidemiological dynamics. Understanding the conditions that promote or hinder the spread of already existing, as well as the invasion of novel, environmentally growing opportunistic diseases is the key for controlling this class of infectious diseases. This chapter aims to give the reader a view on the studies made so far regarding environmentally growing opportunist diseases and present ideas on how the outside-host abiotic and biotic environment could be utilized in pathogen control.

V. Kaitala (✉) • L. Ruokolainen • I. Merikanto • J. Anttila • J. Laakso
Department of Biosciences, University of Helsinki, Helsinki, Finland
e-mail: veijo.kaitala@helsinki.fi; lasse.ruokolainen@helsinki.fi; ilona.merikanto@helsinki.fi; jani.anttila@helsinki.fi; jouni.laakso@helsinki.fi

R.D. Holt
Department of Biology, University of Florida, Gainesville, FL 32611, USA
e-mail: rdholt@ufl.edu

J.K. Blackburn
Department of Geography, Spatial Epidemiology & Ecology Research Laboratory, University of Florida, Gainesville, FL 32611, USA
e-mail: jkblackburn@ufl.edu

C.J. Hurst (ed.), *Modeling the Transmission and Prevention of Infectious Disease*, Advances in Environmental Microbiology 4, DOI 10.1007/978-3-319-60616-3_8

8.1 Introduction

Traditionally, ecological and eco-evolutionary epidemiological models describe the dynamics of infectious diseases by considering susceptible, infected, and recovered hosts with direct host-to-host (Anderson and May 1981) or host-environment-host transmission (Keeling and Rohani 2008), where the "environment" refers to either a passive conduit for pathogen passage or vector-borne transmission. However, this classical framework omits the vast majority of free-living microbes as potential pathogens that reside in a wide range of environmental reservoirs (from soil, to the water column, to commensals in other species). Moreover, it does not fully explain how and why pathogens emerge in the first place. Novel pathogens with already established obligatory life histories frequently emerge through changes in the host range, which can be facilitated by phylogenetic and genetic factors, e.g., by close relatedness and horizontal gene transfer (Antia et al. 2003; Woolhouse et al. 2005). However, the transformation of free-living organisms into opportunistic pathogens, and perhaps in time into obligatory ones, should be given serious consideration as another pathway to pathogenicity. Essentially, for potential pathogens freely living in the environment, a multicellular organism is a rich concentrated patch of resources (Smith 2007).

As a life history strategy for a pathogen, being an environmental opportunist would seem beneficial, as it allows utilization of both within-host and outside-host resources for replication. This gives a competitive advantage for an environmental opportunist against competition between free-living microbes (Wedekind et al. 2010; Merikanto et al. 2014) and permits a kind of bet-hedging among alternative resources (provided there is spillback from the infected host to the environmental reservoirs versus infection being a dead end). It is not thus surprising that many common pathogens have been recognized as environmentally growing opportunistic pathogens that survive and, critically, can reproduce potentially indefinitely in the outside-host environment, e.g., as saprotrophs (Casadevall 2008; Veneault-Fourrey and Martin 2011; Brown et al. 2012). Analysis of disease dynamics has focused on conditions that permit $R_0 > 1$ (i.e., the number of secondary infections per primary infection should exceed unity), because this encapsulates pathogen persistence—for obligate pathogens. If, instead, there are environmental reservoirs, for any given host species, one could have $R_0 \ll 1$ and still observe high prevalence.

In addition to its ecological relevance, environmental opportunism could also be of great evolutionary importance. For instance, one might observe the evolution of high virulence as the outside-host reproduction strategy allows the pathogens to be at least partially free from the transmission-virulence constraints limiting host-specific obligatory pathogens (Read 1994; Levin 1996; Frank 1996; Lipsitch and Moxon 1997; Walther and Ewald 2004). Treating environmentally sustained diseases with traditional methods, e.g., with antibiotics and disinfectants, is not always successful at the population level. This occurs because removal of susceptible hosts or the recovery of infected ones does not eradicate pathogens from their environmental reservoir where they still continue replicating (Rahman et al. 2008; Merikanto et al. 2012; Anttila et al. 2013).

In this chapter we seek to outline the elements of a theoretical framework for environmentally growing opportunist diseases and their emergence. We also discuss several concrete examples of well-known epidemiologically important organisms that fit this framework rather than the classical epidemiological SIR disease transmission compartment model that divides individuals into the categories of susceptible, infectious, and recovered. Furthermore, we also discuss biological control methods, which may be more suitable against these diseases than many traditional methods of disease intervention.

8.2 Environmentally Growing Opportunist Pathogens: Examples

Environmentally growing opportunistic pathogens are plentiful in nature (Casadevall 2008; Veneault-Fourrey and Martin 2011; Brown et al. 2012). These pathogens include bacteria, protozoa, or fungi which variously attack plants and animals including humans. Well-known diseases caused by such pathogens include cholera (*Vibrio cholerae*), various infections by *Pseudomonas aeruginosa*, Legionnaires' disease (*Legionella pneumophila*), listeriosis (*Listeria monocytogenes*), the fungal infection cryptococcosis (*Cryptococcus neoformans*), tuberculosis (*Mycobacterium tuberculosis*), Hansen's disease, also known as leprosy (*Mycobacterium leprae*), bacterial cold-water disease on salmonids (BCWD), rainbow trout fry disease (RTFD), bacterial gill disease on trout (BGD), columnaris disease or cotton-wool disease in cultured and wild freshwater fish (genus *Flavobacterium*), *Burkholderia* in people with cystic fibrosis (Traverse et al. 2013), and urinary tract infections and wound infections in humans (e.g., genus *Serratia*) (Grimont and Grimont 1978; Friedman et al. 2002; Leclerc et al. 2002; Berg et al. 2005; Hall-Stoodley and Stoodley 2005; Soto et al. 2007; Hilbi et al. 2007; Casadevall 2008; Rahman et al. 2008; Freitag et al. 2009; Kunttu et al. 2009; Mahlen 2011; Trivedi et al. 2011; Fisher et al. 2012; Brown et al. 2012). In some case, the term "environmental reservoir" may denote the ability of a potential pathogen to utilize nonliving substrates, while in other cases, the pathogen may be sustained by either unknown or poorly studied alternative hosts (which in some circumstances can be represented by the kinds of models we present in this chapter). Below, we introduce the biology of some of these opportunist pathogens in more detail.

8.2.1 *Flavobacterium columnare*

Columnaris disease causes major economic losses in salmonid and other species in freshwater aquaculture worldwide. The disease agent is a saprotrophic *Flavobacterium columnare* bacteria that can induce massive outbreaks in high-density cultured environments, especially during spring and summer as the water temperatures rise. The *F. columnare* colonize the gills and cause skin erosion often leading to the death of the fish by suffocation, especially in young fish (Declercq et al. 2013). The bacterium can be found in natural waters but is rarely known to cause infection outside fisheries (Pulkkinen et al. 2010; Kunttu et al. 2012). It has been suggested that conditions at fish farms facilitated evolution of virulence in this pathogen due to high availability of susceptible hosts and abundance of saprotrophic resources such as fish food, fish feces, as well as dead fish (Kunttu et al. 2009; Pulkkinen et al. 2010). Given these alternative resources, the pathogen can persist even if very highly virulent on fish hosts.

8.2.2 *Vibrio cholerae*

Cholera is perhaps one of the most well-known environmentally growing opportunist diseases. It is notorious for causing severe epidemics in humans, mostly in developing countries. Although better sanitary conditions have reduced cholera outbreaks, it is still a major problem in Haiti, Southeast Asia, and much of Africa (WHO 2008). The species *V. cholerae* is found naturally in brackish water and estuaries, and the strains vary in virulence from nonpathogenic to highly virulent forms. The most virulent strains in the outside-host environment seem to originate from infected hosts and are released mainly through feces (Faruque et al. 1998; Merrell et al. 2002; Hartley et al. 2006). However, many less virulent or avirulent strains are known to utilize a wide range of other organisms, as well as abiotic surfaces (Lutz et al. 2013). For example, *V. cholerae* resides naturally on the exoskeleton of crustaceans where the existence is facilitated by the bacteria using extracellular chitinase activity to obtain carbon and nitrogen (Mondal et al. 2014).

8.2.3 *Bacillus anthracis*

Anthrax is a bacterial zoonosis of wild and domestic herbivores that secondarily infects humans and some predators and scavengers (Hugh-Jones and de Vos 2002). Outbreaks are documented nearly worldwide, with the distribution of disease limited to environmental conditions that support the causative agent *Bacillus anthracis*, a spore-forming, Gram-positive bacterium (Hugh-Jones and Blackburn 2009). Broadly, outbreaks are concentrated to grassland environments associated with grazing livestock and ungulate wildlife populations (Blackburn et al. 2014b; Turner et al. 2014). Outbreaks can pose serious demographic risks for some vertebrate populations, at least episodically. The microbe can apparently survive in soil [or replicate in the rhizosphere (Saile and Koehler 2006)] for long periods of time under certain environmental conditions (Blackburn et al. 2007; Hugh-Jones and Blackburn 2009). The issue of free-living *B. anthracis* populations is the subject of active investigation, as is how spore generation from infected vertebrates might feed this environmental reservoir. If spores are long-lived, then even small free-living populations could generate large pools of potentially infective spores.

Transmission to vertebrate hosts is not entirely understood, but it is likely a combination of ingesting spores from contaminated soil (particularly in grazers) and consumption of vegetation which may have become contaminated either through "splash" up or by necrophagous flies contaminating vegetation with emesis and feces that are then consumed by browsing species [e.g., white-tailed deer (Blackburn et al. 2010, 2014a)]. Percutaneous infection from biting flies is another potential route of infection (Blackburn et al. 2014b). Inhalation of spores from the environment has limited evidence, though it has been suggested as a possible route of infection in wood bison and cannot be ruled out (WHO 2008). Livestock and wildlife are likely infected through quite similar routes of exposure.

Naturally occurring anthrax (as a disease of vertebrates) can be considered an obligate spillover pathogen with no direct transmission (of quantitative significance) among susceptible hosts (Alexander et al. 2012). Infection occurs in geographic areas where the environment supports pathogen persistence in the environment, and vertebrate hosts coincidentally may be attracted to microhabitats that sustain environmental reservoirs (Blackburn et al. 2014a; Turner et al. 2014). Outside of intentional (nefarious) releases, human anthrax cases are most usually caused by direct contact with infected animals, such as slaughter and meat distribution (Kracalik et al. 2013; Bezymennyi et al. 2014), and can be thought of as secondary spillover onto host B (the humans) from primary spillover into host A (the infected animals). It is plausible that anthrax cases in carnivores and scavengers could occur for comparable reasons. Anthrax outbreaks in the middle latitudes appear to be seasonal across host systems (Blackburn et al. 2014b). For example, deer outbreaks in Texas appear in summer months, with the severity of outbreaks increasing in response to early and intense spring greenup (Blackburn and Goodin 2013). These climatic correlations, coupled with spatial clustering of cases over time, further supports the hypothesis that *B. anthracis* persists in the environment.

8.3 Environmentally Growing Pathogen Dynamics

8.3.1 A General Model of Environmentally Growing Pathogen Dynamics

In general, the dynamics of an environmentally growing pathogen can be divided into two parts: within-host dynamics and outside-host dynamics (Box 8.1). Within-host dynamics can be formulated with the traditional SIR model, describing the reproduction and mortality of the host, becoming infected, and dying or recovering from the disease. In some cases the recovery may occur directly from the infected stage (without acquired immunity), thus leaving the recovery-compartment unnecessary in the disease dynamics model.

The outside-host dynamics are largely determined by the community that resides in the environment and the abiotic resources and conditions there. It includes at least a pathogen capable of either living there or persisting in a dormant state like a spore and being capable of causing infections to certain hosts. The outside-host dynamics are determined by ecological and evolutionary interactions with other pathogens and free-living nonpathogenic taxa. These include competitive, predatory, parasitic, and mutualistic interactions typically occurring in the environmental food webs (Pascual and Dunne 2005).

Several variants of the host dynamics can be appropriate depending, for example, on whether all the S-, I-, and R-individuals participate in the reproduction of the host, or whether all or only a fraction of infected individuals subsequently either die or become sterilized, and whether or not infected hosts have density-dependent interactions with healthy hosts (Merikanto et al. 2012).

The nature of the infection process itself may be crucial in determining the characteristics of the disease dynamics of outbreaks. As given in the anthrax example, transmission may be a result of either direct ingestion from the environment or aided by vector organisms (flies). Infections might be caused by exposure of hosts to the pathogens living in the environment and can also occur at the same time by individual contacts (e.g., direct transmission—in the case of cholera). The functional form of the response to infections (from either source) may not be straightforward to estimate and has profound consequences for epidemiological dynamics (Anttila et al. 2017).

Box 8.1
Here we present a general model of environmentally growing pathogen dynamics, which couples density-dependent growth and community interactions in the outside-host environment with an *SIR* model describing within-host dynamics (but assuming no direct transmission from infected to healthy hosts or vertical transmission):

(continued)

Box 8.1 (continued)

Within-host dynamics:

$$\frac{dS}{dt} = S[g(S,I,R) - Pf(P)] + \theta R \tag{8.1a}$$

$$\frac{dI}{dt} = SPf(P) - I[g'(S,R,I) + \delta + \rho] \tag{8.1b}$$

$$\frac{dR}{dt} = \rho I - R\left[g''(S,R,I) + \theta\right] \tag{8.1c}$$

Outside-host dynamics:

$$\frac{dP}{dt} = P[h(P) + \mathbf{A}_1\mathbf{X}] + \Lambda\delta I \tag{8.1d}$$

$$\frac{d\mathbf{X}}{dt} = \mathbf{X} \bullet (\mathbf{b} + \mathbf{A}_{2\ldots n,2\ldots n}\mathbf{X} + \mathbf{A}_{2:n,1}P) \tag{8.1e}$$

The host consists of susceptible (S), infected (I), and recovered (R) individuals ($H = S + I + R$). Density dependence in the host population is controlled by functions g, g', and g'' describing how individuals in different classes contribute to birth of new susceptibles and mortality of each class. Susceptibles become infected at a rate that depends on the density of pathogen (P) in the environment, controlled by the infectivity function f, which in turn depends on pathogen density. Infected individuals can die either due to competition (g') or disease (δ), or recover (ρ). Finally, recovered individuals lose their immunity at a rate θ and return to the susceptible class (Eq. 8.1a) and die at rate g''. The change in the abundance of recovered individuals depends on the rate of development of an immune response (ρ), and the rate at which immunity is lost (θ) regenerating healthy but susceptible hosts, as well as the effect of density dependence (competition) on mortality, encapsulated by g''.

The community in the outside-host environment consists of a pathogenic species P and a set of nonpathogenic species X_i (contained in vector $\mathbf{X}$) that can be either competitors, predators, or parasites of the pathogen (and each other). Pathogen dynamics in the external environment (Eq. 8.1d) are governed by an intrinsic growth function h and the effect of other species interacting with the pathogen (first row in the interaction matrix $\mathbf{A}$, a_{1i}) and an input Λ of new pathogens released from infected hosts that die directly due to the infection (alternatively, pathogens can also be shed continuously from live infected hosts, in which case the input term would be proportional to I, but not dependent on the mortality rate). For every other, nonpathogenic species in the community, there is a density-independent term b_i

(continued)

Box 8.1 (continued)

(vector **b** in Eq. 8.1e) that differs in sign between producer species and consumer species (Case and Casten 1979). The summation term (in matrix notation, $\mathbf{A}_{2...n,2...n}\mathbf{X}$) contains the effect of interspecific interactions from species other than the pathogen, and the final term gives the effect of the pathogen on the growth of each species X_i. The parameters a_{ij} form a matrix **A**, containing all pair-wise interaction effects in the community (May 1973; Case and Casten 1979) and characterizing the classical Lotka-Volterra model of community interactions. Thus, by modifying the composition of this matrix, one can manipulate the composition of the outside-host species community to ask how patterns in between-species interactions affect the pathogen's role in the system. One could also envision alternative community models, for instance, with either nonlinear functional responses or interference functions. Finally, the • symbol denotes element-vise multiplication between vector X and that resulting from the operations within the parentheses.

8.3.2 *The Role of Pathogen Life Cycle on Epidemiology*

When considering epidemiological dynamics, the specific life cycle of the disease-causing agent can become very important. Here we demonstrate this with a simplified model with a range of different pathogen types. Specifically, we will consider three examples: (1) a classical, directly transmitted obligate pathogen, such as measles (Simpson 1952); (2) an environmentally transmitted obligate pathogen, e.g., many bacteria causing respiratory infections (Walther and Ewald 2004); and (3) an environmental opportunistic pathogen, such as anthrax (Turner et al. 2013). In (1) the pathogen is contained entirely within the infected individuals. The model describes the dynamics of susceptible (S) and infected (I) hosts (1–3) and those of a pathogen (P) (Godfray et al. 1999; Merikanto et al. 2012; Anttila et al. 2013):

$$\frac{dS}{dt} = r_S(1 - f_S S)S - S(\beta P + \gamma I) + \rho I \tag{8.2a}$$

$$\frac{dI}{dt} = S(\beta P + \gamma I) - (\delta + \rho)I \tag{8.2b}$$

$$\frac{dP}{dt} = r_P(1 - f_P P)P + I(\Lambda\delta + \lambda) - \mu_P P. \tag{8.2c}$$

Here the susceptible hosts exhibit logistic renewal governed by growth rate (r_S) and density dependence (f_S). Susceptible hosts can be infected by contacts with pathogens (2, 3) or infected individuals (1), controlled by rate constants β and γ, respectively (Eq. 8.2a). Infected individuals can either return to the susceptible class by recovering from the disease at rate ρ or perish from the disease with rate δ

(Eq. 8.2b). Finally, the pathogen can grow logistically in the environment (3), according to the growth rate r_P and density dependence f_P (Eq. 8.2c) (Merikanto et al. 2012; Anttila et al. 2013). Pathogens can be either released from dead hosts with quantity Λ (3) (also referred to as burst size; Caraco and Wang 2008) or shed continuously from live infected hosts with rate λ (2). Pathogens also undergo density-independent decay in the environment at rate μ_P. We assume that host absorption of pathogens from the environmental pool is quantitatively negligible, compared to other processes governing that population's dynamics. A simplifying assumption here is that infected individuals neither contribute to host reproduction nor to density dependence. This can be taken to mean that infected individuals are too weak either to consume resources or produce offspring or otherwise interfere with the fitness of healthy hosts. Competition between strains with different modes of transmission can lead to complex evolutionary scenarios (Roche et al. 2011).

While the model in Eq. (8.2a–8.2c) can be solved analytically for equilibria and invasibility, the solutions become rather complex except for simple cases, such as a directly transmitted obligate pathogen (case 1,$\beta = 0$). In this case the system has an equilibrium at $S^* = (\delta + \rho)/\gamma$, $I^* = r_S(1 - f_S(\delta + \rho/\gamma))(\delta + \rho/\gamma)/\delta$, which is positive if $\gamma > f_S(\delta + \rho)$.

If $\beta = 0$ in Eq. (8.2a–8.2c), the model becomes isomorphic in form to the Lotka-Volterra predator-prey model with a logistic prey growth term, which is well known to be locally and globally stable. Another simple case is for $\Lambda\delta + \lambda = 0$. This could describe a pathogen, which lives in the environment, but has spillover into dead-end hosts. The free-living pathogen pool will settle to an equilibrium, P^*, and then there will be a constant force of infection imposed on the focal host. This again is analogous to a predator-prey system for which the equilibria are stable. One possible equilibrium is for the host to be excluded by pathogen spillover (a phenomenon related to apparent competition).

The full model described by Eq. (8.2a–8.2c) exhibits a wide range of dynamics when studied numerically. In the simplest case of the *SI* model with host renewal (1), either no infections, host extinction, monotonic convergence to, or dampened oscillations to a stable equilibrium (Fig. 8.1a) can result. In addition to the same qualitative dynamics produced by the directly transmitted disease (1), the environmentally transmitted pathogen (2) can also induce cyclic disease outbreaks (Fig. 8.1b). Finally, the density-dependent outside-host dynamics of the environmental opportunistic pathogen (3) have a stabilizing effect on the dynamics (Merikanto et al. 2012; Anttila et al. 2013). In this case, cyclic outbreaks do not occur (Fig. 8.1c), unless the density-independent decay rate (μ_P) is high enough, such that the environment becomes a sink to the pathogen (Anttila et al. 2013). Due to the environmental growth, this pathogen also has a strong potential to drive the focal host extinct. In the limiting case, where there is no recovery following infection ($\rho = 0$), the condition of host extinction is simply $r_s < \beta P^* = \beta / f_P$. Hosts can become extinct if they have low intrinsic growth rates, and there is a large pool of pathogens in an environmental reservoir.

The behavior of all of these cases depends largely on the balance between infection rates (β, γ) and time spent as infected (depending on the balance between

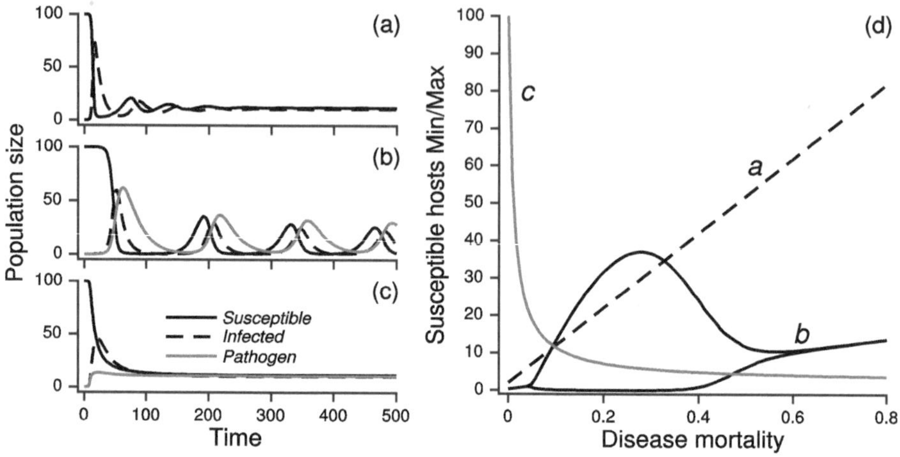

Fig. 8.1 Example dynamics of three different diseases: (**a**) directly transmitted obligate pathogen (1), (**b**) environmentally transmitted obligate pathogen (2), and (**c**) environmental opportunistic pathogen (3). (**d**) The effect of increasing disease mortality (δ) on the severity of the disease in the three different cases (a–c). The minimum and maximum values of the susceptible hosts are shown at the stable or periodic attractor. General parameters: $r_S = 0.1$, $f_S = 0.01$, $f_P = 0.01$, $\mu_P = 0.05$, $\rho = 0.02$, $\delta = 0.1$. Specific parameters: (a) $r_P = 0$, $\lambda = 0$, $\Lambda = 0$, $\beta = 0$, $\gamma = 0.01$; (b) $r_P = 0$, $\Lambda = 0$, $\beta = 0.01$, $\gamma = 0$, $\lambda = 0.1$; (c) $r_P = 1$, $\beta = 0.01$, $\gamma = 0$, $\lambda = 0$, $\Lambda = 1$

ρ and δ). For example, varying the rate of host mortality due to disease infection (δ, which could be modified, e.g., by medical care) can either increase (1) or decrease (3) the density of healthy hosts or lead to disease outbreaks at intermediate mortality rates (2) (Fig. 8.1d). Of course, the dynamics can be affected by other parameters as well, e.g., by the pathogen loss rate μ_P, as explained above.

For simplicity, we have assumed here that infections occur at a constant per capita rate (per susceptible host, per infected host), as is traditionally assumed in epidemiological models of obligate pathogens (Anderson and May 1981; Holt and Pickering 1985). However, assuming other functional forms for infection can have important consequences for host dynamics and potentially cause host exclusion (Regoes et al. 2002; Ben-Ami et al. 2008; Boldin and Kisdi 2012; Anttila et al. 2013). For example, a dose-dependent infection response can introduce an infection threshold (Regoes et al. 2002), a certain level of pathogen exposure that is required for successful invasion of the host—which can protect the host from going extinct when exposed to a virulent pathogen (Anttila et al. 2013). Pathogens vary enormously in the aliquots needed to infect individual hosts, from immeasurably small to astronomical. Holt (1999) provides examples and hypothesizes that analogs of Allee effects (such as saturating host defenses) could explain this pattern.

8.3.3 The Shape of Disease Transmission

As indicated above (Sect. 8.3.2), the shape of disease transmission functions has important effects on epidemiological dynamics. When considering pathogen transmission via the environment, it is logical to think that the rate at which healthy hosts become infected depends on the rate at which they come into contact with infective particles (Anderson and May 1981). While the transmission rate might initially scale linearly with increasing contact rate, it must saturate at some point because (a) population transmission rate decreases when the number of susceptible hosts declines locally and (b) infected individuals become infective after a delay; a constant contact rate is based on the assumptions of infinite, well-mixed populations and an effectively instantaneous infection process. Moreover, at low contact rates, there may be an acceleration in how infection scales with pathogen abundance or infection prevalence.

Indeed, empirical evidence supports a nonlinear relationship between pathogen load and infection probability (Ben-Ami et al. 2008). In their seminal paper, Regoes et al. (2002) showed that a sigmoidal transmission rate that scales with pathogen dose leads to an Allee effect. That is, there is effectively a threshold for pathogen density below which a successful invasion is not possible. At this threshold, pathogen release from infected hosts is balanced by the loss of pathogens from the environment. In general, a sigmoidal infection response arises due to two separate mechanisms. Firstly, the infection threshold results from an over-proportionate increase in infection rate at low pathogen densities. This can arise due to, e.g., pathogen synergism (quorum sensing), biofilm production, toxin production, or host immune system saturation (Regoes et al. 2003; Schmid-Hempel and Frank 2007; Leggett et al. 2012). Secondly, saturation of the infection rate, i.e., an under-proportionate increase in infection rate at high pathogen densities, can arise simply due to a lag between pathogen ingestion and the development of disease symptoms (Regoes et al. 2003). Properly accounting for the latter may require the use of, e.g., delay differential equation models. We will focus, however, on non-linearities that arise because of the biological mechanisms noted above.

In epidemiological studies, the focus tends to be either on the long-term prevalence of the disease or on the ultimate mortality of hosts under a given disease exposure (Ben-Ami et al. 2010). This gives an indication of host disease tolerance (Vale et al. 2014). Such tolerance curves are necessarily sigmoidal in shape, because the response variable (disease prevalence/host mortality) is restricted between zero and one, independent of the details of disease transmission (Regoes et al. 2003; Ben-Ami et al. 2008; Leggett et al. 2012; Anttila et al. 2017).

8.4 Invasion of Environmentally Growing Pathogens

We next consider the disease dynamics of an environmentally growing opportunistic pathogen in competition with a nonpathogenic strain (Merikanto et al. 2014). The competitor is assumed to be superior in the outside-host environment, i.e., we assume that the ability to infect the host carries a cost in terms of outside-host competitive ability (e.g., Sturm et al. 2011). We study the conditions where a new environmentally growing opportunist pathogen strain is able to invade the community and at least coexist with the nonpathogenic strain if not to replace it. We then analyze how outside-host competition affects the long-term disease dynamics of an environmentally growing opportunist.

The disease dynamics given competition are assumed to follow:

$$\frac{dS}{dt} = r_S(1 - S)S - \beta SP - \mu_S S + \rho I \tag{8.3a}$$

$$\frac{dI}{dt} = \beta SP - \delta I - \mu_S I - \rho I \tag{8.3b}$$

$$\frac{dP}{dt} = \Lambda\delta I + r_P(1 - f_{PP}P - f_{BP}B)P - \mu_P P \tag{8.3c}$$

$$\frac{dB}{dt} = r_B(1 - f_{PB}P - f_{BB}B)B - \mu_B B \tag{8.3d}$$

where B denotes the non-infective strain; f_{PB} and f_{BP} are competition coefficients between P and B; r_B and μ_B are the growth rate and mortality of the non-infective strain, respectively; and μ_S is density-independent background mortality for the susceptible and infected hosts due to other causes than the disease.

We use invasion analysis to examine the evolution of pathogenicity. We first assume that the community is at an equilibrium such that the pathogenic population is not present. Thus, given that $r_S > \mu_S$ and $r_B > \mu_B$, the equilibrium of the community is given as

$$(S^*, I^*, P^*, B^*) = \left(\frac{(r_S - \mu_S)}{r_S}, 0, 0, \frac{(r_B - \mu_B)}{r_B f_{BB}}\right), \tag{8.4}$$

where S^* and B^* are the host population and non-infective strain at equilibrium conditions, respectively. Pathogen invasion may begin with a mutant strain P capable of consuming the host as an additional resource. The resource consumption also affects the health of the host. The conditions for the initial invasion can be studied by investigating the growth of the pathogenic strain at this equilibrium. The Jacobian matrix of the community system (8.3a–8.3d) is given as

$$J = \begin{bmatrix} -r_S + \mu_S & \rho & -\beta S^* & 0 \\ 0 & -\delta - \mu_S - \rho & \beta S^* & 0 \\ 0 & \Lambda\delta & r_P - \mu_P - (r_P f_{BP}/r_B f_{BB})(r_B - \mu_B) & 0 \\ 0 & 0 & -f_{PB}(r_B - \mu_B)/f_{BB} & -(r_B - \mu_B) \end{bmatrix} \tag{8.5}$$

It can be shown that two of the four eigenvalues of the above Jacobian matrix are real and negative, given as

$$\lambda_{1,2} = (-(r_S - \mu_S), -(r_B - \mu_B)) \tag{8.6}$$

The two other eigenvalues are obtained from the following submatrix:

$$J = \begin{bmatrix} -\delta - \mu_S - \rho & \beta S^* \\ \Lambda\delta & r_P - \mu_P - (r_P f_{BP}/r_B f_{BB})(r_B - \mu_B) \end{bmatrix} \tag{8.7}$$

The characteristic equation becomes

$$(\lambda + \delta + \mu_S + \rho)(\lambda - [r_P - \mu_P - (r_P f_{BP}/r_B f_{BB})(r_B - \mu_B)]) - \beta\Lambda S^* \delta = 0 \tag{8.8}$$

For a locally stable equilibrium, the real parts of all eigenvalues need to be negative. Thus, invasion becomes possible when the real part of at least one eigenvalue is positive.

When the pathogenicity is developed fully such that $\beta, \Lambda > 0$, then the eigenvalues cannot be solved explicitly. For the sake of argument, we assume that the recovery does not occur, $\rho = 0$, making the analysis transparent. If $\delta > 0$, $\Lambda > 0$, $r_P = r_B$, $\mu_P = \mu_B$, and $f_{PP} = f_{BP} = f_{PB} = f_{BB}$, then the characteristic equation becomes

$$\lambda^2 + (\delta + \mu_S)\lambda - \beta S^* \Lambda\delta = 0 \tag{8.9}$$

This means that one eigenvalue is always real and positive making invasion possible. The pathogen grows faster as compared to the competitor due to the additional resource gained by the disease. This example assumes that there are no costs from evolving infectivity from a nonpathogenic strain. In this case, the nonpathogenic competitor becomes extinct in the external environment.

A more interesting (and possibly biologically more realistic) setting is, however, one where the pathogenic strain will suffer costs from evolving the infectivity trait (Sturm et al. 2011). In other words, transmission within the focal host comes at the cost of some ability to utilize the external environment. The costs can be expressed in different aspects of the life history in the external environment. Natural options would be lowered growth rate (r_P), increase mortality (μ_P), or decreased competitive ability (f_{PB} and f_{BP}). These traits can be traded off with each other or with increased infectivity rate (β) and burst-out rate (Λ). Thus, the trade-off in developing infectivity can range across the ecological parameters, or parameters directly entering into disease dynamics, or both.

A number of phenomena can be illustrated using numerical examples. If the cost of infectivity occurs in the growth rate ($r_P \ll r_B$), then a loss in growth rate can be compensated with a more effective transmission rate (Fig. 8.2a). The effect of the competitor on the pathogen growth rate in the initial stages of invasion is crucial. The stronger the competitor (in terms of f_{BP}), the more efficient the pathogen needs to be in its infection rate of the host. The case $f_{BP} = 0$ means to the pathogen that the competitor is absent. However, even in this case the transmission rate β needs to exceed a positive threshold value. This occurs due to the fact that in this example the growth rate is less than the mortality rate, which must be compensated for.

Equivalent conclusions apply for the burst-out rate Λ, since transmission rate and burst-out rate are exchangeable from the point of view of the pathogen in the invasion process. This can be seen from the characteristic equation (8.8) where these parameters appear together as a product ($\beta\Lambda$).

We next consider a situation where developing the infectivity may result in changes in the pathogen growth rate and its competitive ability. It is expected that increasing the competitive ability of the competitor decreases the possibilities of the invasion of the pathogen. It is also expected that the increasing the growth rate of the pathogen in the environment increases the possibilities of the pathogen to invade. However, the latter prediction may fail.

Figure 8.2b illustrates the invasion results in the parameter space (f_{BP}, r_P). As above, it appears that increasing the strength of the competition decreases the possibilities of the pathogen to invade. However, the results with respect to the pathogen growth rate may appear counterintuitive: increasing the value of the pathogen growth rate will prevent the invasion of the pathogen. This result may be understood by taking a closer look at the competitive dynamics between the competitor and the pathogen. For high values of the interaction strength ($f_{BP} > 1/B^*$), the term $r_P(1 - f_{BP}B)$ in the pathogen dynamics becomes negative. Thus, increasing the value of the pathogen growth rate only intensifies the negative effect of the competitor on the growth of the pathogen at the invasion situation. An increase in the growth rate increases the density dependence, leading to an exclusion of the invading pathogen.

The counterintuitive result above may be based on the strong and biologically unreasonable assumption about the direct relationship of intrinsic growth rate and sensitivity to competition. To address the importance of the density dependence for the purposes of invasion analysis, we revise the model such that density dependence and growth rate can be independent. The corresponding dynamics are revised as follows:

$$\frac{dP}{dt} = \Lambda\delta I + r_P P - d_{PP}P^2 - d_{BP}BP - \mu_P P \qquad (8.10a)$$

$$\frac{dB}{dt} = r_B B - d_{PB}PB - d_{BB}B^2 - \mu_B B \qquad (8.10b)$$

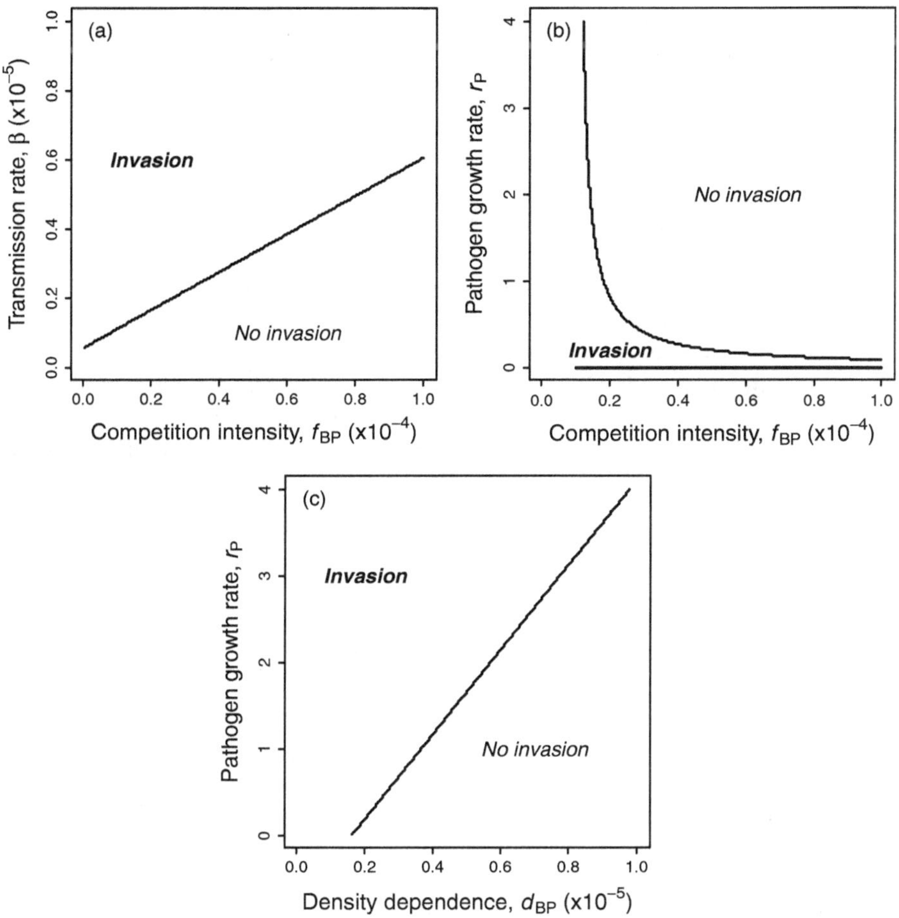

Fig. 8.2 Invasion of a pathogenic mutant is affected by pathogen growth rate and the intensity of outside-host competition. (**a**) Even initially a very low growth rate of the pathogen ($r_P = 0.05$, $r_B = 5$) allows an invasion by the pathogen at low competition pressure with the nonpathogenic competitor. Increased competition between the pathogen and the competitor can be compensated through increased disease transmission rate β. The threshold for the transmission rate at $f_{BP} = 0$ is due to the fact that the mortality of the pathogen is higher than its growth rate ($\mu_P = 0.1$). (**b**) Invasion is possible for low intensities of the effect of the competitor on the pathogen. If the intensity of the competition increases such that $f_{BP} > 1/B^*$ (bolded line on the x-axis), then the term $r_P(1 - f_{BP}B)$ in the pathogen dynamics becomes negative. Thus, increasing the value of the pathogen growth rate only intensifies the negative effect of the competitor on the growth of the pathogen. (**c**) However, if the model is modified such that the density-dependent interaction of the pathogen and its competitor is independent of their growth rates, invasion becomes possible for increasing values of growth rate r_P

where d's are competition coefficients. It becomes clear that the invasion of the pathogen becomes possible for increasing values of the growth rate while increasing competition pressure tends to inhibit it (Fig. 8.2c).

8.5 Coexistence of the Host, Pathogen, and Competitor

Coexistence of the host, pathogen, and competitor is possible for large ranges of the parameters. Thus, an opportunistic environmentally growing strategy is expected to be common in pathogens. Coexistence can show stable dynamics but the dynamics can also fluctuate periodically. Epidemiologically, the latter would correspond to recurrent epidemics in the host. We next look at some examples of the effect of the parameters on the stability of the coexistence dynamics. The examples here are based on model (8.3a–8.3d).

8.5.1 *Pathogen Outside-Host Growth*

We assume here that the competition coefficients f_{PB} and f_{BB} are all equal and the growth rate of the competitor $r_B = 1$. For the studied growth rates of the pathogen, the competitor has a competitive advantage in which case the pathogen would become extinct in the absence of ability to replicate through the disease. For low values of the pathogen growth rate, the disease dynamics are periodic (Fig. 8.3). Increasing values of r_P decreases the amplitudes of the oscillations until the coexistence dynamics of S, P, and B stabilize at $r_P > 0.048$.

8.5.2 *Effect of the Competitor*

We next consider an example where the density effects are otherwise identical except that the density effect (competition strength) of the competitor on the pathogen is varied. The growth rate of the pathogen is small as compared to that of the competitor, which gives the competitor a competitive advantage. When the competition strength of the competitor, f_{BP}, is zero, the situation from the perspective of the pathogen is the same as if the competitor did not exist. In this case the host–pathogen dynamics fluctuate periodically (Fig. 8.4). If the competitor is present, its dynamics also fluctuate periodically as a consequence of the fluctuations of the pathogen. Increasing competition strength of the competitor on the pathogen suppresses the fluctuations until at $f_{BP} = 0.4 \times 10^{-4}$ the dynamics stabilize to an equilibrium.

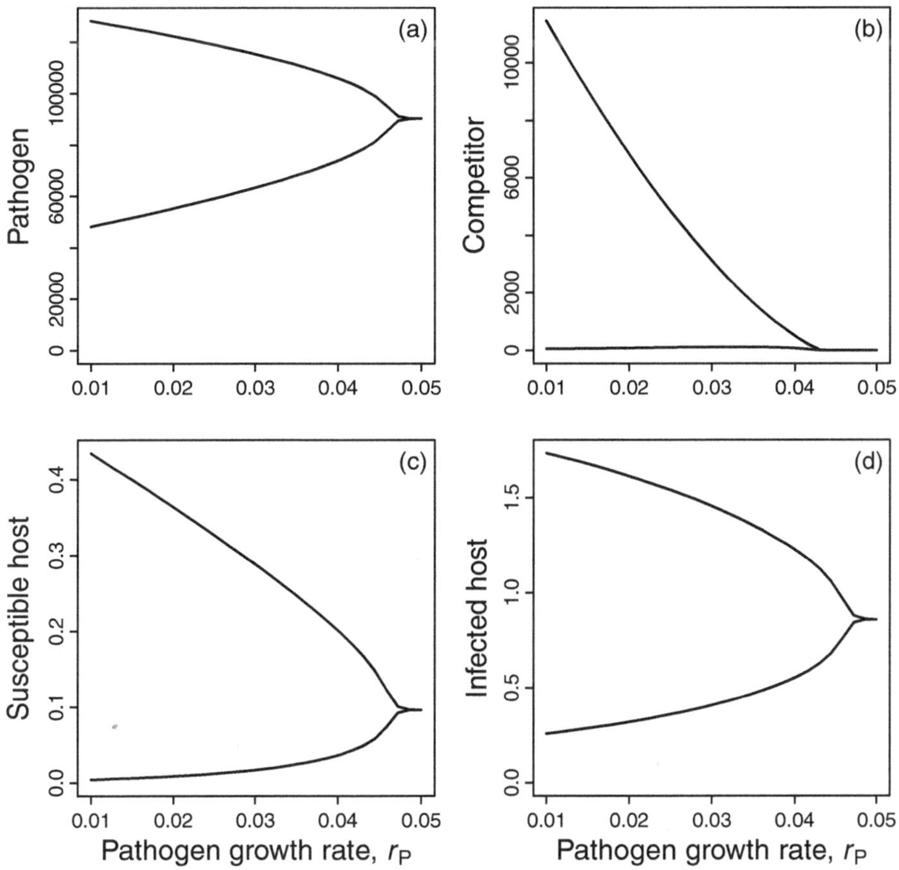

Fig. 8.3 Increasing growth rate r_P of the pathogen in the outside-host environment stabilizes the disease dynamics. Minimum and maximum values of the dynamics of the pathogen (**a**), competitor (**b**), susceptible host (**c**), and infected host (**d**) are shown. For low values of r_P, the dynamics are periodic. A transition from periodic dynamics to stable dynamics (Hopf bifurcation) occurs at $r_P > 0.048$. Parameters as in Merikanto et al. (2014), Table 2, Fig. 1

8.6 Outbreaks in Microbial Communities: Importance of the Infectivity Response and the Microbial Diversity

Interspecific interactions, such as competition, predation, and mutualism, are at the core of ecological research. Such ecological interactions in the outside-host environment may be of great importance in modifying the dynamics of many environmentally transmitted diseases. We next discuss a dynamic model that combines environmental opportunist pathogen-host dynamics and community dynamics outside the host (Anttila et al. 2013). This setting is more realistic in the sense that the pathogen lives among many competitors. We assume that there is a trade-off between virulence and environmental competitive ability. Life-

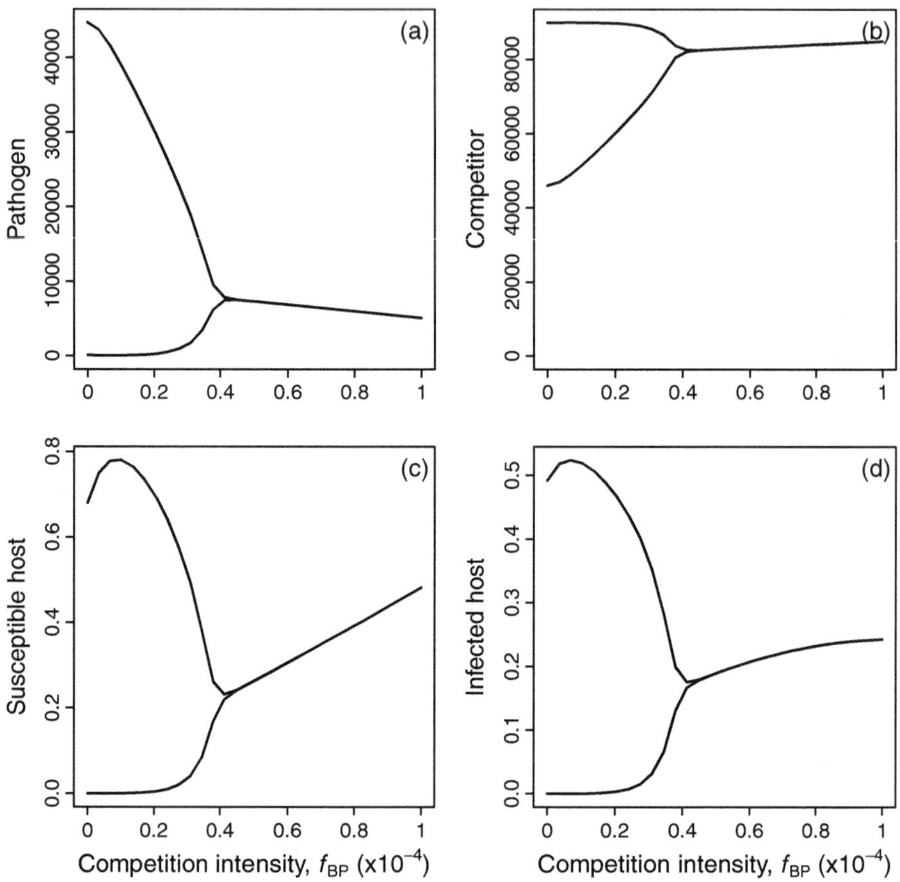

Fig. 8.4 Increasing competition strength f_{BP} of the competitor on the pathogen stabilizes the disease dynamics. The minimum and maximum values of the dynamics of the pathogen (**a**), competitor (**b**), susceptible host (**c**), and infected host (**d**) are shown. When the competition strength of the competitor, f_{BP}, is zero or low, then the host-pathogen dynamics fluctuate periodically. When the competition strength increases, then the dynamics stabilize at $f_{BP} = 0.4 \times 10^{-4}$. Parameters as in Merikanto et al. (2014), Table 2, Fig. 1, except $r_S = 0.1$

history trade-offs are expected to reduce virulence since resource acquisition and defense in the outside- vs. inside-host environments require specialization (Gower and Webster 2005; Sturm et al. 2011; Mikonranta et al. 2012).

The community model is given as follows:

$$\frac{dS}{dt} = r_S\left(1 - \frac{S+I}{K_S}\right)S - \beta S f(P) + \rho I \tag{8.11a}$$

$$\frac{dI}{dt} = \beta Sf(P) - \delta I - \rho I \tag{8.11b}$$

$$\frac{dP}{dt} = \Lambda\delta I + r_P\left(1 - \frac{P + \sum_i(\overline{\alpha} + \alpha)B_i}{K_P}\right)P - \mu_P P \tag{8.11c}$$

$$\frac{dB_i}{dt} = r_P\left(1 - \frac{B_i + (\overline{\alpha} - \alpha)P + \sum_{j\neq i}\overline{\alpha}B_j}{K_P}\right)B_i - \mu_B B_i, i = 1, \ldots, N. \tag{8.11d}$$

K_S is the carrying capacity of the host. Both the pathogen (P) and the competing non-pathogenic strains (B_i) are presumed to grow logistically with rate r_P. Carrying capacities are K_P for all strains. The competition in the outside-host community was modeled as a diffuse competition where the competitive ability of nonpathogens against the pathogen is varied. The intraspecific competition coefficients were set to 1, whereas between-species interaction strength is given by $\overline{\alpha}$. The pathogen faces a trade-off between its ability to cause infections and reduction α in competitive ability. It should be noted that the diffuse competition model used here provides, on average, the same dynamics as a model with randomly generated between-species competition coefficients (this reflects the assumption of linear interactions, on a per capita basis).

The choice of the functional form of the infectivity response can crucially affect the model dynamics. Susceptible hosts are infected with a rate $\beta Sf(P)$ depending on the infectivity response. Two alternative infectivity response functions $f(P)$ were explored. Many recent theoretical studies have for simplicity and algebraic tractability assumed a linear infectivity response, $f(P) = P$. However, this assumption is biologically simplified. Anttila et al. (2013) utilized a more realistic sigmoidal dose-dependent response. The sigmoid infectivity response can be modeled as a Hill function (Regoes et al. 2003) as follows:

$$f(P) = \frac{(P/\mathrm{ID}_{50})^{\kappa}}{1 + (P/\mathrm{ID}_{50})^{\kappa}} \tag{8.12}$$

where the parameters ID_{50} and κ define the shape of the response. The value of the infectivity response approaches 1 as the pathogen density P increases.

The form of the infectivity response function is crucial to the behavior of the epidemics. When the infectivity response is sigmoidal, cyclic dynamics occur at a range of host growth rates (Fig. 8.5a). Cyclic outbreaks occur when the mortality rate (δ) is close the host growth rate. If the host growth rate is much faster than the mortality rate, cycling becomes impossible. When the infectivity response is linear, the pathogen maintains its infectivity even in small doses. The community dynamics remains bounded for the whole range of the host growth rates analyzed (Fig. 8.5b).

There are three crucial differences in pathogen response to increasing competitive pressure that emerge when there is sigmoidal rather than linear transmission (Fig. 8.6). First, cyclic pathogen dynamics appear when the number of competitors

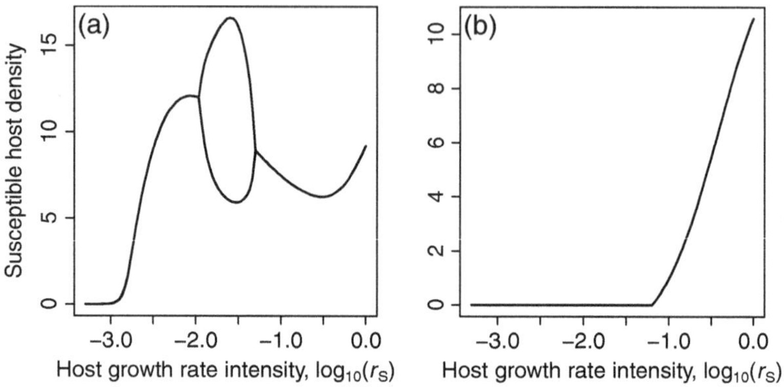

Fig. 8.5 The dynamics of the susceptible host presented as a function of host growth rate under two infectivity scenarios: sigmoidal and linear infectivities. (**a**) Under a sigmoidal pathogen infective response, there is a range with cyclic dynamics. The minima and maxima of population densities are shown. (**b**) When the pathogen infective response is linear, the dynamics are stable. A linear response easily results in local host extinction unless the host growth rate is sufficiently high. In this example, competitive disadvantage (α) is 0.05 and the number of competitor species (N) is 5. $ID_{50} = 3150$ and $\kappa = 4$

is large. The threshold for this transition in dynamics is approximately the number of the competitors where competitive exclusion would occur in the absence of the host. Second, within the range of cyclic dynamics, an alternative attractor appears in which the pathogen becomes extinct. This is due to the Allee effect associated with the sigmoidal response function (Regoes et al. 2003; Anttila et al. 2013). If the initial pathogen density is sufficiently low, it is unable to infect susceptible hosts, and competition outside the host makes it more likely this will occur. Third, increasing the number of the competitors amplifies the cyclic dynamics until the competition becomes strong enough to cause the pathogen to become extinct. This observation can also be expressed as follows: reducing competitor species richness in the outside-host environment can lead to an abrupt emergence of disease outbreaks if the infectivity response of the pathogen is sigmoidal, because there is a positive feedback between raising pathogen levels and nonlinearly increasing infectivity. More generally, decreasing the pressure from competition, e.g., due to use of non-specific disinfectants or antibiotics killing also the competitors, may create fatal disease outbreaks.

8.7 Prevention Strategies Against Environmentally Growing Opportunist Pathogens

Treating infected hosts by using, e.g., disinfectants or antibiotics has often been unsuccessful in eradicating environmentally growing diseases. Firstly, treating the hosts leaves the environmental reservoir population untouched, preventing

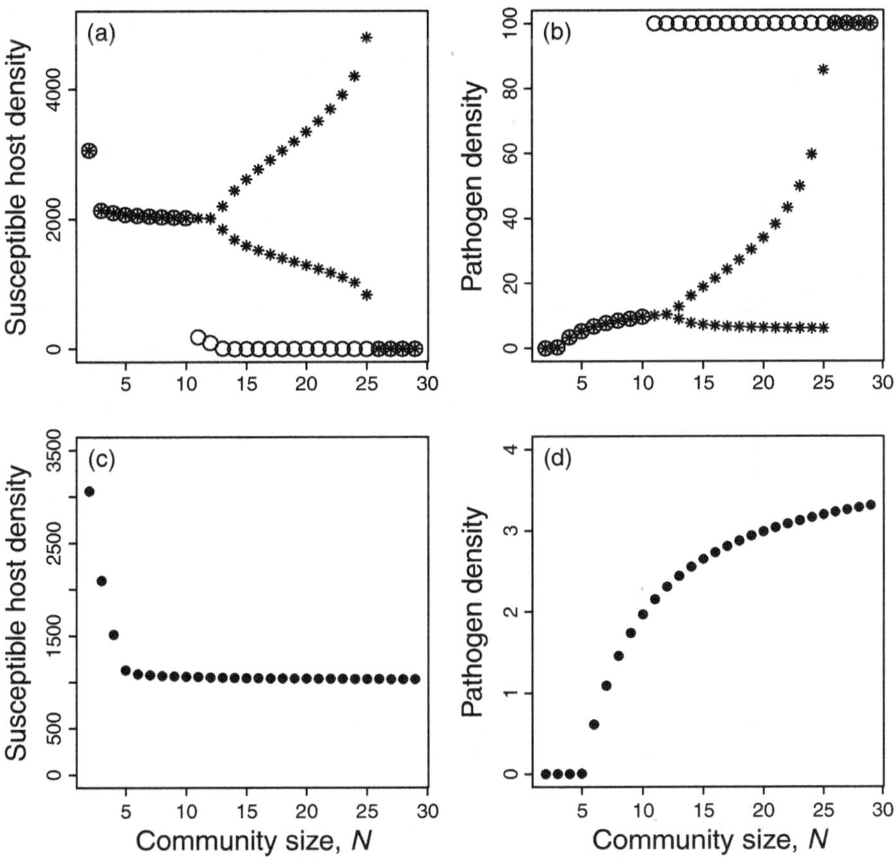

Fig. 8.6 Pathogen and susceptible host dynamics presented as a function of the size of the microbial community (N). Under sigmoidal infectivity response (**a**, **b**), increasing the number of competitors to the pathogen gives rise to cyclic dynamics. For large number of competitors ($N > 25$), the competition drives the pathogen to extinction. Conversely, under linear infectivity response (**c**, **d**), the pathogen is driven to extinction by a relatively small number of competitors $N < 5$. Reduction in competitive ability (α) is 0.04. Figures (**a**) and (**b**) show sigmoidal infectivity response with $ID_{50} = 3150$, $\kappa = 4$. In (**c**) and (**d**) the rate constant β in the linear infectivity response is rescaled as $\beta/2ID_{50}$ for the reasons of comparison with the sigmoidal response (Anttila et al. 2013). The *asterisks* represent minima and maxima of population densities, and the *open circles* show an alternative stable attractor

extirpation of the disease. Secondly, in the worst-case scenario, using nonspecific disinfectants or antibiotics in the outside-host environment may counterintuitively result in an increase in pathogen densities there (Alanis 2005; Merikanto et al. 2012), and therefore the environmental reservoirs can even give rise to disease outbreaks (Anttila et al. 2013). Extensive use of antibiotics against these diseases can also facilitate the spread of antibiotic resistance. For example, high usage of antibiotics has resulted in massive leakages of antibiotics to the environment

(Alanis 2005; Davies and Davies 2010; Oliveira et al. 2012). Novel control methods that lower the pathogen population in the outside-host environment could be a more successful solution in treating these diseases in the future. These methods could consist of reducing available saprotrophic resources in the environment when possible, for instance, by more efficiently removing fish feces and dead fish material in the fisheries, thus directly impacting the reservoir population.

Alternatively, intensifying antagonistic interactions in the outside-host environment could reduce ability of the pathogen to grow there. This includes, e.g., competition between microbes through increasing microbial diversity, increasing bacteria predation by protozoa, or using targeted bacteriophages against pathogenic bacteria strains. This reasoning is based on empirical evidence that there are often trade-offs between ability to defend against competitors, predators, and bacteriophages in the outside-host environment and ability to infect and grow in the within-host environment (Skurnik and Strauch 2006; Friman et al. 2008; Laanto et al. 2012; Mikonranta et al. 2012; Vasanthakrishnan et al. 2015). The need to maintain adaptation to the community interactions outside the host may hamper adaptation to conditions within the host.

8.7.1 Predation

Like competition, predation is also expected to influence disease dynamics. Predation is one of the major factors affecting mortality in bacteria in the outside-host environment (Fenchel 1980; Finlay and Esteban 1998; Sherr and Sherr 2002; Menon et al. 2003). As a means of disease control, predation by protozoa may control pathogen populations effectively. For example, protozoans have been observed to attack and remove *V. cholerae* from environmental water samples (Martínez Pérez et al. 2004). Previous studies have explored how predation can at times reduce disease prevalence when imposed directly on hosts (Packer et al. 2003). Comparable effects can emerge because of predation imposed on pathogens in an environmental reservoir.

8.7.1.1 A Model of Biocontrol by Predation

We next discuss a model of environmentally growing opportunist pathogens that are subject to predation (Merikanto et al. 2017). We first address long-term disease dynamics. We then consider the use of predators as a control method against environmentally growing pathogens.

The disease dynamics under predation are given as

$$\frac{dS}{dt} = r_S S(1 - S) - \beta SP - \mu_S S \tag{8.13a}$$

$$\frac{dI}{dt} = \beta SP - \delta I - \mu_S I \tag{8.13b}$$

$$\frac{dP}{dt} = \Lambda \delta I + r_P(1 - f_{PP}P)P - \frac{aP}{K + cP}Z - \mu_P P \tag{8.13c}$$

$$\frac{dZ}{dt} = m + \frac{ar_Z P}{K + cP}Z - \mu_Z Z \tag{8.13d}$$

where Z denotes the pathogen predators; a is the predator attack rate; K is the predator carrying capacity; c is the predator half saturation constant; r_Z is the conversion rate of the prey resource for predator growth, i.e., predator's efficiency at turning prey into offspring; and μ_Z is the predator death rate. In biological control, predators are fed to the system at an inflow rate m. In some cases, these predators may impose mortality but themselves experience little to no benefit from such consumption (i.e., $r_Z = 0$). In other cases, the predator may be a specialist completely dependent upon the pathogen environmental reservoir ($m = 0$). We first consider the latter case.

8.7.1.2 Disease Dynamics of the Predator–Pathogen System Without Predator Inflow

For low attack rates of the predator ($a < 0.1$), the predator (Z) cannot invade the community. Alternatively, if the predator is present in the community, it will become extinct. The pathogen will remain in the community, and the equilibrium solution ($S^*, I^*, P^* > 0, Z^* = 0$) of the disease dynamics is locally stable. Here the size of the susceptible host population is driven low by the disease (Fig. 8.7).

Higher attack rates ($0.1 < a < 0.6$) allow the predator to survive, and cyclic coexistence of the host, pathogen, and predator becomes possible. Contrary to what one might expect, increasing the attack rate does not suppress the peak densities of the pathogens or of the infected hosts. Neither does it increase the peak densities of the predator itself. However, there is another locally stable equilibrium for the attack rate $a > 0.32$. Thus, for $a > 0.6$ the high peaks of the disease cycles disappear, and the pathogen and the predator levels drop down (Fig. 8.7). In short, a specialist predator is capable of maintaining consistently low pathogen levels only when the attack rate is high enough.

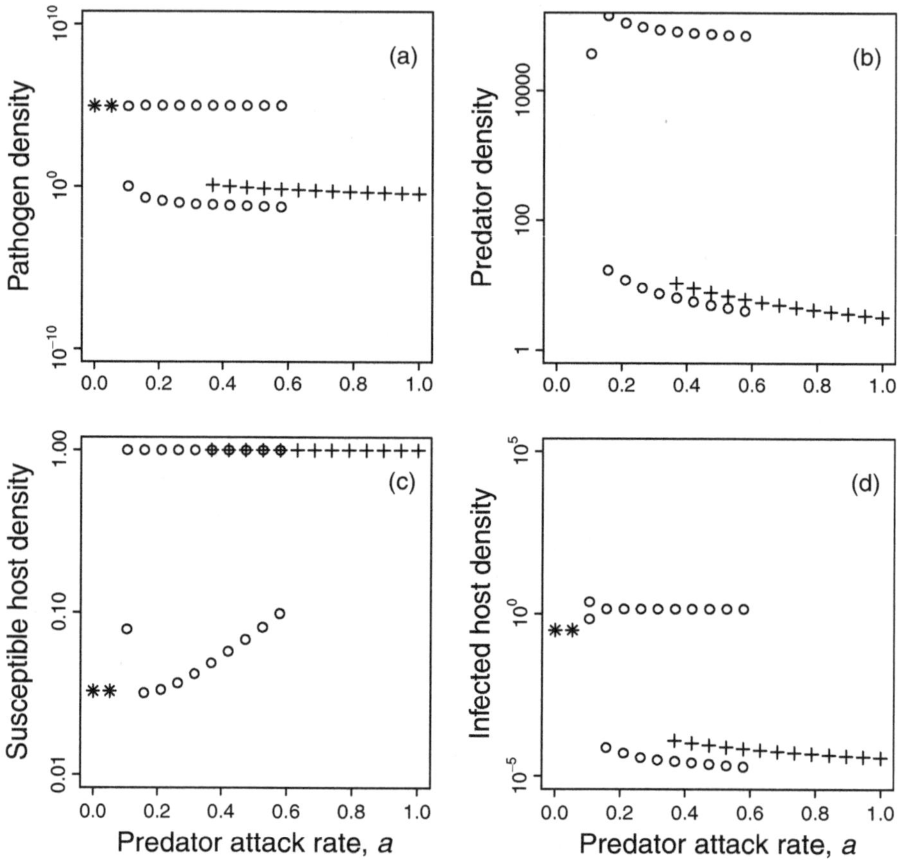

Fig. 8.7 Bifurcation diagrams of the model dynamics of susceptible hosts (S), infected hosts (I), pathogens (P), and predators (Z). The equations used to generate these plots presume that no predator inflow is present ($m = 0$). For the predator attack rates $0.1 < a < 0.6$, there exists a periodic attractor, where the dynamics are cyclic. The maximum and minimum values for population densities are in this area denoted by "*open circle*". For $a > 0.32$ there is an alternative attractor, which is a locally stable equilibrium ($S^*, I^*, P^*, Z^* > 0$), denoted by "*plus*". Thus, the attractor is nonunique for $0.32 < a < 0.6$ allowing different dynamical outcomes. Y-axis is presented on a logarithmic scale (Merikanto et al. 2017)

8.7.1.3 Biocontrol by Predator Inflow

Biological control by a specialist predator can be done in principle in two ways. First, biological control can be processed as a batch process. Here, the predator is introduced into the system of the hosts infected by pathogens. In an ideal case, the predator population will increase, consuming the pathogens until the pathogens become extinct. This case is an extension of the ideas presented above under the title "Disease Dynamics of the Predator–Pathogen System Without Predator Inflow" (though quantifying extinction would require a detailed consideration of transient dynamics and demographic stochasticity). Another option for biological control is to feed the predator into the system continuously. This could also describe a spatially open system with a constant influx into a particular habitat patch from an external source pool. Here the inflow of the predators compensates for the mortality of the predator and decreased growth capacity due to decreased level of pathogens.

A continuous inflow (m) of predators (Z) into the community can either remove the pathogens completely or partially, depending upon the attack rate and the magnitude of the inflow. Even a relatively weak predator (attack rate $a = 0.2$, prey conversion rate $r_Z = 0.3$) is capable of eradicating the pathogen from the system at sufficiently high predator inflow rates m (Fig. 8.8a). For the inflow rate $m < 1.5$, the predator inflow does not seem to have any effect on the fluctuating pathogen levels. (Note, however, that in the absence of the predator the coexistence of the host and pathogen is locally stable, c.f. Fig. 8.7, $a = 0$.) When the inflow rate is increased to $m > 1.5$, the pathogen becomes abruptly extinct.

If the attack rate and the prey conversion rate of the predator are lowered even further ($a = 0.15$, $r_Z = 0.3$), the dynamics stabilize (Fig. 8.8b). The effect of inflow rate becomes visible, however, such that the pathogen level decreases smoothly with increasing inflow rate until an abrupt extinction of the pathogen occurs at $m = 7$. Thus, a relatively high inflow rate of the predator is required to eradicate the pathogen. Interestingly, when $m < 7$, that is, below the threshold for pathogen extinction, the population density of infected host (I) increases with increasing inflow rate m, even if the pathogen population density P decreases. This occurs because the population density of susceptible hosts S increases quicker than the pathogen density decreases.

8.7.2 *Phage Therapy*

To avoid the use of antibiotics in the treatment against bacterial pathogens in aquaculture, the use of bacteriophages has been raised as an alternative option. Phage therapy utilizes lytic bacteriophages that infect the bacteria cell and break it down by lysis (Nakai and Park 2002). The phages often have high specificity to certain bacterial strains, i.e., attacking mainly the targeted bacterial disease agent, unlike antibiotics (Oliveira et al. 2012). This is also the case in columnaris disease (Laanto et al. 2011).

Phage therapy experiments have been promising, but not all experiments to date have been successful in eradicating the disease. Phage therapy mostly treats the hosts in vivo (Skurnik and Strauch 2006; Oliveira et al. 2012). An example of a successful in vivo treatment is catfish infected by columnaris disease agent *F. columnare* (Prasad et al. 2011). Three main problems can be noted about the use of phage therapy. First, because of strong selection on the pathogen, there can be rapid evolution of resistance against bacteriophages (Oliveira et al. 2012). The genetic variation for such selection can arise due either to mutation or horizontal gene transfer among bacterial species. Second, bacteriophages can act as vectors

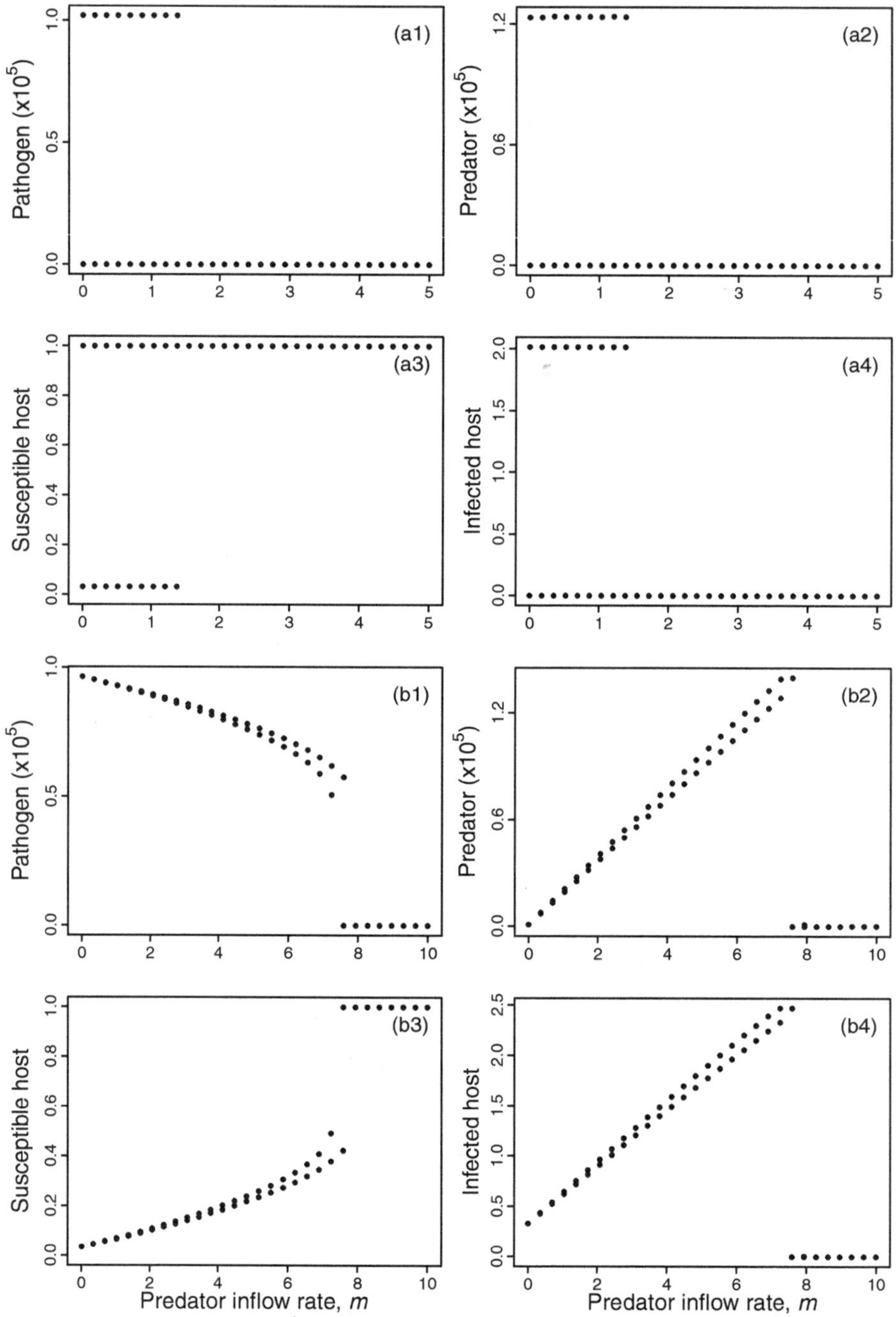

Fig. 8.8 Predator inflow as means of biocontrol in the environmentally growing pathogens. The bifurcation diagrams summarize the resulting community dynamics with predator inflow (m). The minimum and maximum densities of the populations are presented for the cyclic dynamics of

transmitting genes among bacteria allowing the evolution of virulence (Oliveira et al. 2012). Third, treatment of infected hosts solely leaves the outside-host pathogen population untouched and fails to eradicate the disease. Of these, the development of resistance towards phages is however not always a negative outcome in terms of disease control, if there is a clear genetic trade-off between virulence and phage resistance (Skurnik and Strauch 2006). This trade-off has been established for some *F. columnare* strains, but does not apply to all virulent *F. columnare* strains (Laanto et al. 2011).

As columnaris-diseased fish usually carry an infection by multiple strains, the success of phage therapy depends on interactions (e.g., competition) within the infection strain community (Suomalainen et al. 2006). Therefore, it is important to understand the specific disease biology if attempting to use phage therapy as a treatment method against diseases. There are a few theoretical models for phage therapy applied to cholera infections, where phage therapy is targeted on both within-host and outside-host pathogen population (Jensen et al. 2006; Kong et al. 2014). The models by Jensen et al. (2006) and Kong et al. (2014), however, lack the possibility of the evolutionary development of phage resistance and as such are not biologically realistic for studying the long-term disease dynamics expected under phage therapy. Further theoretical as well as experimental work is needed to assess phage therapy as a possible biological control method against environmentally growing opportunist diseases.

8.8 Conclusions

Environmental growth in opportunistic pathogens is an important component in the epidemiological dynamics of many diseases. The life history of an environmentally growing opportunist pathogen can be considered to involve a mixture of alternative reproductive strategies (one within hosts, one outside), compared to non-infective free-living organisms. A crucial difference is how the resources are used in reproducing new individuals. In the simplest scenario, this may be illustrated by considering two bacterial strains. One, the non-infective strain, is able to live in the environment only. The other, the infective strain, also lives in the environment, but it is able to live either in or on a host species. Such an adaptation (i.e., extension in the range of resource use) is often costly and thus implies trade-offs with other life history traits.

Fig. 8.8 (continued) susceptible hosts (S), infected hosts (I), pathogens (P), and predators (Z). Figure (**a**1–**a**4) shows that with a weak predator, the disease dynamics are first cyclic until the pathogen (P), predator (Z), and I go extinct leaving healthy hosts (S) present in its equilibrium density ($a = 0.18$, $r_Z = 0.2$). Figure (**b**1–**b**4) shows the result of presuming a weaker predator ($a = 0.15$). Here, the pathogen population becomes extinct at high values of inflow rate ($m > 8$). At lower values of the inflow rate, the densities of the infected hosts I increase even if the density of the pathogen P decreases as m increases (Merikanto et al. 2017)

Environmental growth serves as one possible mechanism through which pathogen epidemics can arise, devastating hosts, yet without the pathogen self-destructing. We have argued elsewhere by using nonlinear dynamics of ecological interactions with either competitors or predators that disease outbreaks caused by environmental opportunistic pathogens result from interactions involving environmental stochasticity. These involved factors such as the amplitude and autocorrelation of the environmental noise can interact with the details of the infection process, further contributing to the emergence and complexity of outbreak dynamics (Anttila et al. 2015, 2016).

Environmental growth of pathogens is a challenging starting point when considering the evolutionary emergence of novel pathogens. We have addressed here the initial evolution of a disease only in passing in the form of an invasion analysis. However, there is a great need for a better understanding of the costs and benefits of extending resource use of an initially non-infective bacterial organism, from a usually poor environment to resource-rich (but patchy and often short-lived) hosts. From the perspective of our chapter, the main question that arises is articulating the evolutionary pathway of a non-infective bacterial organism that may pass through a phase where it is an environmentally growing opportunistic pathogen to eventually become an obligatory pathogen.

Control of environmentally growing pathogens calls for different approaches as compared to the traditional treatments of obligatory pathogens. The latter can be prevented (and potentially even eradicated, e.g., as in the case of smallpox) by using treatments such as immunization, host medication by antibiotics, and by limiting transmission among hosts. Environmental pathogens by contrast can reside in environmental reservoirs indefinitely, so that even if at one time point all hosts are immunized, in future generations fresh disease cases still can occur. But recognizing the role of environmental pathogens can suggest novel approaches for disease control. In particular, any environmental modification that reduces environmental growth of the pathogen will help to suppress the incidence and prevalence of environmentally opportunistic disease in the host.

We have explored how community interactions such as competition and predation can provide mechanisms for biocontrol. Increasing predator and parasite pressure against the pathogens which cause opportunistic diseases, and facilitating nonpathogenic competitors, reducing resource availability in the reservoir, all can potentially help limit environmentally opportunistic pathogens. Utilizing a deeper understanding of the trade-off between the virulence and environmental growth may provide potentially novel approaches. We emphasize, however, that trade-offs between virulence and the life-history traits can play out in many ways, each providing a new angle to the biocontrol of the diseases.

Acknowledgments This study was funded by the Finnish Academy project #1267541 for VK. Funding for JKB and RDH was provided by the National Institutes of Health Grant 1R01GM117617-01 to JKB and NSF DMS 1515661 for RDH. JKB was also supported by the Emerging Pathogens Institute at the University of Florida and RDH by the University of Florida Foundation. LR was funded by The Academy of Finland.

Compliance with Ethical Standards
Funding: This study was funded by the Academy of Finland to VK (grant number 1267541) and to LR (grant number 286405), by NSF-DMS 1515661 to RDH, and by the National Institutes of Health to JKB and RHD (grant number 1R01GM117617-01).

Conflict of Interest: Veijo Kaitala declares that he has no conflict of interest. Lasse Ruokolainen declares that he has no conflict of interest. Robert D. Holt declares that he has no conflict of interest. Jason Blackburn declares that he has no conflict of interest. Ilona Merikanto declares that she has no conflict of interest. Jani Anttila declares that he has no conflict of interest. Jouni Laakso declares that he has no conflict of interest.

Ethical approval: This article does not contain any studies with human participants or animals performed by any of the authors.

References

Alanis AJ (2005) Resistance to antibiotics: are we in the post-antibiotic era? Arch Med Res 36: 697–705

Alexander KA, Lewis BL, Marathe M et al (2012) Modeling of wildlife-associated zoonoses: applications and caveats. Vector-Borne Zoonotic Dis 12:1005–1018

Anderson RM, May RM (1981) The population dynamics of microparasites and their invertebrate hosts. Philos Trans R Soc B Biol Sci 291:451–524

Antia R, Regoes RR, Koella JC, Bergstrom CT (2003) The role of evolution in the emergence of infectious diseases. Nature 426:658–661

Anttila J, Ruokolainen L, Kaitala V, Laakso J (2013) Loss of competition in the outside host environment generates outbreaks of environmental opportunist pathogens. PLoS One 8:e71621

Anttila J, Kaitala V, Laakso J, Ruokolainen L (2015) Environmental variation generates environmental opportunist pathogen outbreaks. PLoS One 10:e0145511. doi:10.1371/journal.pone.0145511

Anttila J, Laakso J, Kaitala V, Ruokolainen L (2016) Environmental variation enables invasions of environmental opportunist pathogens. Oikos 125:114–1152

Anttila J, Mikonranta L, Ketola T, Kaitala V, Laakso J, Ruokolainen L (2017) A mechanistic underpinning for sigmoid dose-dependent infection. Oikos 126:910–916. doi:10.1111/oik.03242

Ben-Ami F, Regoes RR, Ebert D (2008) A quantitative test of the relationship between parasite dose and infection probability across different host-parasite combinations. Proc R Soc B Biol Sci 275:853–859. doi:10.1098/rspb.2007.1544

Ben-Ami F, Ebert D, Regoes RR (2010) Pathogen dose infectivity curves as a method to analyze the distribution of host susceptibility: a quantitative assessment of maternal effects after food stress and pathogen exposure. Am Nat 175:106–115

Berg G, Eberl L, Hartmann A (2005) The rhizosphere as a reservoir for opportunistic human pathogenic bacteria. Environ Microbiol 7:1673–1685

Bezymennyi M, Bagamian KH, Barro A et al (2014) Spatio-temporal patterns of livestock anthrax in Ukraine during the past century (1913–2012). Appl Geogr 54:129–138

Blackburn JK, Goodin DG (2013) Differentiation of springtime vegetation indices associated with summer anthrax epizootics in west Texas, USA, deer. J Wildl Dis 49:699–703

Blackburn JK, McNyset KM, Curtis A, Hugh-Jones ME (2007) Modeling the geographic distribution of *Bacillus anthracis*, the causative agent of anthrax disease, for the contiguous United States using predictive ecologic niche modeling. Am J Trop Med Hyg 77:1103–1110

Blackburn JK, Curtis A, Hadfield TL et al (2010) Confirmation of *Bacillus anthracis* from flesh-eating flies collected during a West Texas anthrax season. J Wildl Dis 46:918–922

Blackburn JK, Ert MV, Mullins JC et al (2014a) The necrophagous fly anthrax transmission pathway: empirical and genetic evidence from wildlife epizootics. Vector-Borne Zoonotic Dis 14: 576–583

Blackburn JK, Hadfield TL, Curtis AJ, Hugh-Jones ME (2014b) Spatial and temporal patterns of anthrax in white-tailed deer, *Odocoileus virginianus*, and hematophagous flies in West Texas during the summertime anthrax risk period. Ann Assoc Am Geogr 104:939–958

Boldin B, Kisdi E (2012) On the evolutionary dynamics of pathogens with direct and environmental transmission. Evolution 66:2514–2527

Brown SP, Cornforth DM, Mideo N (2012) Evolution of virulence in opportunistic pathogens: generalism, plasticity, and control. Trends Microbiol 20:336–342. doi:10.1016/j.tim.2012.04.005

Caraco T, Wang I-N (2008) Free-living pathogens: life-history constraints and strain competition. J Theor Biol 250:569–579. doi:10.1016/j.jtbi.2007.10.029

Casadevall A (2008) Evolution of intracellular pathogens. Annu Rev Microbiol 62:19–33. doi:10.1146/annurev.micro.61.080706.093305

Case TJ, Casten RG (1979) Global stability and multiple domains of attraction in ecological systems. Am Nat 113:705–714

Davies J, Davies D (2010) Origins and evolution of antibiotic resistance. Microbiol Mol Biol Rev 74:417–433. doi:10.1128/MMBR.00016-10

Declercq AM, Haesebrouck F, Van den Broeck W et al (2013) Columnaris disease in fish: a review with emphasis on bacterium-host interactions. Vet Res 44:27. doi:10.1186/1297-9716-44-27

Faruque SM, Albert JM, Mekalanos JJ (1998) Epidemiology, genetics, and ecology of toxigenic *Vibrio cholerae*. Microbiol Mol Biol Rev 62:1301–1314

Fenchel T (1980) Suspension feeding in ciliated protozoa: functional response and particle size selection. Microb Ecol 6:1–11. doi:10.1007/BF02020370

Finlay B, Esteban G (1998) Freshwater protozoa: biodiversity and ecological function. Biodivers Conserv 7:1163–1186

Fisher MC, Henk DA, Briggs CJ et al (2012) Emerging fungal threats to animal, plant and ecosystem health. Nature 484:186–194. doi:10.1038/nature10947

Frank SA (1996) Models of parasite virulence. Q Rev Biol 71:37. doi:10.1086/419267

Freitag NE, Port GC, Miner MD (2009) *Listeria monocytogenes* – from saprophyte to intracellular pathogen. Nat Rev Microbiol 7:623–628

Friedman H, Yamamoto Y, Klein TW (2002) *Legionella pneumophila* pathogenesis and immunity. Semin Pediatr Infect Dis 13:273–279

Friman V-P, Lindstedt C, Hiltunen T et al (2008) Predation on multiple trophic levels shapes the evolution of pathogen virulence. PLoS One 4:e6761

Godfray HCJ, Briggs CJ, Barlow ND et al (1999) A model of insect–pathogen dynamics in which a pathogenic bacterium can also reproduce saprophytically. Proc R Soc B Biol Sci 266:233–240. doi:10.1098/rspb.1999.0627

Gower CM, Webster JP (2005) Intraspecific competition and the evolution of virulence in a parasitic nematode. Evolution 59:544–553

Grimont PAD, Grimont F (1978) The genus *Serratia*. Annu Rev Microbiol 32:221–248

Hall-Stoodley L, Stoodley P (2005) Biofilm formation and dispersal and the transmission of human pathogens. Trends Microbiol 13:7–10. doi:10.1016/j.tim.2004.11.004

Hartley DM, Morris JG, Smith DL (2006) Hyperinfectivity: a critical element in the ability of *V. cholerae* to cause epidemics? PLoS Med 3:e7

Hilbi H, Weber SS, Ragaz C et al (2007) Environmental predators as models for bacterial pathogenesis. Environ Microbiol 9:563–575. doi:10.1111/j.1462-2920.2007.01238.x

Holt RD (1999) A biogeographical and landscape perspective on within-host infection dynamics. In: Bell CR, Brylinsky M, Johnson-Green P (eds) Proceedings of the 8th international symposium of microbial ecology. Atlantic Canada Society for Microbial Ecology, Halifax, Canada, pp 583–588

Holt RD, Pickering J (1985) Infectious disease and species coexistence: a model of Lotka-Volterra form. Am Nat 126:196–211

Hugh-Jones ME, Blackburn J (2009) The ecology of *Bacillus anthracis*. Mol Asp Med 30:356–367

Hugh-Jones ME, de Vos V (2002) Anthrax and wildlife. Rev Sci Tech Int Off Epizoot 21:359–383

Jensen MA, Faruque SM, Mekalanos JJ, Levin BR (2006) Modeling the role of bacteriophage in the control of cholera outbreaks. Proc Natl Acad Sci 103:4652–4657. doi:10.1073/pnas.0600166103

Keeling MJ, Rohani P (2008) Modeling infectious diseases in humans and animals. Princeton University Press, Princeton

Kong JD, Davis W, Wang H (2014) Dynamics of a cholera transmission model with immunological threshold and natural phage control in reservoir. Bull Math Biol 76:2025–2051

Kracalik IT, Malania L, Tsertsvadze N et al (2013) Evidence of local persistence of human anthrax in the country of Georgia associated with environmental and anthropogenic factors. PLoS Negl Trop Dis 7:e2388

Kunttu HMT, Valtonen ET, Jokinen EI, Suomalainen L-R (2009) Saprophytism of a fish pathogen as a transmission strategy. Epidemics 1:96–100. doi:10.1016/j.epidem.2009.04.003

Kunttu HMT, Sundberg L-R, Pulkkinen K, Valtonen ET (2012) Environment may be the source of *Flavobacterium columnare* outbreaks at fish farms. Environ Microbiol Rep 4:398–402

Laanto E, Sundberg L-R, Bamford JKH (2011) Phage specificity of the freshwater fish pathogen *Flavobacterium columnare*. Appl Environ Microbiol 77:7868–7872

Laanto E, Bamford JKH, Laakso J, Sundberg L-R (2012) Phage-driven loss of virulence in a fish pathogenic bacterium. PLoS One 7:e53157

Leclerc H, Schwartzbrod L, Dei-Cas E (2002) Microbial agents associated with waterborne diseases. Crit Rev Microbiol 28:371–409

Leggett HC, Cornwallis CK, West SA (2012) Mechanisms of pathogenesis, infective dose and virulence in human parasites. PLoS Pathog 8:e1002512

Levin BR (1996) The evolution and maintenance of virulence in microparasites. Emerg Infect Dis 2:93–102

Lipsitch M, Moxon E (1997) Virulence and transmissibility of pathogens: what is the relationship? Trends Microbiol 5:31–37. doi:10.1016/S0966-842X(97)81772-6

Lutz C, Erken M, Noorian P et al (2013) Environmental reservoirs and mechanisms of persistence of *Vibrio cholerae*. Front Microbiol 4:375

Mahlen SD (2011) *Serratia* infections: from military experiments to current practice. Clin Microbiol Rev 24:755–791

Martínez Pérez ME, Macek M, Galván MTC (2004) Do protozoa control the elimination of *Vibrio cholerae* in brackish water? Int Rev Hydrobiol 89:215–227

May RM (1973) Stability and complexity in model ecosystems. Princeton University Press, Princeton

Menon P, Billen G, Servais P (2003) Mortality rates of autochthonous and fecal bacteria in natural aquatic ecosystems. Water Res 37:4151–4158. doi:10.1016/S0043-1354(03)00349-X

Merikanto I, Laakso J, Kaitala V (2012) Outside-host growth of pathogens attenuates epidemiological outbreaks. PLoS One 7:e50158

Merikanto I, Laakso J, Kaitala V (2014) Invasion ability and disease dynamics of environmentally growing opportunistic pathogens under outside-host competition. PLoS One 9:e113436

Merikanto I, Laakso J, Kaitala V (2017) Outside-host predation as a biological control against an environmental opportunist disease. Ecol Mod 347:85–93

Merrell DS, Butler SM, Qadri F et al (2002) Host-induced epidemic spread of the cholera bacterium. Nature 417:642–645. doi:10.1038/nature00778

Mikonranta L, Friman V-P, Laakso J (2012) Life history trade-offs and relaxed selection can decrease bacterial virulence in environmental reservoirs. PLoS One 7:e43801

Mondal M, Nag D, Koley H et al (2014) The *Vibrio cholerae* extracellular chitinase ChiA2 is important for survival and pathogenesis in the host intestine. PLoS One. doi:10.1371/journal.pone.0103119

Nakai T, Park SC (2002) Bacteriophage therapy of infectious diseases in aquaculture. Res Microbiol 153:13–18. doi:10.1016/S0923-2508(01)01280-3

Oliveira J, Castilho F, Cunha A, Pereira MJ (2012) Bacteriophage therapy as a bacterial control strategy in aquaculture. Aquac Int 20:879–910. doi:10.1007/s10499-012-9515-7

Packer C, Holt R, Dobson A, Hudson P (2003) Keeping the herds healthy and alert: impacts of predation upon prey with specialist pathogens. Ecol Lett 6:797–802

Pascual M, Dunne JA (2005) Ecological networks: linking structure to dynamics in food webs. Oxford University Press, New York

Prasad Y, Kumar D, Sharma AK (2011) Lytic bacteriophages specific to *Flavobacterium columnare* rescue catfish, *Clarias batrachus* (Linn.) from columnaris disease. J Environ Biol Acad Environ Biol India 32:161–168

Pulkkinen K, Suomalainen L-R, Read AF et al (2010) Intensive fish farming and the evolution of pathogen virulence: the case of columnaris disease in Finland. Proc R Soc B Biol Sci 277: 593–600. doi:10.1098/rspb.2009.1659

Rahman MH, Biswas K, Hossain MA et al (2008) Distribution of genes for virulence and ecological fitness among diverse *Vibrio cholerae* population in a cholera endemic area: tracking the evolution of pathogenic strains. DNA Cell Biol 27:347–355

Read AF (1994) The evolution of virulence. Trends Microbiol 2:73–76. doi:10.1016/0966-842X(94)90537-1

Regoes RR, Ebert D, Bonhoeffer S (2002) Dose-dependent infection rates of parasites produce the Allee effect in epidemiology. Proc R Soc B Biol Sci 269:271–279. doi:10.1098/rspb.2001.1816

Regoes RR, Hottinger JW, Sygnarski L, Ebert D (2003) The infection rate of *Daphnia magna* by *Pasteuria ramosa* conforms with the mass-action principle. Epidemiol Infect 131:957–966

Roche B, Drake JM, Rohani P (2011) The curse of the pharaoh revisited: evolutionary bi-stability in environmentally transmitted pathogens: bi-stability arises from environmental transmission. Ecol Lett 14:569–575. doi:10.1111/j.1461-0248.2011.01619.x

Saile E, Koehler TM (2006) *Bacillus anthracis* multiplication, persistence, and genetic exchange in the rhizosphere of grass plants. Appl Environ Microbiol 72:3168–3174

Schmid-Hempel P, Frank SA (2007) Pathogenesis, virulence, and infective dose. PLoS Pathog 3:e147

Sherr EB, Sherr BF (2002) Significance of predation by protists in aquatic microbial food webs. Antonie Van Leeuwenhoek 81:293–308

Simpson RH (1952) Infectiousness of communicable diseases in the household: (measles, chickenpox, and mumps). Lancet 260:549–554

Skurnik M, Strauch E (2006) Phage therapy: facts and fiction. Int J Med Microbiol IJMM 296: 5–14. doi:10.1016/j.ijmm.2005.09.002

Smith V (2007) Host resource supplies influence the dynamics and outcome of infectious disease. Integr Comp Biol 47:310–316

Soto E, Mauel MJ, Karsi A, Lawrence ML (2007) Genetic and virulence characterization of *Flavobacterium columnare* from channel catfish (*Ictalurus punctatus*). J Appl Microbiol 104: 1302–1310

Sturm A, Heinemann M, Arnoldini M et al (2011) The cost of virulence: retarded growth of *Salmonella typhimurium* cells expressing type III secretion system 1. PLoS Pathog 7: e1002143. doi:10.1371/journal.ppat.1002143

Suomalainen L-R, Kunttu H, Valtonen ET et al (2006) Molecular diversity and growth features of *Flavobacterium columnare* strains isolated in Finland. Dis Aquat Org 70:55–61

Traverse CC, Mayo-Smith LM, Poltak SR, Cooper VS (2013) Tangled bank of experimentally evolved *Burkholderia* biofilms reflects selection during chronic infections. Proc Natl Acad Sci 110:E250–E259

Trivedi SR, Malik R, Meyer W, Sykes JE (2011) Feline cryptococcosis impact of current research on clinical management. J Feline Med Surg 13:163–172. doi:10.1016/j.jfms.2011.01.009

Turner WC, Imologhome P, Havarua Z et al (2013) Soil ingestion, nutrition and the seasonality of anthrax in herbivores of Etosha National Park. Ecosphere 4:1–19

Turner WC, Kausrud KL, Krishnappa YS et al (2014) Fatal attraction: vegetation responses to nutrient inputs attract herbivores to infectious anthrax carcass sites. Proc R Soc Lond B Biol Sci 281:20141785

Vale PF, Fenton A, Brown SP (2014) Limiting damage during infection: lessons from infection tolerance for novel therapeutics. PLoS Biol 12:e1001769

Vasanthakrishnan RB, de las Heras A, Scortti M et al (2015) PrfA regulation offsets the cost of *Listeria* virulence outside the host: cost of *Listeria* virulence. Environ Microbiol 17(11): 4566–4579. doi:10.1111/1462-2920.12980

Veneault-Fourrey C, Martin F (2011) Mutualistic interactions on a knife-edge between saprotrophy and pathogenesis. Curr Opin Plant Biol 14:444–450. doi:10.1016/j.pbi.2011.03.022

Walther BA, Ewald PW (2004) Pathogen survival in the external environment and the evolution of virulence. Biol Rev 79:849–869. doi:10.1017/S1464793104006475

Wedekind C, Gessner MO, Vazquez F et al (2010) Elevated resource availability sufficient to turn opportunistic into virulent fish pathogens. Ecology 91:1251–1256. doi:10.1890/09-1067.1

WHO (2008) Anthrax in humans and animals, 4th edn. WHO, Geneva

Woolhouse MEJ, Haydon DT, Antia R (2005) Emerging pathogens: the epidemiology and evolution of species jumps. Trends Ecol Evol 20:238–244

Chapter 9
Of Ducks and Men: Ecology and Evolution of a Zoonotic Pathogen in a Wild Reservoir Host

Michelle Wille, Neus Latorre-Margalef, and Jonas Waldenström

Abstract A hallmark of disease is that most pathogens are able to infect more than one host species. However, for most pathogens, we still have a limited understanding of how this affects epidemiology, persistence and virulence of infections—including several zoonotic pathogens that reside in wild animal reservoirs and spillover into humans. In this chapter, we review the current knowledge of mallard (*Anas platyrhynchos*) as host for pathogens. This species is widely distributed, often occupying habitats close to humans and livestock, and is an important game bird species and the ancestor to domestic ducks—thereby being an excellent model species to highlight aspects of the wildlife, domestic animal interface and the relevance for human health. We discuss mallard as host for a range of pathogens but focus more in depth of it as a reservoir host for influenza A virus (IAV). Over the last decades, IAV research has surged, prompted in part to the genesis and spread of highly pathogenic virus variants that have been devastating to domestic poultry and caused a number of human spillover infections. The aim of this chapter is to synthesise and review the intricate interactions of virus, host and environmental factors governing IAV epidemiology and evolution.

M. Wille
Zoonosis Science Center, Department of Medical Biochemistry and Microbiology, Uppsala University, Uppsala, Sweden

Centre for Ecology and Evolution in Microbial Model Systems, Linnaeus University, SE 391 82 Kalmar, Sweden
e-mail: Michelle.Wille@influenzacentre.org

N. Latorre-Margalef
Functional Zoology/Molecular Ecology and Evolution Lab, Department of Biology, Lund University, SE 223 62 Lund, Sweden
e-mail: neus.latorre-margalef@biol.lu.se

J. Waldenström (✉)
Centre for Ecology and Evolution in Microbial Model Systems, Linnaeus University, SE 391 82 Kalmar, Sweden
e-mail: jonas.waldenstrom@lnu.se

C.J. Hurst (ed.), *Modeling the Transmission and Prevention of Infectious Disease*, Advances in Environmental Microbiology 4, DOI 10.1007/978-3-319-60616-3_9

9.1 Introduction

Pathogenic microorganisms are a reality for all organisms, with profound impacts on the ecology and evolution of species, populations and individuals (Schmid Hempel 2011). The outcome of infection is variable, ranging from asymptomatic conditions to severe disease and death, depending on factors relating to the host (such as age, immune functions, condition and genetics), the pathogen (such as factors relating to pathogenicity, virulence or transmission) and the environment (e.g. climate, food availability, competition and predation) and their interactions—sometimes described as the pathogen-host-environment matrix (Schmid Hempel 2011). In recent years, a growing understanding has emerged where disease dynamics needs to incorporate the realism of multihost-multipathogen systems—the majority of pathogens can infect more than one host, and most hosts can be infected with several different pathogens (Bordes and Morand 2011).

For humans, it has been estimated that the majority of infectious diseases are zoonotic, e.g. shared with at least one other animal host (Woolhouse and Gowtage-Sequeria 2005). However, we know very little about the dynamics of diseases outside the clinic, even for well-known human diseases such as campylobacteriosis, influenza or rabies, and even less for pathogens that only occasionally spillover into humans. As a response, calls for joint forces, such as the EcoHealth Alliance and One Health initiatives, have been launched, with the aim of bridging the gaps between professionals in human and veterinary medicine and ecology (Roger et al. 2016).

In this chapter, we will focus on the mallard (*Anas platyrhynchos*), the world's most numerous wild duck species and the ancestor to domestic ducks, and its role for maintaining pathogens of relevance for humans and domestic animals. Of particular interest is the influenza A virus (IAV), for which the mallard and related waterfowl species are the key reservoir in nature. Because of its great abundance, its expansive distribution and its preference for human-influenced environments, it is a potential bridge species between wild animals, domestic animals and humans, specifically for pathogens to reach domestic ducks and gallinaceous poultry. Moreover, it is an important game species across its distribution, with large bag limits. Most importantly, however, it is one of very few wildlife species from which there is sufficient data to discuss how host ecology affects the disease dynamics in humans and domestic animals. As such, mallard is an important model species for multihost and zoonotic diseases.

9.2 Waterfowl and Mallard Biology

Waterfowl (order Anseriformes) include ducks, geese and swans and are well known due to their historical importance for hunting, domestication and aviculture. They have a nearly cosmopolitan distribution, except for Antarctica, occupying virtually every habitat associated with water. The approximately 150 species alive today include forms ranging from those occupying highly specialised niches to generalist species able to successfully exploit any wetland habitat, artificial or native. To occupy this wide range of environments, waterfowl are diverse in their anatomy, behaviour and physiology.

Among waterfowl, the mallard is perhaps the archetypical and most recognisable anatid species. Mallards are native to the Holarctic and, however, are thriving globally due to successful introductions to New Zealand, Australia, Peru, Brazil, Uruguay, Argentina, Chile, the Falkland Islands, South Africa and Hawaii (Cramp and Simmons 1977; Drilling et al. 2002). The success of this species is owed to its adaptability; it thrives not only in the wild but is also extremely tolerant of human presence or disturbance and utilises wetlands of all sizes in and around human settlements.

The species is mostly migratory in its native range, but some populations are sedentary with low levels of dispersal. Within migratory populations, there is large flexibility in migratory behaviour, and it is often referred to as being a partial migrant, where some individuals (or populations) are strictly migratory, some strictly residents and others switching between these states depending on specific conditions, such as food availability or cold spells (Cramp et al. 1985; Drilling et al. 2002). It is believed, but rarely tested, that females exhibit a higher degree of philopatry to the natal site than males. As pair formation occurs during late winter at shared wintering sites, this means that a mixing of populations will occur over time. Indeed, this is observed at the genetic level where hardly any population subdivision is evident in either mitochondrial or nuclear genes across the natural breeding distribution of the species (Cramp and Simmons 1977; Drilling et al. 2002; Kraus et al. 2011).

The global mallard population is large, approximately 19 million individuals, of which 7.5 million breed in Europe (Wetlands International 2012). Mortality rates are high, especially during the first year of life, and the turnover rate of mallards in Northern Europe has been estimated to be roughly 1/3, meaning a substantial recruitment rate of young individuals into the population each year (Bentz 1985; Munster and Fouchier 2009). Additionally, large numbers are reared and released for hunting purposes, locally contributing to high densities. It is estimated that 270,000 individuals are released in North America (USFWS 2003, 2011) and greater than 2,850,000 released into Europe annually (Delany and Scott 2006; Birdlife-International 2004; Champagnon 2011). In Eurasia and North America, mallard breeding is mainly restricted to spring and summer resulting in a marked increase in immunologically naïve juveniles in autumn (see Sect. 9.3.2.2).

Nearly all domestic ducks are derived from the mallard, with the exception of Muscovy ducks (*Cairina moschata*), and domestication dates back to at least the twelfth century (Drilling et al. 2002). Approximately 600 million ducks are farmed in China: 60% of the world population of domestic ducks. As a result, there are more domesticated ducks in China than there are wild mallards across the globe. These large numbers, specifically the large input of young birds, are imperative in the maintenance of infection diseases. Thus, it is a combination of large distribution range, large population size and turnover rate in wild mallards and enormous population size of farmed ducks, most of which are free-range, that make mallard one of the largest aquatic reservoir hosts for diseases. Association with other wild waterfowl, association with farming and domestic poultry and association with humans in urbanised areas make this species extremely important for zoonotic transmission and spillover events.

9.3 Mallards as Hosts for Pathogenic Microorganisms

Our understanding of avian diseases, particularly in wild birds, is mainly the result of intensive study of poultry diseases, where wildlife can contribute as a source of infection, such as avian influenza (see more below), Newcastle disease (e.g. Kim et al. 2012) or salmonellosis and campylobacteriosis (e.g. Hald et al. 2016; Berglund 2014). Other examples include diseases that cause large mortality events in the wild, such as avian cholera (Samuel et al. 2007) or Wellfleet Bay virus (Allison et al. 2014). It is only more recently, with the onset of molecular methods, that we have started to assess the presence of infectious agents in asymptomatic hosts. For example, recently the tufted duck (*Aythya fuligula*) microbiome (Strong et al. 2013) and domestic duck viral metagenome (Fawaz et al. 2016) were described. However, the agents we know of are just the tip of the iceberg; for instance, it is predicted that there are 32,000 virus species to be discovered in mammals (Anthony et al. 2013), and it is reasonable to assume that diversity in avian species is in line with that of mammalian hosts. Furthermore, given that we have identified an infectious agent in avian wild birds, we may still know very little about the ecology, epidemiology or even basic biology of these agents.

Wild animal hosts have variable importance to the epidemiology of infectious agents, ranging from optimal or major hosts to minor hosts and to accidental hosts following spillover infections. This is reflected by the adaptation of particular pathogens to their hosts, the efficiency of the parasite to exploit the host effectively for replication and, crucially, the transmission to new hosts. In order to ascertain the specific role of each host species in pathogen epidemiology, it is imperative to combine large screening efforts, molecular-based phylogeny approaches and infection experiments, which unfortunately is rarely met. Compounded with this, we have limited understanding of the role of mallards as hosts for pathogens: are these birds the central reservoir, important but not central to the epidemiology or merely permissive to spillover infection? Moreover, variation in pathogen phenotypic characteristics is usually unknown, such as variation in virulence, pathogenicity, survival in the environment and duration of infection. These properties will depend on both host and pathogen and likely are variable among genetic variants and/or strains of the pathogen. Regardless, our current catalogue of disease-causing agents is probably an underestimation, and with the advent of deep sequencing and more sensitive screening tools, we will likely uncover numerous new disease-causing agents such as picornaviruses (Woo et al. 2010) or disentangle complex epidemiology of known pathogens in new hosts such as coronaviruses (Jackwood et al. 2012).

9.3.1 Mallards as Hosts for Spillover Infections

An important feature of spillover infections is that the disease typically is not further maintained in the population, because it is not likely transmitted by the spillover host to other hosts, and hence, the disease does not become established within the new population. There are a number of viruses that are shared between wild waterfowl and gallinaceous poultry, of which several are believed to spillover from wild birds into poultry and are of animal health concern, including Newcastle disease virus and infectious bronchitis virus (Table 9.1). It is important to note, however, that transmission can occur also in the other direction, from poultry to wild birds, sometimes associated with wildlife mortality events, as, for instance, noted with duck plague (Converse and Kidd 2001) and avian cholera (Botzler 1991; Gordus 1993) in North America.

Two major sources exist for spillover infections to mallards: spillover from poultry or spillover from other wild birds. Spillover infections from other non-avian hosts are also possible but occur much less frequently or may be underappreciated. First, due to high genetic similarity and sharing habitat with domesticated conspecifics or utilising habitat surrounding intensive poultry farms, there is a high risk for spillover infections from both gallinaceous poultry and domesticated ducks (e.g. Christensen et al. 1998; Wang et al. 2008; Shin et al. 2000, 2002). Indeed, pathogens more frequently found in poultry are identified in wild mallards but very infrequently (Table 9.1). These pathogens are usually only detected in wild mallards utilising habitat surrounding poultry farms as they are used as sentinels. For example, avian pneumovirus has been detected in mallard sentinels and wild mallards in the vicinity of poultry operations (Shin et al. 2000, 2002). Given the large population sizes of birds reared for meat production, one can hypothesise that the occurrence of spillover of poultry-associated pathogens into mallard and other wild bird populations is underestimated. Although modern poultry production units enforce barrier protection, no system can truly be regarded closed—and in many parts of the world, poultry are reared in open units or are let free to roam the environments exposing a large wildlife/domestic animal interface. The extent of this can be exemplified with repeated isolation of bacteria in wild birds with antibiotic resistance profiles suggesting origin in anthropogenic environments (e.g. Stedt et al. 2014; Hasan et al. 2014; Hernandez et al. 2013; Bonnedahl et al. 2014).

An example of spillover from other wild bird species into mallards is West Nile virus (WNV). With the rapid spread of WNV across North America in 2005 and onwards, resulting in wild bird mortality, there was intensive surveillance in numerous bird species (George et al. 2015; LaDeau et al. 2007). Passerine birds were identified as main avian host with spillover into other bird families. These

Table 9.1 Some examples of RNA viruses documented in the poultry industry (chickens, turkey, domestic duck) which are found in wild waterfowl, either as spillover hosts or as natural reservoirs

Genome	Family	Genus	Poultry (chickens, turkey, domestic duck)		Wild waterfowl			References[a]
			Virus	Disease, signs and symptoms	Virus[b]	Host type	Signs and symptoms	
dsRNA	*Birnaviridae*	*Avibirnavirus*	Infectious bursal disease virus	Egg drop, bursal disease		Documented	Clinically asymptomatic	Jeon et al. (2008), Kibenge et al. (1988)
	Reoviridae	*Orthoreovirus*	Avian orthoreovirus (ARV)	Tenosynovitis, viral arthritis	Avian orthoreovirus	Documented		Yu et al. (2014), Jones and Guneratne (1984), Zhang et al. (2007), Yun et al. (2013)
ssRNA	*Astroviridae*	*Astrovirus*	Avian nephritis virus (ANV), chicken astrovirus (CAstV), turkey astrovirus type 1 (TAstV1), turkey astrovirus type 2 (TAstV2), duck astrovirus (DuAstV) (formerly named duck hepatitis virus 2), duck hepatitis virus 3 (DuHV3)	Runting-stunting syndrome (RSS) of broiler chickens, poult enteritis complex (PEC) and poult enteritis mortality syndrome (PEMS) of turkeys, hepatitis in ducks	Avian astrovirus Group 1, Group 2, Group 3	Evidence of closely related low pathogenicity viruses to all highly pathogenic forms	Clinically asymptomatic	Wobeser (1997), Chu et al. (2012), Todd et al. (2009), Pantin-Jackwood et al. (2011), Fu et al. (2009)
	Bornaviridae	*Bornavirus*	Psittacine birds	Proventricular dilatation disease (PDD)	Avian bornavirus	Circulation, putative reservoir	Clinically asymptomatic forms Pathogenic forms	Delnatte et al. (2014), Guo et al. (2014), Berg et al. (2001), Delnatte et al. (2013)
	Coronaviridae	*Coronavirus*	Infectious bronchitis virus (IBV), turkey coronavirus (TCoV)	Infectious bronchitis in chickens; turkey coronavirus enteritis; decreased egg production, morbidity, mortality	Low pathogenic infectious bronchitis and related avian low pathogenic viruses	Circulation, putative reservoir	Clinically asymptomatic	Cavanagh (2005), Chu et al. (2011), Hughes et al. (2009), Jonassen et al. (2005), Muradrasoli et al. (2009, 2010), Woo et al. (2009), Guy (2000)

	Orthomyxoviridae	*Influenza A virus*	Influenza A virus subtype (IAV) H5, H7, H9	Avian influenza, respiratory infections, high mortality	Influenza A virus subtype H1–H12, H14, H15	Natural reservoir	Clinically asymptomatic H5 may cause mortality in some species	Olsen et al. (2006), Alexander (2000b, 2007), Bröjer et al. (2009), Chen et al. (2005), Feare (2010), Hatchette et al. (2004), Latorre-Margalef et al. (2014), Wilcox et al. (2011)
	Paramyxoviridae		Avian paramyxovirus type 1 (AMPV-1; Newcastle disease virus)	Decreased egg production, morbidity, mortality	Avian paramyxovirus types 1, 4, 6, 8, 9	Putative reservoir for type 1	Clinically asymptomatic	Leighton and Heckert (2007), Alexander (2000a), Rosseel et al. (2011), Kuiken et al. (1998), White et al. (2014), Tolf et al. (2013b), Alexander et al. (1999), Hoque et al. 2012), Jindal et al. (2009), Ramey et al. (2013)
	Paramyxoviridae	*Metapneumovirus*	Avian pneumoviruses	Upper respiratory tract infections, turkey rhinotracheitis		Documented, spillover	Clinically asymptomatic	de Graaf et al. (2008), Bennett et al. (2004)
	Picornaviridae	*Picornavirus*	Duck hepatitis virus types 1 and 3	Hepatitis, morbidity		Putative circulation	Clinically asymptomatic	Wobeser (1997), Koci and Schultz-Cherry (2002), De Benedictis et al. (2011), Lambert et al. (1991)

[a]Not an exhaustive list of reference

[b]Often wild bird viruses that are related to poultry viruses are not described beyond a short sequence fragment, so no official virus name is listed

species acquired infection from mosquito vectors but with large interspecies variation in reservoir competence, i.e. in how well the virus could replicate in the birds to achieve sufficient viremia to allow transmission to the vector, and severity of infection. Mallards utilise wetland habitats, so it is not surprising that WNV was detected in this species too (Grard et al. 2007; Lindh et al. 2008; Lobo et al. 2009), but it is not considered a reservoir host. Mallards are also accidentally infected with fungal pathogens due to consuming food laced with fungal spores, such as *Aspergillus fumigatus*, most notably at times with food shortenings and inclement weather (Adrian et al. 1978; USGS 1999).

The concept of spillover infection, minor host and major host is useful for discussions on the various roles different species can have in the epidemiology of a disease. However, it should be noted that the distinctions are not always clear-cut; rather host type is a continuum starting with occasional individual infections in a new host, continuing to stuttered transmission chains and multihost disease dynamics with increasing specialisation in the different hosts and ending with host-specific transmission and disease (Wolfe et al. 2007; Morse 1995, 2004; Church 2004; Fenton and Pedersen 2005). Importantly, this transgression depends on both the frequency of interspecies transmission and the intraspecies transmission in the novel host given successful interspecies transmission (Fenton and Pedersen 2005). Mallards may act as minor host for many waterfowl diseases due to their association with other waterfowl. Even if repeatedly infected, mallards might not have a central role in the epidemiology of a particular disease. For example, numerous mallards die in outbreaks of the bacterially associated disease avian cholera, also called fowl cholera (Blanchong et al. 2006; Botzler 1991, 2002), but currently these mortality events are occurring in high arctic breeding areas of eiders (*Somateria mollissima*) (Descamps et al. 2012), Ross's geese (*Chen rossii*) and snow geese (*Chen caerulescens*) (Samuel et al. 2005a, b), and it is geese that are being implicated as long-term carriers of the bacterium (Samuel et al. 2005a, b).

9.3.2 *Mallards as the Main Reservoir: Influenza A Viruses*

9.3.2.1 Influenza A Viruses (IAVs) as a Multihost Pathogen

Influenza A viruses (IAVs) are probably best known for their ability to cause seasonal epidemics and pandemics in humans, such as the pandemic of 1918 Spanish influenza, or the circulating seasonal influenza. They are, however, to the largest extent viruses associated with wild birds, especially those that occupy wetlands and in particular waterfowl (Alexander 2000b; Olsen et al. 2006). It is in waterfowl, and particularly mallards, that the largest genetic and antigenic variation of IAVs occurs (Fig. 9.1). In addition to wild birds and seasonal influenza in humans (Rambaut et al. 2008b), IAVs also circulate in pigs (Vincent et al. 2014), horses (Daly et al. 2011), marine mammals (Groth et al. 2014), bats (Wu et al. 2014b) and domestic birds (Olsen et al. 2006; Webster et al. 1992; Alexander

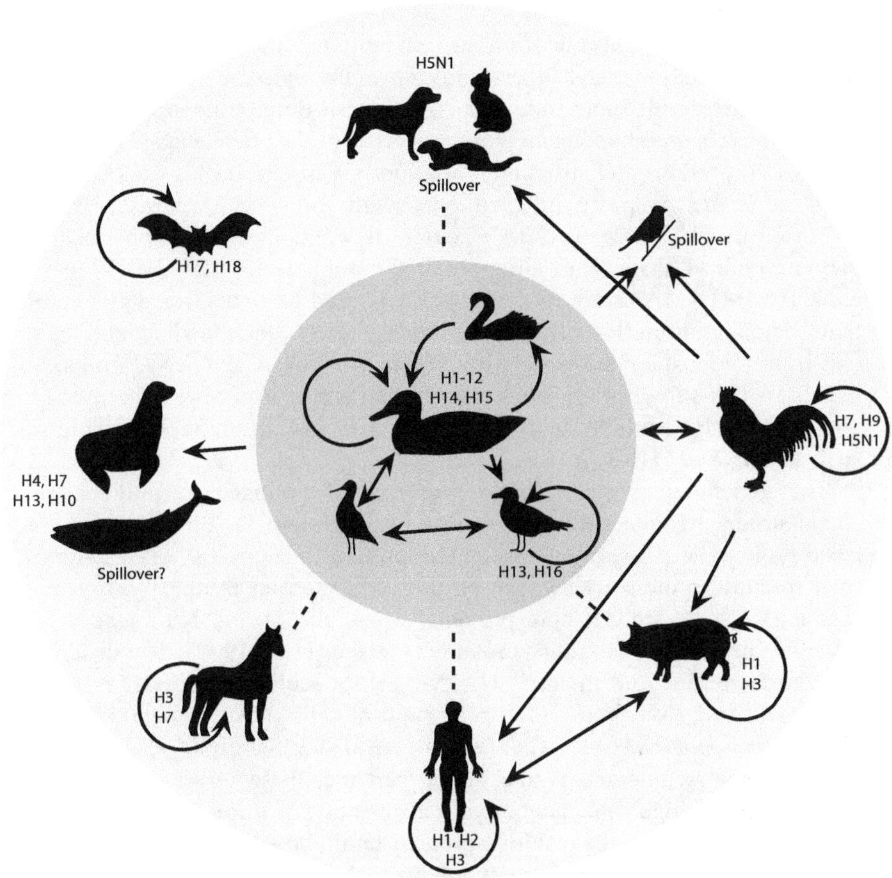

Fig. 9.1 Host range and transmission of IAV. The wild bird reservoir comprises of waterfowl and gulls (dark grey), with direct spillover to other avian species such as passerines and poultry. H5N1, which is amplified in poultry, has subsequently spilled over to wild bird species, mammalian hosts such as cats and dogs and humans. The relationship between bats and other host groups is unknown. Solid lines represent known routes of transmission, dashed lines are infrequent routes of transmission, and semicircles demonstrate circulation of IAV in that host group [Reproduced with permission, Wille (2015) LNU PRESS]

2000b). These viruses are subtyped based on the two surface proteins hemagglutinin and neuraminidase and are further classified as either highly pathogenic (HPIAV) or low pathogenic (LPIAV) based on their virulence in poultry (see Boxes 9.1 and 9.2).

Box 9.1 Influenza Classification and Structure

Influenza viruses belong to the family *Orthomyxoviridae* (Kawaoka et al. 2005) and are divided into influenza A, influenza B and influenza C viruses. This division is based upon antigenic properties of the nucleocapsid (NP) and matrix (M) proteins and structural variations (Webster and Kawaoka 1988). Wild birds are naturally infected only with influenza A viruses (IAV) (Webster et al. 1992) The IAV virion is enveloped and spherical or pleiomorphic in shape with an approximate diameter of 120 nm (Webster et al. 1992). The IAVs are further classified based on two surface glycoproteins: hemagglutinin (HA) and neuraminidase (NA) which mediate entry and exit from the host cells, respectively. There are 18 HA and 9 NA forms, of which 16 HA subtypes are present in birds (Wu et al. 2014b; Olsen et al. 2006). These HA and NA subtypes can occur in 144 different combinations, such as H5N1 or H1N1.

The genome consists of eight segments of unlinked, negative-sense, single-stranded RNA: PB2 (polymerase basic protein 2), PB1 (polymerase basic protein 1), PA (polymerase acidic protein), HA, NP, NA, M and NS (nonstructural protein) (Kawaoka et al. 2005; Webster et al. 1992). These segments encode for ten core proteins, where the M and NS encode two proteins, and several auxiliary proteins (Webster et al. 1992). The different proteins have functions in entry (HA, M2), RNA replication (PB2, PB1, PA, NP), packaging (M1, NS2), exit from the host cells (NA, M1) and immune system evasion (NS, HA, NA) (Webster et al. 1992; Samji 2009).

Due to the segmented nature of the genome, these viruses are able to dramatically change their genotype (and phenotype) through reassortment. Following coinfection the resulting progeny could be any one of 256 possible combinations of the parental genotypes due to the process of virion packaging (Steel and Lowen 2014). Due to the error-prone RNA-dependent RNA polymerase that lacks proofreading ability, these viruses have a high mutation rate (3.4×10^{-3} sub/site/year) (Chen and Holmes 2006), which is about a million times that of vertebrates (Pybus and Rambaut 2009). This rapid rate of change allows for the continued immune evasion of antigenically important segments, such as the HA and NA.

Box 9.2 Determinants of Pathogenicity of Avian IAV

Avian IAVs are categorised into two groups: low pathogenic influenza A viruses (LPIAVs) and highly pathogenic influenza A viruses (HPIAVs). The pathogenicity trait is based on virulence of the virus in chickens and is an important consideration in prevention, control and eradication strategies in

(continued)

Box 9.2 (continued)

commercial fowl (Swayne and Suarez 2000). Wild birds infected with LPIAV generally show no clinical signs of infection (Olsen et al. 2006; Webster et al. 1992). However, it has been demonstrated that LPIAV infections may induce fever (Jourdain et al. 2010) and affect body mass and migratory ability (van Gils et al. 2007; Latorre-Margalef et al. 2009b; Jourdain et al. 2010), but overall the effects of LPIAV infection on wild birds are still poorly understood. In birds, LPIAVs preferentially infect the epithelium of the lower gastrointestinal tract and are shed predominantly through the feces (Webster et al. 1976, 1978; Slemons and Easterday 1978; Engering et al. 2013; Daoust et al. 2011). These viruses are thought to be transmitted mainly by faecal-oral route through bird-bird contact (Webster et al. 1992) and water-borne transmission (Webster et al. 1992; Roche et al. 2009).

In contrast, HPIAV preferentially infects the epithelium of the respiratory tract, including the trachea, lungs and air sacs (Bröjer et al. 2009; Keawcharoen et al. 2008; Worobey et al. 2014). However, lesions associated with HPIAV have been found throughout birds; these viruses are organ promiscuous (Bröjer et al. 2009) As a result, HPIAV infection normally results in significant morbidity and mortality of the infected bird host (Webster and Rott 1987; Alexander 2007). Mechanistically, the switch from LPIAV to HPIAV follows changes in the HA protein. The hemagglutinin protein is produced as a precursor, HA0, which is cleaved into HA1 and HA2 during virus maturation by host tissue-restricted proteases. The introduction of basic amino acid residues to the cleavage site allows for increased HA cleavability by more ubiquitous proteases, which, in turn, allows for enhanced replication outside the gastrointestinal tract (Alexander 2000a). The subtypes H5 and H7 have accounted for most HPIAV isolations in wild birds (Alexander 2007; Olsen et al. 2006). The switch from low to high pathogenicity forms occurs most often after the introduction of these LPIAV H5 and H7 into poultry (Alexander 2000b), and has never been documented in wild bird hosts (Alexander 2000b, 2007). HPIAV has been isolated predominantly from domestic gallinaceous birds (chickens, turkeys, quail) (Alexander 2000b; Perkins and Swayne 2001; Wobeser 1992; Chen et al. 2005), but spillover outbreaks have occurred.

Despite the broad host range, wild birds are the natural reservoir for LPIAV and exhibit no clinical symptoms of infection. Within this reservoir, LPIAVs have been isolated from at least 105 species in 26 different families, though this number has undoubtedly increased since the last substantial reviews of bird hosts in 2006 and 2007 (Olsen et al. 2006; Stallknecht and Brown 2007). However, all bird species are not equally permissive hosts, and different groups play different roles in the epidemiology and maintenance of IAV. Viruses are most frequently detected in

Anseriformes (ducks, geese, swans), where H1–H12 are routinely detected (Munster et al. 2007; Alexander 2007; Olsen et al. 2006). Some Charadriiformes (shorebirds, gulls) are also important: most notable are gulls, which are the natural reservoir for H13 and H16 viruses and in which other subtypes are infrequently detected (Wille et al. 2011a; Arnal et al. 2014). These two subtypes are restricted to gulls, and ducks are not permissive to infection with H13 (Brown et al. 2012) or H16 (Fereidouni et al. 2014). Shorebirds play an interesting role, in that IAVs appear to be rare in this group, except in some species such as ruddy turnstone (*Arenaria interpres*) and red knot (*Calidris canutus*) in one geographic location: Delaware Bay, USA. This site is recognised as an important stopover location, and many different IAV subtypes are detected in spring-staging shorebirds; however, it is believed that shorebirds amplify LPIAV circulating in local waterfowl and the dynamics are the result of a unique ecological event: the spawning of horseshoe crabs (*Limulus polyphemus*) (Pearce et al. 2009; Winker et al. 2008; Maxted et al. 2012; Krauss et al. 2010). Other bird orders are believed to be spillover hosts, including the species-rich Passeriformes (Slusher et al. 2014) (Fig. 9.1). Within the Anseriformes, dabbling ducks, and particularly mallards, have accounted for most LPIAV isolations globally (Olsen et al. 2006). This may in part be due to sampling bias (Hoye et al. 2010), but it is highly likely that dabbling ducks, such as mallards, do actually have higher infection rates than other species.

The importance of waterfowl IAVs, and more recently poultry-adapted IAVs, in the context of emerging disease, is when they occasionally transmit to other species, particularly to mammals (humans, pigs, horses). Genetic barriers between host groups limit the free transmission of IAVs; however, spillovers do occur. These spillover events may result in isolated outbreaks with little or no onward transmission, such as spillover of HPIAV H5N1 to dogs (Songserm et al. 2006b), cats (Songserm et al. 2006a) and tigers (Mushtaq et al. 2008) or LPIAV H10 into seals (Bodewes et al. 2015; Zohari et al. 2014). The continued spillover of LPIAV H7N9 to humans (Gao et al. 2013; Kageyama et al. 2013), and subsequent adaptation to mammalian hosts, is of further concern. Rather than the spillover of entirely avian viruses, it is the incorporation of avian or swine gene segments that is of high concern as at least three major human pandemics of IAV were caused by viruses containing gene segments of avian origin (Lindstrom et al. 2004; Rabadan et al. 2006; Scholtissek et al. 1978; Taubenberger et al. 2005), and swine viruses played a role in the most recent H1N1 pandemic. Indeed, in an analysis of cross species transmission, wild birds, domestic birds and swine showed the highest connectivity, and further, swine and wild birds were the dominant species for global virus delivery (Ren et al. 2016). Further, Worobey et al. (2014) proposed that the Western Hemisphere panzootic of equine influenza in 1872–1873 may have resulted in the introduction of equine origin segments into human and avian IAV, particularly of the internal genes into avian IAV lineages. This combined with the first records of highly pathogenic avian influenza in poultry, which coincide with the transition to industrial animal production, may have been imperative in the successful emergence of novel avian viruses (Worobey et al. 2014). As such, reassortment is the driving factor in the ability of IAV to successfully emerge in multiple host species and remerge in populations.

Phylogeography of IAV in birds is shaped by host species movement and migration patterns.

Many dabbling duck species are migratory, or partial migrants, in the Northern Hemisphere, generally displaying a higher propensity for migration the further north the breeding distribution is located. In tropical regions, ducks are either resident or migrate in relation to rain and dry seasons, sometimes with irruptive movements. However, compared to other bird groups like gulls, terns or shorebirds, they show less long-distance migrations across substantial geographic barriers, such as oceans or deserts. As a result, they tend to migrate within the Old World (Europe, Africa and Asia) and the New World (North and South America) (Olsen et al. 2006). Due to the geographically segregated nature of their waterfowl hosts, avian IAVs can be divided into two main phylogenetic clades: Eurasian and North American (Olsen et al. 2006). More recently, it has been proposed that there may be a distinct IAV lineage in South America as well (Pereda et al. 2008; Nelson et al. 2016), perhaps reflected by limited waterfowl migration across the Gulf of Mexico or the Isthmus of Panama. Indeed, more recent work in blue-winged teals (*Anas discors*) in two different parts of their migratory routes, in the southern USA and Guatemala, has demonstrated viral phylogenetic signal from North America, rather than from South America (Ramey et al. 2014; Gonzalez-Reiche et al. 2012).

This pattern of hemispheric signal due to independently evolving major lineages is conserved across all eight RNA segments of the IAV genome; however, due to occasional introductions and subsequent competitive exclusion, some of these broad geographic lineages are replaced (Bahl et al. 2009, 2013). Within waterfowl hosts, it is extremely rare to find a virus with a geographic mosaic of segment origin. Winker and Gibson quantified avian movement between Asia and Alaska and demonstrated a large influx of birds between these continents (Winker and Gibson 2010). More targeted work in species such as northern pintails (*Anas acuta*), a species that breeds on both sides of the Bering Strait, has demonstrated movement of viral segments from Asia into North America (Koehler et al. 2008; Ramey et al. 2010). However, despite sharing habitats with these pintails, detection of IAV with differing geographic origins within Alaskan mallards is infrequent (Pearce et al. 2011), suggesting that there may be some type of host species barrier or fitness consequences for these mosaic viruses. It had been hypothesised that this concept of a natural host species barrier prevented Asian viruses from entering North America. However, more recently, it has been proposed that wild birds migrating between Asia and Alaska were the conduit for the introduction of highly pathogenic H5N8 to North America (highly pathogenic IAV is further discussed in Box 9.2 and Sect. 9.3.2.5) (Ramey et al. 2016; Lee et al. 2015). Unlike waterfowl, geographic mosaic viruses are more common in gulls (Wille et al. 2011a, b; Huang et al. 2014b; Dusek et al. 2014), which is in part driven by gulls having different migration and movement patterns as compared to ducks. For instance, great black-backed gulls (*Larus marinus*) banded in eastern Canada have been recorded in Western Europe (Wille et al. 2011b). Similarly, studies of common murres (*Uria aalge*), which may interact with other seabirds and gulls in overwintering areas, have detected virus genomes with geographic mosaicism (Huang et al. 2014a; Lang et al. 2016). Genetically, there does not seem to be host species segregation in avian IAV, with the exception of gulls, wherein there are gull-specific lineages for the NA, M, NP and NS segments and the HA subtypes H13 and H16 are gull specific (Chen and Holmes 2009; Wille et al. 2011a).

The large genetic diversity of IAVs is a result of two mechanisms: genetic drift and genetic shift. Genetic drift occurs due to an error-prone RNA-dependant RNA polymerase, which lacks proofreading ability (Gething et al. 1980; Both et al. 1983; Webster et al. 1992). An early concept in IAV evolution was that avian IAVs are in "evolutionary stasis" in that the evolutionary arms race between host and virus is less intense in avian systems resulting in little selective requirement to repeatedly fix amino acid changes that evade the immune response (Chen and Holmes 2006; Suarez 2000). This hypothesis has been refuted using genetic studies demonstrating high rates of mutation due to genetic drift. Avian IAVs have been demonstrated to have rapid rates of evolutionary change, characterised by accumulations of synonymous and non-synonymous mutations (Chen and Holmes 2006, 2010; Bahl et al. 2009). A synonymous mutation is one which changes the nucleic acid sequence without changing the amino acid sequence of the encoded protein. Non-synonymous nucleic acid mutations do change the amino acid sequence. The rate of mutations varies across segments, with an average rate of 3.41×10^{-3} substitutions/site/year. To put this in context, the rate of IAV mutation is a million times greater than that of vertebrate genomic DNA, and this allows for the rapid adaptation of these viruses to new environments (Pybus and Rambaut 2009). A result is that there are a number of forward evolving lineages for all RNA segments.

The second mechanism through which IAV can diversify is genetic shift, which occurs due to coinfection and reassortment. Reassortment occurs due to the unlinked nature of the eight RNA segments, and thus, if a cell is infected by more than one virus, the progeny virions may contain various combinations of segments from the different parental viruses (Webster et al. 1992; Gething et al. 1980). Thus, given a coinfection with two IAVs, each with 8 segments, 256 different genetic progenies are possible, generating significant viral diversity (Ma et al. 2016). Due to the frequent reassortment in avian IAVs, virus genotypes, or genome constellations, are rarely isolated across consecutive days at the same location (Dugan et al. 2008). Given this, it is unsurprising that across an autumn season, over 50% of viruses from mallards are reassorted, across a number of different subtypes (Wille et al. 2013). Furthermore, this is likely driven by seasonal dynamics of subtype presence and virus load in the population (Wille et al. 2013). Thus, IAVs do not circulate as fixed genome constellations, but rather as transient constellations that rapidly change, even within the same host species, location and time period (see Box 9.3).

Box 9.3 Influenza A Virus Evolutionary Genetics

Genetic drift and shift do not necessarily correspond to antigenic change or change in phenotype. However, given change in phenotype, the progeny viruses may have a selective advantage due to host immune system evasion. Studies of human IAV H3N2 have demonstrated that genetic drift is a gradual and continuous process, resulting in a ladder-shaped phylogeny (Rambaut et al. 2008a). Antigenic shift, however, is more punctuated in that the accumulation of a number of mutations at specific positions will result in viruses occupying a new phenotype (Koel et al. 2013; Smith et al. 2004). Our comparatively less knowledge regarding antigenic change and inter-lineage evolution in avian IAVs is partly due to large concurrent genetic variation and insufficient sampling but also compounded by a possible long-term tenacity of viruses in the abiotic environment where "old" viruses have been hypothesised to reappear in the population of birds after some time (Roche et al. 2014).

Genetic drift and genetic shift allow for IAV to rapidly diversify; however, current genetic structure of IAV is due to an interplay between diversification and selective sweeps in the population. Viruses with specific genome constellations that attain a much higher fitness will rapidly increase in frequency. These constellations may drive competing lineages to extinction and may be driven to fixation, thus eliminating circulating diversity—known as a selective sweep. Although it is only the antigenic segments that may be selected for, as these are the ones that interact directly with the immune system, the other segments in that successful constellation will be "carried along", demonstrating a "hitchhiking" mechanism (Chen and Holmes 2010). As a result, there will be a selective sweep across not only antigenic segments but all segments of IAV.

Despite a proposed ancient co-evolution of birds and IAVs, dating of current lineages suggests these are of recent origin. The time of origin of the circulating PB2, PB1, PA, NP and M segments is only approximately 100–130 years ago. The most recent common ancestor for the more divergent HA, NA and NS segments is more ancient; however, intra-subtype radiation occurred more recently as well (Chen and Holmes 2006; Worobey et al. 2014). Coincidentally, during that time period when the first descriptions of HPIAV in domestic chickens occurred, there was a transition to more intensive chicken farming. Additionally, the time period 1872–1873 corresponds with a severe panzootic of equine influenza, coupled with reports of influenza in domestic birds following local equine outbreaks (McDonald et al. 2009). Thus, it is hypothesised that these events resulted in a global sweep of avian IAV resulting in these shallow divergence times (Worobey et al. 2014). While this global sweep has had large implications in the genetic structure of IAV, numerous local sweeps have occurred as well, driven by the introduction of a novel segment or segments following reassortment (e.g. Bahl et al. 2009).

9.3.2.2 Dynamics of IAV in Mallards: All Birds Are Not Equal

Mallard has high IAV prevalence across years and locations, and the largest number of viruses has been isolated from this species and with a high diversity of subtypes (Olsen et al. 2006, 2014). This species is also a dominant component in species composition of many IAV surveillance studies. A review by Olsen et al. (2006) demonstrated that nearly 50% of all waterfowl samples analysed for IAV were from mallard, with a global viral prevalence of 12.9%. The number of collected samples has risen dramatically since the review; a search of the Influenza Research Database (IRD; http://www.fludb.org) indicates 64,194 samples have been collected from mallards with 3271 HA sequences generated. In our own study site in southeast Sweden, 22,229 cloacal/faecal samples were collected in 2002–2009, generating 1081 isolated IAVs across 74 HA-NA subtypes (Latorre-Margalef et al. 2014) (Fig. 9.3b).

The prevalence of IAV in mallard follows a seasonal pattern, whereby it is low during the late winter, spring and summer, followed by a peak in viral prevalence during the autumn migration (Latorre-Margalef et al. 2014; Wilcox et al. 2011; Ito et al. 1995; Hatchette et al. 2004; Olsen et al. 2006). This pattern has been observed at a number of study sites across the Northern Hemisphere, including Sweden (Latorre-Margalef et al. 2014), the Netherlands (Munster et al. 2007; van Dijk et al. 2014), Canada (Alberta; Hatchette et al. 2004; Sharp et al. 1997), the USA (Minnesota; Wilcox et al. 2011), California (Hill et al. 2012) and Alaska (Ip et al. 2008; Runstadler et al. 2007), and prevalence can be up to 30% during the autumnal peak.

This prevalence pattern is largely driven by the ecology of the waterfowl hosts. In the autumn months, widely dispersed breeding individuals congregate during migration, which increases host density. Further, during the summer months, breeding has led to the production of young, which are immunologically naïve. The high turnover rate of mallards results in a substantial recruitment of immunologically naïve individuals into the population (Munster and Fouchier 2009; Bentz 1985). Not only do hatch-year birds account for more infections than all other age classes (Wilcox et al. 2011; Ip et al. 2008; Webster et al. 1992), but it also has been demonstrated that young birds and migrants are important drivers in IAV dynamics (van Dijk et al. 2014; Avril et al. 2016).

Compounded with an increase in immunologically naïve individuals, there is a decline in anti-IAV antibodies in second-year mallards during the summer months, suggesting a decrease in general herd immunity during this period allowing for reinfections with IAV the next autumn (Tolf et al. 2013a). Thus, it is a combination of mallard phenology, ecology and biology that are drivers for the seasonal pattern of IAV prevalence (Fig. 9.2).

Within the autumnal prevalence peak, there are many different HA-NA subtypes co-circulating. To date, 102 of the possible 144 HA-NA subtype combinations have been detected in wild birds, globally (Olson et al. 2014) (Fig. 9.3a). Mallards represent a substantial proportion of this figure, whereby 74 HA-NA subtype combinations have been detected in mallards from a Swedish study site alone (2002–2009), and most globally detected subtypes have been found in mallard and other Anseriformes elsewhere (Figs. 9.1 and 9.3). Within mallards, some IAV subtypes are very common, and others are either rare or absent. Common subtypes are usually isolated every year, such as H6 and H4 in Europe (Latorre-Margalef et al. 2014) (Fig. 9.3b) or H3 in North America (Bahl et al. 2013; Bahl et al. 2009; Wilcox et al. 2011). Some subtypes exhibit a more outbreak-like pattern, whereby they are common in some years and absent in others (Thangavel et al. 2011; Wilcox

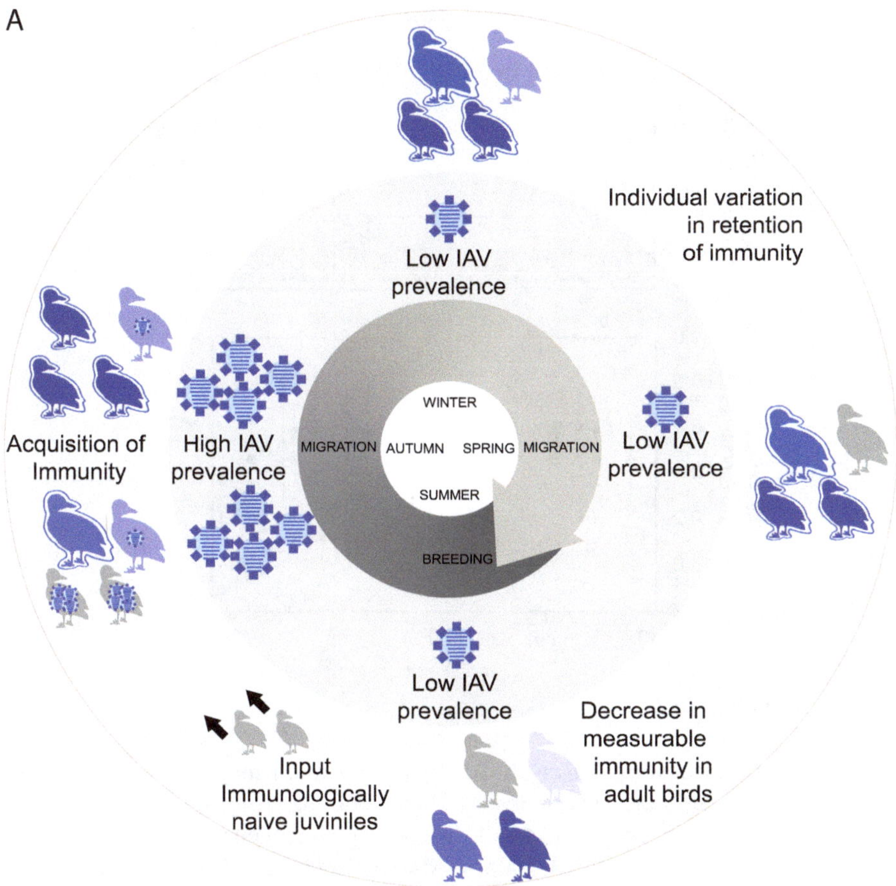

Fig. 9.2 (**a**) Seasonal dynamics of IAV in mallard is influenced by an input of immunologically naïve individuals and a decrease in immunity. (**b**) Seasonal prevalence of IAV, also illustrated by the second concentric circle of the schematic. (**c**) Number and proportion of newly ringed mallards at Ottenby demonstrating an increase in young or newly ringed individuals in the summer prior to the prevalence peak. (**d**) Seasonal levels of anti-NP antibodies of second-year birds living in a duck trap demonstrating individual variation in retention of immunity, long-term immunity following infection in the previous autumn and a marked drop in antibodies during the summer months followed by an increase following reinfection in the autumn [Panel A is modified from Latorre-Margalef (2012) LNU Press, Panels B and C were reproduced with permission from the Proceedings of the Royal Society B, and Panel D was modified from Tolf et al. (2013a) PLoS One]

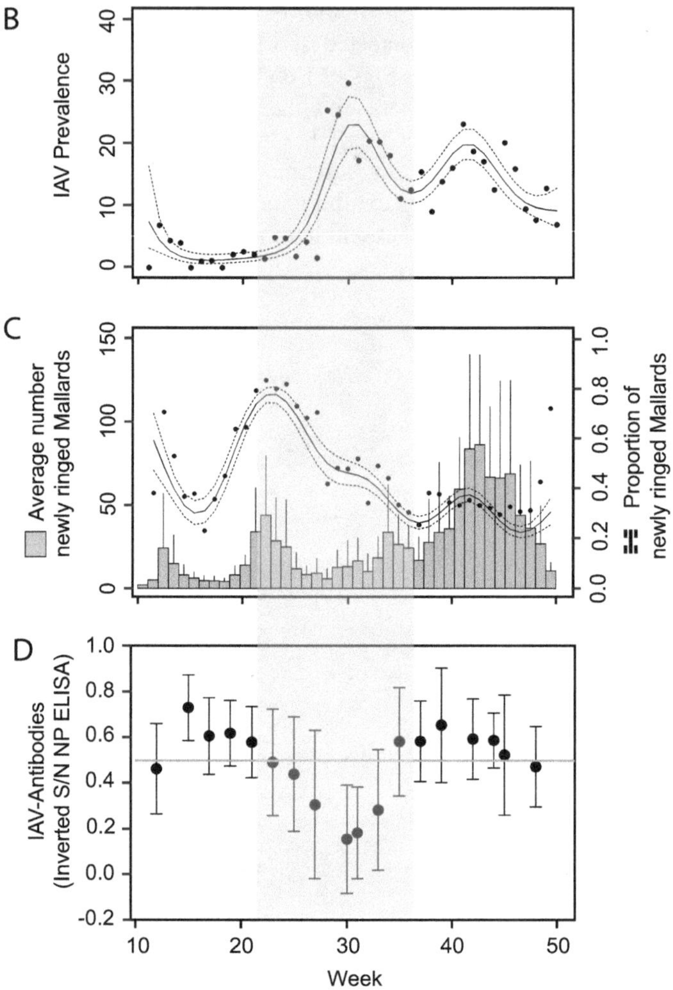

Fig. 9.2 (continued)

et al. 2011; Latorre-Margalef et al. 2014). Rare viruses may be isolated in low numbers every year or only sporadically. This overall pattern is observed across the Northern Hemisphere; however, small differences also occur between study sites and continents (Bahl et al. 2013; Latorre-Margalef et al. 2014; Olson et al. 2014). Furthermore, some HA-NA combinations are overrepresented, such as H4N6, H6N2 and H3N8, which are consistently isolated. Alternatively, some HA subtypes may be paired with any NA subtypes, suggesting fitness differences between HA-NA subtypes (Latorre-Margalef et al. 2014; Dugan et al. 2008). Surprisingly, when challenging ducks with combinations that are common (e.g. H3N8) compared to uncommon (e.g. H4N8), there appear to be no fitness differences such as virus

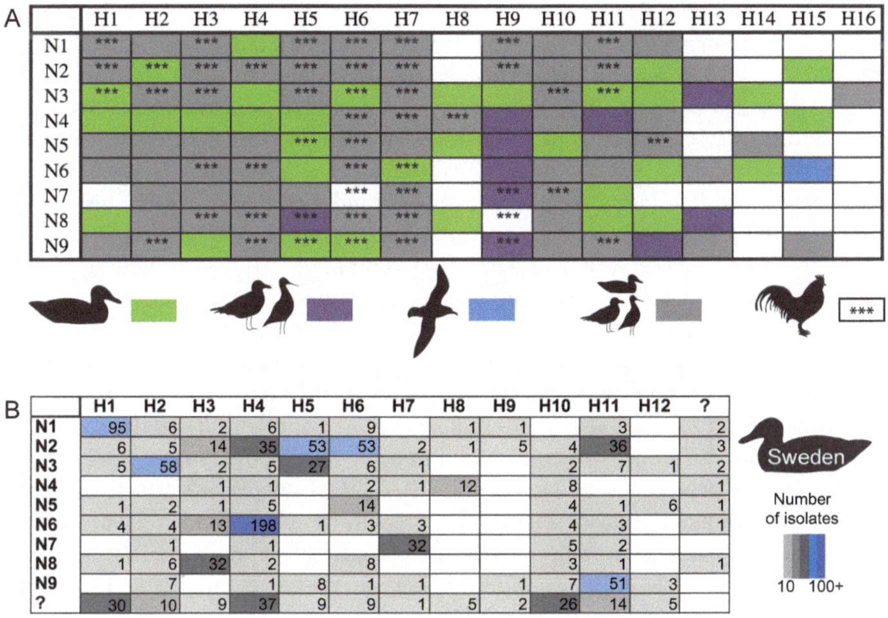

Fig. 9.3 Subtype diversity of IAV. (**a**) Global subtype distribution and detection across different avian host groups: Anseriformes, Charadriiformes, Procellariiformes, more than one order and poultry (Modified from Olson et al. 2014). (**b**) Subtype distribution and frequency in mallards utilising Ottenby Bird Observatory in Sweden as a stopover side, 2002–2009 (Modified from Latorre-Margalef et al. 2014)

load or duration of shedding (Lebarbenchon et al. 2012); thus, the mechanisms that drive subtype abundance patterns are still unknown.

Furthermore, some subtypes or group of subtypes appear early in the season, and others appear late in the season. This is attributed to HA subtype-specific immunity against specific subtypes (homosubtypic immunity) or closely related subtypes (heterosubtypic immunity) (Latorre-Margalef et al. 2013). For example, in our long-term monitoring of migratory mallards at a stopover site in SE Sweden, the first viruses to appear in the season are normally H3 class viruses, which include phylogenetically related H3, H4, H7 and H10 subtypes. This is in contrast to the later-arriving H1 class viruses (H1 clade (H1, H2, H5 and H6) and H9 clade (H8, H9 and H12)) (Latorre-Margalef et al. 2014). This is a relatively new way of characterising IAV subtype dynamics and therefore has not been assessed at other study sites.

Most field studies conducted on LPIAV in mallard focus on describing population-level parameters, such as virus prevalence or seroprevalence. Although important, population-level data can mask processes occurring at the individual level and fail to acknowledge individual variation in infection patterns. The biggest hurdle for conducting individual-level disease ecology studies is to follow individuals and their disease states over time. One approach used by Tolf et al. (2013a) was to introduce immunologically naïve commercially reared mallards in a duck trap used for attracting wild birds. These ducks became naturally infected by their wild conspecifics as they were sharing water and separated only by some mesh, and by daily sampling of these birds, it was possible to create individual disease histories of 1.5 years of length. Although these birds shared overall trends, there were considerable differences between individuals in which they were infected with LPIAV subtypes, coinfection patterns, lengths of shedding, clearance of infection and immune responses (Fig. 9.4) (Tolf et al. 2013a). Another approach is to use data from wild birds that are captured repeatedly over time. The resulting disease histories can be analysed by capture-mark-recapture (CMR) modelling techniques to estimate how infection parameters are affected by host categories (such as age and sex) and seasonal factors. Using multistate-CMR models on 3500 individual mallards across seven autumn seasons, Avril et al. (2016) demonstrated individual-level differences in both infection force and recovery rate. Specifically, for most years, prevalence and risk of LPIAV infection peaked at a single time during the autumn migration season, but the timing, shape and intensity of the infection curve showed strong annual heterogeneity. In contrast, the seasonal pattern of recovery rate only varied in intensity across years. Adults and juveniles displayed similar seasonal patterns of infection and recovery each year. However, juveniles experienced twice the risk of becoming infected as compared to adults, whereas recovery rates were similar across age categories (Avril et al. 2016).

9.3.2.3 Mallard Immunity to Influenza A Virus

Most microorganisms are immediately and non-specifically detected and cleared by the innate immune system, and it may alone succeed in repelling the pathogen while allowing time for the adaptive immune response to be mounted. Infection by IAV in ducks is initially combatted by components of the innate immune such as interferon-induced proteins. The adaptive immunity then develops neutralising antibodies, both subtype specific and those to conserved epitopes across subtypes (Magor 2011; Vanderven et al. 2012; Lundqvist et al. 2006). Taken together, this means that IAV infection in mallards is acute and of short duration—the average length of infection is 1 week, depending on host type, age and previous infection history. The complexity of the mallard immune system and response is still being disentangled. Following infection, it has been demonstrated that *RIG-I* (retinoic acid-inducible gene 1) is highly upregulated at the site of infection: the gastrointestinal tract or lungs of ducks infected with LPIAV and HPIAV, respectively (Barber et al. 2008; Vanderven et al. 2012). The *RIG-I* gene is absent in chickens and may explain why chickens display severe morbidity and mortality following infection, whereas mallards may display no clinical signs of disease (Barber et al. 2008; Vanderven et al. 2012). Other important innate immune genes are effectors of the interferon (IFN) pathways. Interestingly, the major histocompatibility complex

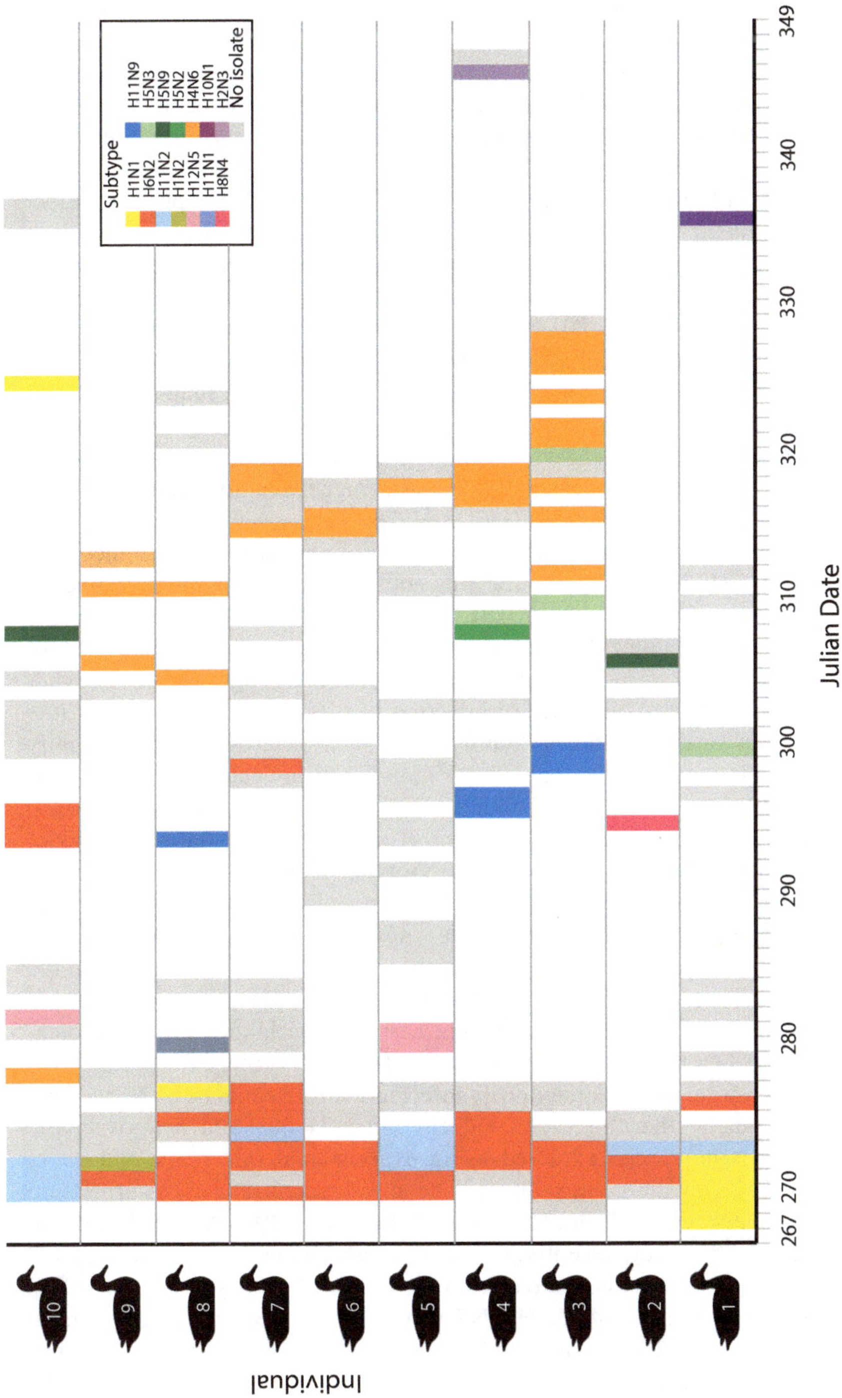

Fig. 9.4 Ten mallards residing in a duck trap were sampled daily across autumn 2009. Different individuals show different patterns of infection, despite following the expected seasonal trend. Individuals are plotted on the y-axis. *White* space indicates birds were negative for influenza, and colours refer to different HA subtypes with the exception of *grey*, which are samples positive by real-time PCR but negative by culture and therefore not subtyped (Modified from Wille et al. 2013, Virology)

(MHC), a part of the acquired immune response, appears to be important both early and late in IAV infection (Vanderven et al. 2012).

Despite combatting LPIAV infection rapidly, ducks may have poor long-term immune memory (Magor 2011), illustrated by a pattern of seroconversion and seroreversion (Tolf et al. 2013a), hypothesised to be due to the structure of the immunoglobulins, such as the translocation of the IgA (Magor 2011; Magor et al. 1998, 1999) and a truncation of the IgY (Warr et al. 1995) which appear to affect antibody functionality. Truncated IgY is able to neutralise viruses but is not involved in agglutination, complement fixation or opsonisation (Lundqvist et al. 2006). However, infection experiments (Fereidouni et al. 2010) and natural experimental infections (Tolf et al. 2013a) have demonstrated the presence of anti-NP antibodies, which are not neutralising, for months following infection. In long-lived species, such as shearwater, geese or swans, antibodies may be long-lived—antibodies against a Newcastle disease virus were detectable for a number of years in Cory's shearwater (*Calonectris borealis*) (Ramos et al. 2014), and in swans and geese, antibody prevalence increases with age, suggesting long-term antibody retention and accumulation with age (Hill et al. 2016).

Hill et al. further demonstrated that the breadth of antibody response increases with age, that is, individuals have neutralising antibodies against a larger number of HA subtypes with age (Hill et al. 2016). In chickens, protection against IAV is primarily through antibodies directed at the HA (Kapczynski and Swayne 2009), which is likely also true in mallards. Antibodies directed against NA or other proteins may contribute to clearance of infection (e.g. Nayak et al. 2010), but it is unclear. As previously mentioned, both homo- and heterosubtypic immunity develop following natural infections, where individuals infected with a particular subtype are unlikely to be reinfected with that same HA subtype later in the season (homosubtypic immunity) and across seasons (Tolf et al. 2013a; Latorre-Margalef et al. 2013). That is, ducks infected with H3 viruses should not be reinfected with H3 viruses. Homosubtypic immunity, however, is not always complete, and a field study utilising vaccines demonstrated the escape of an H3 virus from H3 neutralising antibodies which was hypothesised to be due to antigenic shift in the field viruses (Wille et al. 2016). This phenomenon is also seen in escape of HPIAV H5 viruses from the H5 vaccine in birds. Furthermore, partial or complete protection is apparent when reinfected with a closely related HA subtype (heterosubtypic immunity). Therefore, the duck previously infected with H3 could be protected, and thus not infected, by the closely related H4 virus. Homo- and heterosubtypic immunity have largely been explored using experimental infections, and many have been done so in the context of vaccine development and cross-protection against highly pathogenic H5 and H7 viruses (Costa et al. 2010, 2011; Fereidouni et al. 2009, 2010). But, more recently, studies have begun to explore protection and immunity patterns in low pathogenic infections, for example, Segovia et al. (2017) which investigated H3N8, H4N6, H10N7 and H14N5 infections in a balanced design (Segovia et al. 2017). Latorre-Margalef et al. (2016a, b), which assessed protection of H3 antibodies against an array of other virus subtypes, showed that the degree of protection was correlated with phylogenetic relatedness between viruses, where highest protection was induced to closely related HAs (Latorre-Margalef et al. 2016a, b). This acquired immunity shapes the dynamics of many diseases and is likely a driver for the continuing divergence of HA types. Given these findings, one can hypothesise that it is this immunity that drives the order and patterns of subtypes that occur in a population of birds across an autumn season. However, this premise warrants further study, especially tests of how HA-specific immunity and cross immunity affect future infection probability and virus load.

While there is great interest in IAV ecoimmunology, we are still largely using proxies for the mallard immune response, with a focus on the acquired immune response. In order to better understand host response to infection, continued work assessing the response of the innate response is imperative, as this response is coupled to the acquired response and may explain some of the patterns we see at this level.

9.3.2.4 Impact of LPIAV Infection on Hosts and Host Ecology

There is limited and contentious knowledge regarding the effect of LPIAV infection on wild birds, on short- and long-term impacts on host fitness, either at the individual bird level or bird population level. The current dogma is that birds, especially dabbling ducks, infected with LPIAV exhibit no clinical disease signs, despite being infected and reinfected with a virus and shedding these viruses at high viral loads in the gastrointestinal tract. These viruses replicate in the surface epithelium of the respiratory tract and gastrointestinal tract, and gross lesions are absent at the site of infection in natural infections (Kuiken 2013) (Box 9.2). Furthermore, there does not appear to be any increase (or decrease) in immune parameters of mallards naturally infected with LPIAV (van Dijk et al. 2015b). Granted there is a limited physiological response of individuals, this may still translate into short-term ecological effects as infections tend to be acute and short. Wild birds can experience physiological stress as a result of limited nutritional resources and variable energy expenditure during the year, which could have an effect on the course of disease within the host and therefore the host population. Interestingly, poor body condition due to food limitation in mallards in the context of IAV infection has indicated limited viral shedding compared to individuals in good condition (Arsnoe et al. 2011). Latorre-Margalef et al. (2009b) demonstrated a negative impact of LPIAV infection on body mass, and the amount of virus shed by infected juveniles was negatively correlated with body mass. This has been countered, wherein it is unclear if LPIAVs affect the body mass of individuals or whether birds in poor physical condition are more susceptible to acquiring infection (Flint and Franson 2009; Latorre-Margalef et al. 2009a). In a study of white-fronted geese (*Anser albifrons*), individuals with a lower body weight had a higher probability of infection but only for 1 of 4 years (Kleijn et al. 2010). In turn, during a study on Bewick's swans (*Cygnus bewickii*), it was found that birds experiencing their first infection (naïve-infected) had a reduced foraging rate but had similar body stores to reinfected and uninfected individuals (Hoye et al. 2016). This study reflects the reduced refuelling and feeding rates detected in an earlier study (van Gils et al. 2007).

Latorre-Margalef et al. (2009b) further found no effects of overall staging time or the speed and distance of subsequent migration. van Dijk et al. (2015a) found a weak negative association between LPIAV infection in mallard and regional movements (>100 m) on the final days of tracking, being exacerbated by poor weather conditions (van Dijk et al. 2015a), but a recent tracking study in Sweden found no differences in activity or movement between infected and uninfected mallards during stopover in autumn (Bengtsson et al. 2016). In naturally infected Bewick's swans, which demonstrated reduced refuelling and feeding rates, the birds also had delayed and protracted migration distances as when infected with LPIAV; however, the sample size of this study was only two infected birds (van Gils et al. 2007). This trend was not observed when applying a more experimental set-up wherein birds were infected and released (Hoye 2011). Interestingly, a follow-up study on Bewick's swans detected a potential difference in survival, where naïve-infected swans were unlikely to be resighted 1 year after infection, compared to uninfected or reinfected individuals (Hoye et al. 2016). This study also illustrates [a difference in response] between individuals infected for the first time and those uninfected and reinfected, whereby [birds that have been infected] and reinfected [have similar responses to birds that have never been infected] (Hoye et al. 2016). This is perhaps not surprising as immunologically naïve individuals have a much higher risk of infection (Avril et al. 2016).

Despite limited physiological signs of infection, it has been hypothesised that LPIAV infection may be affecting digestive tract functioning. Wild birds delicately balance energy intake and energy output, and decreased gastrointestinal functioning could translate into reduced body mass, delayed staging or decreased movements of individuals (Kuiken 2013). As of yet, there are few studies using natural systems, due to the difficulty in carrying out such experiments and disentangling all the confounding factors during data analysis. Experimental infections may provide insight; however, these studies rarely reflect natural conditions, and the results are dependent upon mode of inoculation, strains and conditions (Kuiken 2013). Low virulence and limited clinical signs have been interpreted as a long-standing co-evolutionary relationship between IAV and the host (van Dijk et al. 2015b), but further research addressing this is warranted.

9.3.2.5 LPIAV, HPIAV and the Interface with Poultry

In the sections above, we have mainly addressed wild birds as carriers of low pathogenic viruses (for extended definition, see Box 9.2). However, viruses with a highly pathogenic phenotype can be detected, either as spillover infections or in sustained transmission among wild waterfowl. Actually, the very first record of IAV in wild birds was an outbreak in common terns (*Sterna hirundo*) in South Africa, 1961, resulting in the mortality of at least 1300 individuals (Becker

1966). This record remains unusual as it is the only recorded case of an outbreak of HPIAV in wild birds with no direct link to outbreaks in poultry.

The HPIAV H5N1, colloquially referred to as "bird flu", was first identified in 1996; however, it wasn't until 2005 that it resulted in the mass mortality of wild and domestic birds alike (Gauthier-Clerc et al. 2007; Feare 2010; Chen et al. 2005). It has since spread to countries in Asia, the Middle East, Africa and Europe, resulting in the culling of 400 billion chickens, turkeys and ducks and over 600 human cases (FAO 2012). Despite many years of research and vaccine development, this virus continues to cause outbreaks in Asia and Africa (FAO 2012). In November and December 2014, there were new incursions of HPIAV H5 into Europe and North America, the latter of which is a geographic range expansion. A novel HPIAV H5N8 resulted in the culling of poultry in Asia, Europe and North America. This strain was first reported in Chinese duck farms in 2010 (Wu et al. 2014a) and was detected in both poultry and wild birds in Korea, following an outbreak in 2014 (Lee et al. 2014). North America had not previously been affected by HPIAV H5N1, and the proposed conduit for entrance of this virus into North America is Beringia or from Asia into Alaska with migrating wild birds (Lee et al. 2015; Ramey et al. 2016). Unlike HPIAV H5N1, the HPIAV H5N8 doesn't appear to cause widespread morbidity or mortality in wild birds; hence, it entered North America and Europe virtually undetected in wild birds; that is, following mortality events in poultry, it was detected in wild birds from surveillance studies that were retrospectively screened (e.g. Ramey et al. 2016). Further, in North America, there is evidence that HPIAV H5N8 has reassorted with low pathogenic avian viruses resulting in HPIAV H5N1, H5N2 and H5N8 (e.g. Pasick et al. 2015). Similar to HPIAV H5N1, this virus has been detected in wild ducks in Asia, Europe and North America, suggesting wild birds as contributors in the long-distance dispersal (Gauthier-Clerc et al. 2007; Feare 2010; European Food Safety Authority 2014; Verhagen et al. 2015; Lee et al. 2015). Intriguingly, H5N8 seemed to disappear from North America following massive expansion in 2014 prompting questions about the role of wild birds in perpetuating this virus (Krauss et al. 2016). However, there have been severe outbreaks of H5N8 in 2016/2017 in North America and further across the globe.

The HPIAV infection experiments conducted on waterfowl have shown large variation in disease severity depending on the host species (e.g. Perkins and Swayne 2001; Perkins and Swayne 2002; Ellis et al. 2004; Keawcharoen et al. 2008; Brown et al. 2006, 2008; Pasick et al. 2007; Liu et al. 2005). Generally, dabbling ducks show fewer and less severe symptoms—and are sometimes asymptomatic despite shedding virus—than other duck species such as diving ducks (Bröjer et al. 2009; Pantin-Jackwood and Swayne 2007). This may be explained by intrinsic factors of the host such as the composition of immune branches and type/severity of the immune response. For example, RIG-I seems to be important in clearing IAV infection and is present in mallard but absent in chickens (Barber et al. 2008). Difference in response may also be partially explained by previous exposure to LPIAV, which reduces disease symptoms. For example, in an experimental study, birds that were first exposed to LPIAV had a less severe response to HPIAV after being reinfected (Fereidouni et al. 2009). The intrinsic features of wild birds that result in differing levels of infection are largely unknown, which is perhaps the result of our limited understanding of ducks (and wild bird) immune responses. Furthermore, despite numerous surveillance schemes, we are unable to predict the emergence and expansion of highly pathogenic IAV, as clearly illustrated by the recent emergence and range expansion of HPIAV H5N8.

9.4 Future Directions

What can we learn from the mallard in terms of ecology and evolution of disease? Not surprisingly, the first thing to note is how little we yet know of diseases in wildlife, especially for those diseases that can infect multiple host species and display large strain/antigenic variation. With the exception of IAV, the current knowledge on basic parameters such as host range, prevalence and distribution is sketchy at most for avian diseases. Interest in avian pathogens has primarily been driven by unexpected events, such as the introduction of West Nile virus in the USA or the spread of HPIAV H5N1 in Europe, resulting with intense surveillance activity for a few years and then winding down again with the entrance of another attention-grabbing disease on the scene. A more systematic sampling approach is needed, preferably representing a long-term focus coupled with large-scale efforts to study pathogens from their wild hosts. Fortunately, with the development of molecular methods and decreasing sequencing costs, we are better equipped for conducting these types of studies, and it is expected that the available information will increase substantially the coming years. Although important, molecular detection is the starting point, not the goal; in order to address ecological and evolutionary questions more accurately, they need to be complemented with efforts to isolate and characterise pathogens (Latorre-Margalef et al. 2016a, b; McClintock et al. 2010). This allows for functional analyses of pathogenicity and virulence, either in vitro or in animal models.

For IAV, long-term monitoring studies are available representing Europe and North America, and studies are emerging from Asia, Africa and South America, too. Collectively, these studies have provided genome data across the range of the virus enabling studies of evolutionary questions. However, for phylodynamic studies, even these large datasets are often insufficient as the global diversity of IAV is so large, leading to undersampling issues and, often, limited and biased spatial-temporal resolution. We do, however, have a basic understanding on the natural dynamics of LPIAV in wild mallards, including how virus prevalence varies between age classes and over time. Although most of the current literature focuses on host population-level data, an increasing trend for analyses conducted at the individual level is evident. This includes approaches to study movements and stopover behaviour in relation to infection, as well as capturing individual-based epidemiological parameters of disease dynamics (e.g. Avril et al. 2016; Tolf et al. 2013a; Latorre-Margalef et al. 2014). These in-depth, long-term studies are extremely valuable, and the continuation of such series (although expensive and logistically challenging) will be an important part of future research.

Building on the advances made during the last 50 years, the IAV research field is well suited to combine ecology and epidemiology for disease studies. Of particular interest would be to use the mallard-IAV system, for which we have a lot of "baseline" data, and to focus more on physiology such as the effect of infection on hosts, ecoimmunology or the interplay between the immune response following infection and host life history traits—characterising the immune systems and general host immunological responses to infections—and the interplay between IAV and other members of the virome and microbiome, to illuminate interspecies transmission and reveal dynamics within the Anseriform reservoir beyond mallards or to tackle questions pertaining to basic epidemiological and disease ecology theory such as host range, resistance vs tolerance, etc. Surveillance and characterisation studies are imperative, however, as future advances almost certainly will hinge on multidisciplinary work.

Compliance with Ethical Standards

Funding: This study was funded by the Swedish Research Council (2013-7510, 2015-03877).

Conflict of Interest: Michelle Wille declares that he/she has no conflict of interest. Neus Latorre-Margalef declares that he/she has no conflict of interest. Jonas Waldenström declares that he/she has no conflict of interest.

Ethical approval: All applicable international, national and/or institutional guidelines for the care and use of animals were followed.

References

Adrian WJ, Spraker TR, Davies RB (1978) Epornitics of aspergillosis in mallards (*Anas platyrhynchos*) in north central Colorado. J Wildl Dis 14(2):212–217

Alexander DJ (2000a) Newcastle disease and other avian paramyxoviruses. Rev Sci Tech OIE 19:443–462

Alexander DJ (2000b) A review of avian influenza in different bird species. Vet Microbiol 74:3–13

Alexander DJ (2007) An overview of the epidemiology of avian influenza. Vaccine 25:5637–5644

Alexander DJ, Banks J, Collins MS, Manvell RJ, Frost KM, Speidel EC, Aldous EW (1999) Antigenic and genetic characterisation of Newcastle disease viruses isolated from outbreaks in domestic fowl and turkeys in Great Britain during 1997. Vet Rec 145(15):417–421

Allison AB, Ballard JR, Tesh RB, Brown JD, Ruder MG, Keel MK, Munk BA, Mickley RM, Gibbs SE, Travassos da Rosa AP, Ellis JC, Ip HS, Shern-Bochsler VI, Rogers MB, Ghedin E, Holmes EC, Parrish CR, Dwyer C (2014) Cyclic avian mass mortality in the northeastern United States is associated with a novel orthomyxovirus. J Virol 89(2):1389–1403. doi:10.1128/JVI.02019-14

Anthony SJ, Epstein JH, Murray KA, Navarrete-Macias I, Zambrana-Torrelio CM, Solovyov A, Ojeda-Flores R, Arrigo NC, Islam A, Ali Khan S, Hosseini P, Bogich TL, Olival KJ, Sanchez-Leon MD, Karesh WB, Goldstein T, Luby SP, Morse SS, Mazet JA, Daszak P, Lipkin WI (2013) A strategy to estimate unknown viral diversity in mammals. mBio 4(5):e00598–e00513. doi:10.1128/mBio.00598-13

Arnal A, Vittecoq M, Pearce-Duvet J, Gauthier-Clerc M, Boulinier T, Jourdain E (2014) Laridae: a neglected reservoir that could play a major role in avian influenza virus epidemiological dynamics. Crit Rev Microbiol 41(4):508–519. doi:10.3109/1040841X.1042013.1870967

Arsnoe DM, Ip HS, Owen JC (2011) Influence of body condition on influenza A virus infection in mallard ducks: experimental infection data. PloS one 6(8):e22633. doi:10.1371/journal.pone.0022633

Avril A, Grosbois V, Latorre-Margalef N, Gaidet N, Tolf C, Olsen B, Waldenström J (2016) Capturing individual-level parameters of influenza A virus dynamics in wild ducks using multistate models. J Appl Ecol 53(4):1289–1297. doi:10.1111/1365-2664.12699

Bahl J, Vijaykrishna D, Holmes EC, Smith GJD, Guan Y (2009) Gene flow and competitive exclusion of avian influenza A virus in natual resevoir hosts. Virology 390:289–297

Bahl J, Krauss S, Kuhnert D, Fourment M, Raven G, Pryor SP, Niles LJ, Danner A, Walker D, Mendenhall IH, Su YC, Dugan VG, Halpin RA, Stockwell TB, Webby RJ, Wentworth DE, Drummond AJ, Smith GJ, Webster RG (2013) Influenza A virus migration and persistence in North American wild birds. PLoS Pathog 9(8):e1003570. doi:10.1371/journal.ppat.1003570

Barber MRW, Aldridge JR, Webster RG, Magor KE (2008) Association of RIG-I with innate immunity of ducks to influenza. Proc Natl Acad Sci USA 107:5913–5918

Becker WB (1966) The isolation and classification of Tern virus: influenza virus A/Tern/South Africa/1961. J Hyg 64:309–320

Bengtsson D, Safi K, Avril A, Fiedler W, Wikelski M, Gunnarsson G, Elmberg J, Tolf C, Olsen B, Waldenström J (2016) Does influenza A virus infection affect movement behaviour during stopover in its wild reservoir host? R Soc Open Sci 3(2):150633. doi:10.1098/rsos.150633

Bennett RS, Nezworski J, Velayudhan BT, Nagaraja KV, Zeman DH, Dyer N, Graham T, Lauer DC, Njenga MK, Halvorson DA (2004) Evidence of avian pneumovirus spread beyond Minnesota among wild and domestic birds in central North America. Avian Dis 48 (4):902–908. doi:10.1637/7208-051804r

Bentz P-G (1985) Studies on some urban mallard Anas platyrhynchos populations in Scandinavia. Part I: Causes of death, mortality and longevity among Malmö mallards as shown by ringing recoveries. Fauna Norv Ser C Cinclus 8:44–56

Berg M, Johansson M, Montell H, Berg AL (2001) Wild birds as a possible natural reservoir of Borna disease virus. Epidemiol Infect 127(1):173–178

Berglund PG (2014) Exploring the epidemiology and population structure of Campylobacter jejuni in humans, broilers and wild birds. PhD Thesis, Linnaeus University, Kalmar, Sweden

Birdlife-International (2004) Birds in Europe: population estimates, trends and conservation status. Birdlife International, Cambridge, UK

Blanchong JA, Samuel MD, Mack G (2006) Multi-species patterns of avian cholera mortality in Nebraska's Rainwater Basin. J wildl Dis 42(1):81–91

Bodewes R, Bestebroer TM, van der Vries E, Verhagen JH, Herfst S, Koopmans MP, Fouchier RA, Pfankuche VM, Wohlsein P, Siebert U, Baumgartner W, Osterhaus AD (2015) Avian influenza A(H10N7) virus-associated mass deaths among harbor seals. Emerg Infect Dis 21 (4):720–722. doi:10.3201/eid2104.141675

Bonnedahl J, Hernandez J, Stedt J, Waldenstrom J, Olsen B, Drobni M (2014) Extended-spectrum beta-lactamases in Escherichia coli and Klebsiella pneumoniae in Gulls, Alaska, USA. Emerg Infect Dis 20(5):897–899. doi:10.3201/eid2005.130325

Bordes F, Morand S (2011) The impact of multiple infections on wild animal hosts: a review. Infect Ecol Epidemiol 1. doi:10.3402/iee.v1i0.7346

Both GW, Sleigh MJ, Cox NJ, Kendal AP (1983) Antigenic drift in influenza virus H3 hemagglutinin from 1968 to 1980: multiple evolutionary pathways and sequential amino acid changes at key antigenic sites. J Virol 48:52–60

Botzler RG (1991) Epizootiology of avian cholera in wildfowl. J wildl Dis 27(3):367–395

Botzler RG (2002) Avian cholera on north coast California: distinctive epizootiological features. Ann N Y Acad Sci 969:224–228

Bröjer C, Ågren EO, Uhlhorn H, Bernodt K, Mörner T, Jansson DS, Mattsson R, Zohari S, Thoren P, Berg M, Gavier-Widen D (2009) Pathology of natural highly pathogenic avian influenza H5N1 infection in wild tufted ducks (*Aythya fuligula*). J Vet Diagn Invest 21:579–587

Brown JD, Stallknecht DE, Beck JR, Suarez DL, Swayne DE (2006) Susceptibility of North American ducks and gulls to H5N1 highly pathogenic avian influenza viruses. Emerg Infect Dis 12:1663–1670

Brown JD, Stallknecht DE, Swayne DE (2008) Experimental infection of swans and geese with highly pathogenic avian influenza virus (H5N1) of Asian lineage. Emerg Infect Dis 14 (1):136–142. doi:10.3201/eid1401.070740

Brown J, Poulson R, Carter D, Lebarbenchon C, Pantin-Jackwood M, Spackman E, Shepherd E, Killian M, Stallknecht D (2012) Susceptibility of avian species to North American H13 low pathogenic avian influenza viruses. Avian Dis 56(4 Suppl):969–975. doi:10.1637/10158-040912-Reg.1

Cavanagh D (2005) Coronaviruses in poultry and other birds. Avian Pathol 34(6):439–448. doi:10.1080/03079450500367682

Champagnon J (2011) Consequences of the introduction of individuals within harvested populations: the case of the mallard *Anas platyrhynchos*. PhD Thesis, University of Monpellier, Montpellier, France

Chen R, Holmes EC (2006) Avian influenza virus exhibits rapid evolutionary dynamics. Mol Biol Evol 23:2336–2341

Chen R, Holmes EC (2009) Frequent inter-species transmission and geographic subdivision in avian influenza viruses from wild birds. Virology 383:156–161

Chen R, Holmes EC (2010) Hitchhiking and the population genetic structure of avian influenza virus. J Mol Evol 70:98–105

Chen H, Smith GJD, Zhang SY, Qin K, Wang J, Li KS, Webster RG, Peiris JSM, Guan Y (2005) H5N1 virus outbreak in migratory waterfowl. Nature 436:191–192

Christensen JP, Dietz HH, Bisgaard M (1998) Phenotypic and genotypic characters of isolates of Pasteurella multocida obtained from back-yard poultry and from two outbreaks of avian cholera in avifauna in Denmark. Avian Pathol 27(4):373–381. doi:10.1080/03079459808419354

Chu DK, Leung CY, Gilbert M, Joyner PH, Ng EM, Tse TM, Guan Y, Peiris JS, Poon LL (2011) Avian coronavirus in wild aquatic birds. J Virol 85(23):12815–12820. doi:10.1128/JVI.05838-11

Chu DK, Leung CY, Perera HK, Ng EM, Gilbert M, Joyner PH, Grioni A, Ades G, Guan Y, Peiris JS, Poon LL (2012) A novel group of avian astroviruses in wild aquatic birds. J Virol 86 (24):13772–13778. doi:10.1128/JVI.02105-12

Church DL (2004) Major factors affecting the emergence and re-emergence of infectious diseases. Clin Lab Med 24(3):559–586. doi:10.1016/j.cll.2004.05.008

Converse KA, Kidd GA (2001) Duck plague epizootics in the United States, 1967–1995. J Wildl Dis 37:347–357

Costa TP, Brown JD, Howerth EW, Stallknecht DE (2010) Effect of a prior exposure to a low pathogenic avian influenza virus in the outcome of a heterosubtypic low pathogenic avian influenza infection in mallards (*Anas platyrhynchos*). Avian Dis 54:1286–1291

Costa TP, Brown JD, Howerth EW, Stallknecht DE, Swayne DE (2011) Homo- and heterosubtypic low pathogenic avian influenza exposure on H5N1 highly pathogenic avian influenza virus infection in wood ducks (*Aix sponsa*). PLoS One 6(1):e15987. doi:10.1371/journal.pone.0015987

Cramp S, Simmons KEL (1977) Mallard (*Anas platyrhynchos*). In: Cramp S, Simmons KEL, Ferguson-Lees IJ et al (eds) Birds of the Western Palearctic, vol 1. Oxford University Press, London, pp 505–519

Cramp S, Brooks DJ, Dunn E, Gillmor R, Hollom PAD, Hudson R, Nicholson EM, Ogilvie MA, Olney PJS, Roselaar CS, Simmons KEL, Voos KH, Wallace DIM, Wattel J, Wilson MG (1985) Handbook of the birds of Europe, the Middle East and North Africa. The birds of the Western Palearctic, vol 4. Oxford University Press, Oxford

Daly JM, MacRae S, Newton JR, Wattrang E, Elton DM (2011) Equine influenza: a review of an unpredictable virus. Vet J 189(1):7–14. doi:10.1016/j.tvjl.2010.06.026

Daoust P-Y, Kibenge FSB, Fouchier RAM, van de Bildt MWG, Kuiken T (2011) Replication of low pathogenic avian influenza virus in naturally infected mallard ducks (*Anas platyrhynchos*) causes no morphologic lesions. J Wildl Dis 47:401–409

De Benedictis P, Schultz-Cherry S, Burnham A, Cattoli G (2011) Astrovirus infections in humans and animals: molecular biology, genetic diversity, and interspecies transmissions. Infect Genet Evol 11(7):1529–1544. doi:10.1016/j.meegid.2011.07.24

de Graaf M, Osterhaus AD, Fouchier RA, Holmes EC (2008) Evolutionary dynamics of human and avian metapneumoviruses. J Gen Virol 89(Pt 12):2933–2942. doi:10.1099/vir.0.2008/006957-0

Delany S, Scott D (2006) Waterbird population estimates, 4th edn. Wetlands International, Netherlands

Delnatte P, Ojkic D, Delay J, Campbell D, Crawshaw G, Smith DA (2013) Pathology and diagnosis of avian bornavirus infection in wild Canada geese (*Branta canadensis*), trumpeter swans (*Cygnus buccinator*) and mute swans (*Cygnus olor*) in Canada: a retrospective study. Avian Pathol 42(2):114–128. doi:10.1080/03079457.2013.769669

Delnatte P, Nagy E, Ojkic D, Leishman D, Crawshaw G, Elias K, Smith DA (2014) Avian bornavirus in free-ranging waterfowl: prevalence of antibodies and cloacal shedding of viral RNA. J Wildl Dis 50(3):512–523. doi:10.7589/2013-08-218

Descamps S, Jenouvrier S, Gilchrist HG, Forbes MR (2012) Avian cholera, a threat to the viability of an Arctic seabird colony? PloS one 7(2):e29659. doi:10.1371/journal.pone.0029659

Drilling N, Titman R, Mckinney F (2002) Mallard (Anas platyrhynchos). In: Poole A (ed) Birds of North America Online. Cornell Lab of Ornithology, Ithaca, doi: 10.2173/bna.658. http://bna.birds.cornell.edu/bna/species/658

Dugan VG, Chen R, Spiro DJ, Sengamalay N, Zaborsky J, Ghedin E, Nolting J, Swayne DE, Runstadler JA, Happ GM, Senne DA, Wang R, Slemons RD, Holmes EC, Taubenberger JK (2008) The evolutionary genetics and emergence of avian influenza A viruses in wild birds. PLoS Pathog 4:e1000076. doi:10.1371/journal/ppat/1000076

Dusek RJ, Hallgrimsson GT, Ip HS, Jonsson JE, Sreevatsan S, Nashold SW, TeSlaa JL, Enomoto S, Halpin RA, Lin X, Fedorova N, Stockwell TB, Dugan VG, Wentworth DE, Hall JS (2014) North Atlantic migratory bird flyways provide routes for intercontinental movement of avian influenza viruses. PLoS One 9(3):e92075. doi:10.1371/journal.pone.0092075

Ellis TM, Bousfield RB, Bissett LA, Dyrting KC, Luk GS, Tsim ST, Sturm-Ramirez K, Webster RG, Guan Y, Malik Peiris JS (2004) Investigation of outbreaks of highly pathogenic H5N1 avian influenza in waterfowl and wild birds in Hong Kong in late 2002. Avian Pathol 33 (5):492–505. doi:10.1080/03079450400003601

Engering A, Hogerwerf L, Slingenbergh J (2013) Pathogen-host-environment interplay and disease emergence. Emerg Microbes Infect 2(2):e5. doi:10.1038/emi.2013.5

European Food Safety Authority (2014) Highly pathogenic avian influenza A subtype H5N8. EFSA J 12:3941

FAO (2012) H5N1 Highly pathogenic avian influenza global review. EMPRES/GLEW Report, FAO Issue 31. http://www.fao.org/docrep/015/an388e/an388e.pdf

Fawaz M, Vijayakumar P, Mishra A, Gandhale PN, Dutta R, Kamble NM, Sudhakar SB, Roychoudhary P, Kumar H, Kulkarni DD, Raut AA (2016) Duck gut viral metagenome analysis captures snapshot of viral diversity. Gut Pathog 8:30. doi:10.1186/s13099-016-0113-5

Feare CJ (2010) Role of wild birds in the spread of highly pathogenic avian influenza virus H5N1 and implications for global surveillance. Avian Dis 54:201–212

Fenton A, Pedersen AB (2005) Community epidemiology framework for classifying disease threats. Emerg Infect Dis 11(12):1815–1821. doi:10.3201/eid1112.050306

Fereidouni SR, Starick E, Beer M, Wilking H, Kalthoff D, Grund C, Häuslaigner R, Breithaupt A, Lange E, Harder TC (2009) Highly pathogenic avian influenza virus infection of mallards with homo- and heterosubtypic immunity induced by low pathogenic avian influenza viruses. PLoS One 4:e6706. doi:10.1371/journal.pone.0006706

Fereidouni SR, Grund C, Häuslaigner R, Lange E, Wilking H, Harder TC, Beer M, Starick E (2010) Dynamics of specific antibody responses induced in mallards after infection by or immunization with low pathogenic avian influenza viruses. Avian Dis 54:79–85

Fereidouni SR, Harder TC, Globig A, Starick E (2014) Failure of productive infection of mallards (Anas platyrhynchos) with H16 subtype of avian influenza viruses. Influenza Other Respir Virus 8(6):613–616. doi:10.1111/irv.12275

Flint PL, Franson JC (2009) Does influenza A affect body condition of wild mallard ducks, or vice versa? Proc R Soc B 276:2345–2346

Fu Y, Pan M, Wang X, Xu Y, Xie X, Knowles NJ, Yang H, Zhang D (2009) Complete sequence of a duck astrovirus associated with fatal hepatitis in ducklings. J Gen Virol 90(Pt 5):1104–1108. doi:10.1099/vir.0.008599-0

Gao R, Cao B, Hu Y, Feng Z, Wang D, Hu W, Chen J, Jie Z, Qiu H, Xu K, Xu X, Lu H, Zhu W, Gao Z, Xiang N, Shen Y, He Z, Gu Y, Zhang Z, Yang Y, Zhao X, Zhou L, Li X, Zou S, Zhang Y, Yang L, Guo J, Dong J, Li Q, Dong L, Zhu Y, Bai T, Wang S, Hao P, Yang W, Han J, Yu H, Li D, Gao GF, Wu G, Wang Y, Yuan Z, Shu Y (2013) Human infection with a novel avian-origin influenza A (H7N9) virus. N Engl J Med 368(20):1888–1897. doi:10.1056/NEJMoa1304459

Gauthier-Clerc M, Lebarbenchon C, Thomas F (2007) Recent expansion of highly pathogenic avian influenza H5N1: a critical review. Ibis 149:202–214

George TL, Harrigan RJ, LaManna JA, DeSante DF, Saracco JF, Smith TB (2015) Persistent impacts of West Nile virus on North American bird populations. Proc Natl Acad Sci U S A 112 (46):14290–14294. doi:10.1073/pnas.1507747112

Gething MJ, Bye J, Skehel JJ, Wakefield M (1980) Cloning and DNA sequence of double-stranded copies of hemagglutinin genes from H2 and H3 strains elucidates antigenic shift and drift in human influenza virus. Nature 287:310–306

Gonzalez-Reiche AS, Morales-Betoulle ME, Alvarez D, Betoulle JL, Muller ML, Sosa SM, Perez DR (2012) Influenza A viruses from wild birds in Guatemala belong to the North American lineage. Plos One 7(3):e32873. doi:10.1371/journal.pone.0032873

Gordus AG (1993) Notes on the first known avian cholera epizootic in wildfowl in North America. J Wildl Dis 29(2):367

Grard G, Moureau G, Charrel RN, Lemasson JJ, Gonzalez JP, Gallian P, Gritsun TS, Holmes EC, Gould EA, de Lamballerie X (2007) Genetic characterization of tick-borne flaviviruses: new insights into evolution, pathogenetic determinants and taxonomy. Virology 361(1):80–92. doi:10.1016/j.virol.2006.09.015

Groth M, Lange J, Kanrai P, Pleschka S, Scholtissek C, Krumbholz A, Platzer M, Sauerbrei A, Zell R (2014) The genome of an influenza virus from a pilot whale: relation to influenza viruses of gulls and marine mammals. Infect Genet Evol 24C:183–186. doi:10.1016/j.meegid.2014.03.026

Guo J, Shivaprasad HL, Rech RR, Heatley JJ, Tizard I, Payne S (2014) Characterization of a new genotype of avian bornavirus from wild ducks. Virology J 11(1):197. doi:10.1186/s12985-014-0197-9

Guy JS (2000) Turkey coronavirus is more closely related to avian infectious bronchitis virus than to mammalian coronaviruses: a review. Avian Pathol 29(3):207–212. doi:10.1080/03079450050045459

Hald B, Skov MN, Nielsen EM, Rahbek C, Madsen JJ, Waino M, Chriel M, Nordentoft S, Baggesen DL, Madsen M (2016) *Campylobacter jejuni* and *Campylobacter coli* in wild birds on Danish livestock farms. Acta Vet Scand 58:11. doi:10.1186/s13028-016-0192-9

Hasan B, Melhus A, Sandegren L, Alam M, Olsen B (2014) The gull (*Chroicocephalus brunnicephalus*) as an environmental bioindicator and reservoir for antibiotic resistance on the coastlines of the Bay of Bengal. Microb Drug Resist 20(5):466–471. doi:10.1089/mdr.2013.0233

Hatchette TF, Walker D, Johnson C, Baker A, Pryor SP, Webster RG (2004) Influenza A viruses in feral Canadian ducks: extensive reassortment in nature. J Gen Virol 85:2327–2337

Hernandez J, Johansson A, Stedt J, Bengtsson S, Porczak A, Granholm S, Gonzalez-Acuna D, Olsen B, Bonnedahl J, Drobni M (2013) Characterization and comparison of extended-spectrum beta-lactamase (ESBL) resistance genotypes and population structure of Escherichia

coli isolated from Franklin's gulls (*Leucophaeus pipixcan*) and humans in Chile. PLoS One 8 (9):e76150. doi:10.1371/journal.pone.0076150
Hill NJ, Takekawa JY, Ackerman JT, Hobson KA, Herring G, Cardona CJ, Runstadler JA, Boyce WM (2012) Migration strategy affects avian influenza dynamics in mallards (*Anas platyrhynchos*). Mol Ecol 21(24):5986–5999. doi:10.1111/j.1365-294X.2012.05735.x
Hill SC, Manvell RJ, Schulenburg B, Shell W, Wikramaratna PS, Perrins CM, Sheldon BC, Brown IH, Pybus OG (2016) Antibody responses to avian influenza viruses in wild birds broadens with age. Proc R Soc B 283. doi:10.1098/rspb.2016.2159
Hoque MA, Burgess GW, Karo-Karo D, Cheam AL, Skerratt LF (2012) Monitoring of wild birds for Newcastle disease virus in north Queensland, Australia. Prev Vet Med 103(1):49–62. doi:10.1016/j.prevetmed.2011.08.013
Hoye B (2011) Host–pathogen interactions on the move: migratory waterfowl and avian influenza viruses. Utrecht University, The Netherlands
Hoye BJ, Munster VJ, Nishiura H, Klaassen M, Fouchier RAM (2010) Surveillance of wild birds for avian influenza virus. Emerg Infect Dis 16(12):1827–1834. doi:10.3201/Eid1612.100589
Hoye B, Munster VJ, Huig N, de Vries P, Oosterbeek K, Tijsen W, Klaassen M, Fouchier RAM, van Gils JA (2016) Hampered performance of migratory swans: intra- and inter-seasonal effects of avian influenza virus. Integr Comp Biol 56:317–329
Huang Y, Robertson GJ, Ojkic D, Whitney H, Lang AS (2014a) Diverse inter-continental and host lineage reassortant avian influenza A viruses in pelagic seabirds. Infect Genet Evol 22:103–111. doi:10.1016/j.meegid.2014.01.014
Huang Y, Wille M, Benkaroun J, Munro H, Bond AL, Fifield DA, Robertson GJ, Ojkic D, Whitney H, Lang AS (2014b) Perpetuation and reassortment of gull influenza A viruses in Atlantic North America. Virology 456-457:353–363. doi:10.1016/j.virol.2014.04.009
Hughes LA, Savage C, Naylor C, Bennett M, Chantrey J, Jones R (2009) Genetically diverse coronaviruses in wild bird populations of northern England. Emerg Infect Dis 15 (7):1091–1094. doi:10.3201/eid1507.090067
Ip HS, Flint PL, Franson JC, Dusek RJ, Derkson DV, Gill RE Jr, Ely CE, Pearce JM, Lanctot RB, Matsuoka SM, Irons DB, Fischer JB, Oates RM, Petersen MR, Fondell TF, Rocque DA, Pedersen JC, Rothe TC (2008) Prevalence of influenza A viruses in wild migratory birds in Alaska: patterns of variation in detection at a crossroads of intercontinental flyways. Virol J 5:71–81
Ito T, Okazaki K, Kawaoka Y, Takada A, Webster RG, Kida H (1995) Perpetuation of influenza A viruses in Alaskan waterfowl reservoirs. Arch Virol 140:1163–1172
Jackwood MW, Hall D, Handel A (2012) Molecular evolution and emergence of avian gammacoronaviruses. Infect Genet Evol 12(6):1305–1311. doi:10.1016/j.meegid.2012.05.003
Jeon WJ, Lee EK, Joh SJ, Kwon JH, Yang CB, Yoon YS, Choi KS (2008) Very virulent infectious bursal disease virus isolated from wild birds in Korea: epidemiological implications. Virus Res 137(1):153–156. doi:10.1016/j.virusres.2008.06.013
Jindal N, Chander Y, Chockalingam AK, de Abin M, Redig PT, Goyal SM (2009) Phylogenetic analysis of Newcastle disease viruses isolated from waterfowl in the upper midwest region of the United States. Virol J 6:191. doi:10.1186/1743-422X-6-191
Jonassen CM, Kofstad T, Larsen IL, Lovland A, Handeland K, Follestad A, Lillehaug A (2005) Molecular identification and characterization of novel coronaviruses infecting graylag geese (*Anser anser*), feral pigeons (*Columbia livia*) and mallards (*Anas platyrhynchos*). J Gen Virol 86(6):1597–1607. doi:10.1099/vir.0.80927-0
Jones R, Guneratne J (1984) The pathogenicity of some avian reoviruses with particular reference to tenosynovitis. Avian Pathol 13:173–189
Jourdain E, Gunnarsson G, Wahlgren J, Latorre-Margalef N, Bröjer C, Sahlin S, Svensson L, Waldenström J, Lundkvist Å, Olsen B (2010) Influenza virus in a natural host, the mallard: experimental infection data. PLoS One 5:e8935. doi:10.1371/journal.pone.0008935

Kageyama T, Fujisaki S, Takashita E, Xu H, Yamada S, Uchida Y, Neumann G, Saito T, Kawaoka Y, Tashiro M (2013) Genetic analysis of novel avian A(H7N9) influenza viruses isolated from patients in China, February to April 2013. Eurosurveillance 18(15):20453

Kapczynski DR, Swayne DE (2009) Influenza vaccines for avian species. Curr Top Microbiol Immunol 333:133–152. doi:10.1007/978-3-540-92165-3_6

Kawaoka Y, Cox NJ, Haller O, Hongo S, Klenk H-D, Lamb RA, McCauley J, Palese P, Rimstad E, Webster RG (2005) Orthomyxoviridae. In: Fauquet CM, Mayo MA, Maniloff J, Desselberger U, Ball LA (eds) Virus taxonomy: eighth report of the international committee for the taxonomy of viruses. Elsevier Academic Press, San Diego, USA, pp 681–693

Keawcharoen J, van Riel G, van Amerongen G, Bestebroer TM, Beyer WEP, van Lavieren R, Osterhaus ADME, Fouchier RAM, Kuiken T (2008) Wild ducks as long-distance vectors of highly pathogenic avian influenza virus (H5N1). Emerg Infect Dis 14:600–606

Kibenge FS, Dhillon AS, Russell RG (1988) Biochemistry and immunology of infectious bursal disease virus. J Gen Virol 69:1757–1775

Kim BY, Lee DH, Kim MS, Jang JH, Lee YN, Park JK, Yuk SS, Lee JB, Park SY, Choi IS, Song CS (2012) Exchange of Newcastle disease viruses in Korea: the relatedness of isolates between wild birds, live bird markets, poultry farms and neighboring countries. Infect Genet Evol 12 (2):478–482. doi:10.1016/j.meegid.2011.12.004

Kleijn D, Munster VJ, Ebbinge BS, Jonkers DA, Müskens GJDM, van Randen Y, Fouchier RAM (2010) Dynamics and ecological consequences of avian influenza virus infection in greater white-fronted geese in thier winter staging areas. Proc R Soc B 277:2041–2048

Koci D, Schultz-Cherry S (2002) Avian astroviruses. Avian Pathol 31(3):213–227. doi:10.1080/03079450220136521

Koehler AV, Pearce JM, Flint PL, Franson JC, Ip HS (2008) Genetic evidence of interncontinental movement of avian influenza in a migratory bird: the Northern Pintail (*Anas acuta*). Mol Ecol 17:4754–4762

Koel BF, Burke DF, Bestebroer TM, van der Vliet S, Zondag GC, Vervaet G, Skepner E, Lewis NS, Spronken MI, Russell CA, Eropkin MY, Hurt AC, Barr IG, de Jong JC, Rimmelzwaan GF, Osterhaus AD, Fouchier RA, Smith DJ (2013) Substitutions near the receptor binding site determine major antigenic change during influenza virus evolution. Science 342 (6161):976–979. doi:10.1126/science.1244730

Kraus RH, Zeddeman A, van Hooft P, Sartakov D, Soloviev SA, Ydenberg RC, Prins HH (2011) Evolution and connectivity in the world-wide migration system of the mallard: inferences from mitochondrial DNA. BMC Genet 12:99. doi:10.1186/1471-2156-12-99

Krauss S, Stallknecht DE, Negovetich NJ, Niles LJ, Webby RJ, Webster RG (2010) Coincident ruddy turnstone migration and horseshoe crab spawning creates an ecological 'hot spot' for influenza viruses. Proc Biol Sci 277(1699):3373–3379. doi:10.1098/rspb.2010.1090

Krauss S, Stallknecht DE, Slemons RD, Bowman AS, Poulson RL, Nolting JM, Knowles JP, Webster RG (2016) The engima of the apparent dissapearance of Eurasian highly pathogenic H5 clade 2.3.4.4 influenza A viruses in North American waterfowl. Proc Natl Acad Sci USA 113(32):9033–9038

Kuiken T (2013) Is low pathogenic avian influenza virus virulent for wild waterbirds? Proc Biol Sci 280(1763):20130990. doi:10.1098/rspb.2013.0990

Kuiken T, Frandsen D, Clavijo A (1998) Newcastle disease in cormorants. Can Vet J La revue veterinaire canadienne 39(5):299

LaDeau SL, Kilpatrick AM, Marra PP (2007) West Nile virus emergence and large-scale declines of North American bird populations. Nature 447(7145):710–713. doi:10.1038/nature05829

Lambert V, Cova L, Chevallier P, Mehrotra R, Trepo C (1991) Natural and experimental infection of wild mallard ducks with duck hepatitis B virus. J Gen Virol 72:417–420

Lang AS, Lebarbenchon C, Ramey AM, Robertson GJ, Waldenström J, Wille M (2016) Assessing the role of seabirds in the ecology of influenza A viruses. Avian Dis 60:378–386

Latorre-Margalef N, Gunnarsson G, Munster VJ, Fouchier RAM, Osterhaus ADME, Elmberg J, Olsen B, Wallensten A, Fransson T, Brudin L, Waldenström J (2009a) Does influenza A affect body condition of wild mallard ducks, or vice versa? A reply to flint and franson. Proc R Soc B 276:2347–2349

Latorre-Margalef N, Gunnarsson G, Munster VJ, Fouchier RAM, Osterhaus ADME, Elmberg J, Olsen B, Wallensten A, Haemig PD, Fransson T, Brudin L, Waldenström J (2009b) Effects of influenza A virus infection on migrating mallard ducks. Proc R Soc B 276:1029–1036

Latorre-Margalef N (2012) Ecology and evolution of influenza A virus in Mallards Anas platyrhynchos. Doctoral thesis, Linnaeus University Press. ISBN: 978-91-86983-61-1

Latorre-Margalef N, Grosbois V, Wahlgren J, Munster VJ, Tolf C, Fouchier RAM, Osterhaus ADME, Waldenström J, Olsen B (2013) Heterosubtypic immunity to influenza A virus infections in mallards may explain existence of multiple virus subtypes. PLoS Pathog 9(6): e1003443. doi:10.1371/journal.ppat.1003443

Latorre-Margalef N, Tolf C, Grosbois V, Avril A, Bengtsson D, Wille M, Osterhaus ADME, Fouchier RAM, Olsen B, Waldenström J (2014) Long-term variation in influenza A virus prevalence and subtype diversity in a migratory mallards in Northern Europe. Proc R Soc B 281:20140098. doi:10.1098/rspb.2014.0098

Latorre-Margalef N, Avril A, Tolf C, Olsen B, Waldenstrom J (2016a) How does sampling methodology influenza molecular detection and isolation success in influenza virus field samples? Appl Environ Microbiol 82:1147–1153

Latorre-Margalef N, Brown JD, Fojtik A, Poulson RL, Carter D, Franca M, Stallknecht DE (2016b) Competition between IAV subtypes through heterosubtypic immunity modulates re-infection and antibody dynamics in the mallard reservoir. bioRxiv Preprint

Lebarbenchon C, Sreevatsan S, Lefèvre T, Yang M, Ramakrishnan MA, Brown JD, Stallknecht DE (2012) Reassortment influenza A viruses in wild duck populations: effects on viral shedding and persistence in water. Proc R Soc B. doi:10.1098/rspb.2012.1271

Lee YJ, Kang HM, Lee EK, Song BM, Jeong J, Kwon YK, Kim HR, Lee KJ, Hong MS, Jang I, Choi KS, Kim JY, Lee HJ, Kang MS, Jeong OM, Baek JH, Joo YS, Park YH, Lee HS (2014) Novel reassortant influenza A(H5N8) viruses, South Korea, 2014. Emerg Infect Dis 20 (6):1087–1089. doi:10.3201/eid2006.140233

Lee DH, Torchetti MK, Winker K, Ip HS, Song CS, Swayne DE (2015) Intercontinental spread of Asian-origin H5N8 to North America through Beringia by migratory birds. J Virol 89 (12):6521–6524. doi:10.1128/Jvi.00728-15

Leighton FA, Heckert RA (2007) Newcastle disease and related avian paramyxoviruses. In: Thomas NJ, Hunter BD, Atkinson CT (eds) Infectious diseases of wild birds. Blackwell, Ames, pp 3–16

Lindh E, Huovilainen A, Ratti O, Ek-Kommonen C, Sironen T, Huhtamo E, Poysa H, Vaheri A, Vapalahti O (2008) Orthomyxo-, paramyxo- and flavivirus infections in wild waterfowl in Finland. Virol J 5:35. doi:10.1186/1743-422X-5-35

Lindstrom SE, Cox NJ, Klimov AI (2004) Genetic analysis of human H2N2 and early H3N2 influenza viruses, 1957–1972: evidence for genetic divergence and multiple reassortment events. Virology 328:101–119

Liu J, Xiao H, Lei F, Zhu Q, Qin K, Zhang XW, Zhang XL, Zhao D, Wang G, Feng Y, Ma J, Liu W, Wang J, Gao GF (2005) Highly pathogenic H5N1 influenza virus infection in migratory birds. Science 309(5738):1206. doi:10.1126/science.1115273

Lobo FP, Mota BE, Pena SD, Azevedo V, Macedo AM, Tauch A, Machado CR, Franco GR (2009) Virus-host coevolution: common patterns of nucleotide motif usage in Flaviviridae and their hosts. PLoS One 4(7):e6282. doi:10.1371/journal.pone.0006282

Lundqvist ML, Middleton DL, Radford C, Warr GW, Magor KE (2006) Immunoglobulins of the non-galliform birds: antibody expression and repertoire in the duck. Dev Comp Immunol 30 (1–2):93–100. doi:10.1016/j.dci.2005.06.019

Ma EJ, Hill NJ, Zabilansky J, Yuan K, Runstadler JA (2016) Reticulate evolution is favored in influenza niche switching. Proc Natl Acad Sci USA 113(19):5335–5339. doi:10.1073/pnas.1522921113

Magor KE (2011) Immunoglobulin genetics and antibody responses to influenza in ducks. Dev Comp Immunol 35:1008–1016

Magor KE, Warr GW, Bando Y, Middleton DL, Higgins DA (1998) Secretory immune system of the duck (*Anas platyrhynchos*). Identification and expression of the genes encoding IgA and IgM heavy chains. Eur J Immunol 28:1063–1068

Magor KE, Higgins DA, Middleton DL, Warr GW (1999) Opposite orientation of the alpha- and upsilon-chain constant region genes in the immunoglobulin heavy chain locus of the duck. Immunogenetics 49:692–695

Maxted AM, Luttrell MP, Goekjian VH, Brown JD, Niles LJ, Dey AD, Kalasz KS, Swayne DE, Stallknecht DE (2012) Avian influenza virus infection dynamics in shorebird hosts. J Wildl Dis 48(2):322–334. doi:10.7589/0090-3558-48.2.322

McClintock BT, Nichols JD, Bailey LL, MacKenzie DI, Kendall WL, Franklin AB (2010) Seeking a second opinion: uncertainty in disease ecology. Ecol Lett 13:659–674

McDonald SM, Matthijnssens J, McAllen JK, Hine E, Overton L, Wang S, Lemey P, Zeller M, Van Ranst M, Sprio DJ, Patton JT (2009) Evolutionary dynamics of human rotaviruses: balancing reassortment with preferred genome constellations. PLoS Pathog 5:e1000634. doi:10.1371/journal.ppat.1000634

Morse SS (1995) Factors in the emergence of infectious diseases. Emerg Infect Dis 1(1):7–15. doi:10.3201/eid0101.950102

Morse SS (2004) Factors and determinants of disease emergence. Rev Sci Tech OIE 23:443–451

Munster V, Fouchier RAM (2009) Avian influenza virus: of virus and bird ecology. Vaccine 27:6340–6344

Munster VJ, Baas C, Lexmond P, Waldenström J, Wallensten A, Fransson T, Rimmelzwaan GF, Beyer WEP, Schutten M, Olsen B, Osterhaus ADME, Fouchier RAM (2007) Spatial, temporal, and species variation in prevalence of influenza A viruses in wild migratory birds. PLoS Pathog 3:e61. doi:10.1371/journal.ppat.0030061

Muradrasoli S, Mohamed N, Hornyak A, Fohlman J, Olsen B, Belak S, Blomberg J (2009) Broadly targeted multiprobe QPCR for detection of coronaviruses: coronavirus is common among mallard ducks (*Anas platyrhynchos*). J Virol Methods 159(2):277–287. doi:10.1016/j.jviromet.2009.04.022

Muradrasoli S, Balint A, Wahlgren J, Waldenstrom J, Belak S, Blomberg J, Olsen B (2010) Prevalence and phylogeny of coronaviruses in wild birds from the Bering Strait area (Beringia). PLoS One 5(10):e13640. doi:10.1371/journal.pone.0013640

Mushtaq MH, Juan H, Jiang P, Li Y, Li T, Du Y, Mukhtar MM (2008) Complete genome analysis of a highly pathogenic H5N1 influenza A virus isolated from a tiger in China. Arch Virol 153 (8):1569–1574. doi:10.1007/s00705-008-0145-3

Nayak B, Kumar S, DiNapoli JM, Paldurai A, Perez DR, Collins PL, Samal SK (2010) Contributions of the avian influenza virus HA, NA, and M2 surface proteins to the induction of neutralizing antibodies and protective immunity. J Virol 84(5):2408–2420. doi:10.1128/JVI.02135-09

Nelson MI, Pollett S, Ghersi B, Silva M, Simons MP, Icochea E, Gonzalez AE, Segovia K, Kasper MR, Montgomery JM, Bausch DG (2016) The genetic diversity of influenza A viruses in wild birds in Peru. PLoS One 11(1):e0146059. doi:10.1371/journal.pone.0146059

Olsen B, Munster VJ, Wallensten A, Waldenström J, Osterhaus ADME, Fouchier RAM (2006) Global patterns of influenza A virus in wild birds. Science 312:384–388

Olson SH, Parmley J, Soos C, Gilbert M, Latorre-Margalef N, Hall JS, Hansbro PM, Leighton F, Munster V, Joly D (2014) Sampling strategies and biodiversity of influenza A subtypes in wild birds. PLoS One 9(3):e90826. doi:10.1371/journal.pone.0090826

Pantin-Jackwood MJ, Swayne DE (2007) Pathobiology of asian highly pathogenic avian influenza H5N1 virus infections in ducks. Avian Dis 50:250–259

Pantin-Jackwood MJ, Strother KO, Mundt E, Zsak L, Day JM, Spackman E (2011) Molecular characterization of avian astroviruses. Arch Virol 156(2):235–244. doi:10.1007/s00705-010-0849-z

Pasick J, Berhane Y, Embury-Hyatt C, Copps J, Kehler H, Handel K, Babiuk S, Hooper-McGrevy K, Li Y, Mai Le Q, Lien Phuong S (2007) Susceptibility of Canada Geese (*Branta canadensis*) to highly pathogenic avian influenza virus (H5N1). Emerg Infect Dis 13(12):1821–1827. doi:10.3201/eid1312.070502

Pasick J, Berhane Y, Joseph T, Bowes V, Hisanaga T, Handel K, Alexandersen S (2015) Reassortant highly pathogenic influenza A H5N2 virus containing gene segments related to Eurasian H5N8 in British Columbia, Canada, 2014. Sci Rep 5:9484. doi:10.1038/srep09484

Pearce JM, Ramey AM, Ip HS, Gill REJ (2009) Limited evidence of trans-hemispheric movement of avian influenza viruses among contemporary North American shorebird isolates. Virus Res 148:44–50

Pearce JM, Reeves AB, Ramey AM, Hupp JW, Ip HS, Bertram M, Petrula MJ, Scotton BD, Trust KA, Meixell BW, Runstadler JA (2011) Interspecific exchange of avian influenza virus genes in Alaska: the influenza of trans-hemispheric migratory tendency and breeding ground sympatry. Mol Ecol 20:1015–1025

Pereda AJ, Uhart M, Perez AA, Zaccagnini ME, La Sala L, Decarre J, Goijman A, Solari L, Suarez R, Craig MI, Vagnozzi A, Rimondi A, Konig G, Terrera MV, Kaloghlian A, Song H, Sorrell EM, Perez DR (2008) Avian influenza virus isolated in wild waterfowl in Argentina: evience of a potentially unqie phylogenetic lineage in South America. Virology 378:363–370

Perkins LE, Swayne DE (2001) Pathobiology of A/chicken/Hong Kong/220/97(H5N1) avian influenza virus in seven gallinaceous species. Vet Pathol 38:149–164

Perkins LEL, Swayne DE (2002) Susceptibility of Laughing Gulls (*Larus atricilla*) to H5N1 and H5N2 highly pathogenic avian influenza viruses. Avian Dis 46:877–885

Pybus OG, Rambaut A (2009) Evolutionary analysis of the dynamics of viral infectious disease. Nat Rev Genet 10(8):540–550. doi:10.1038/nrg2583

Rabadan R, Levine AJ, Robins H (2006) Comparison of avian and human influenza A viruses reveals a mutational bias on the viral genomes. J Virol 80:11887–11891

Rambaut A, Pybus OG, Nelson MI, Viboud C, Taubenberger JK, Holmes EC (2008a) The genomic and epidemiological dynamics of human influenza A virus. Nature 453:615–619

Rambaut A, Pybus OG, Nelson MI, Viboud C, Taubenberger JK, Holmes EC (2008b) The genomic and epidemiological dynamics of human influenza A virus. Nature 453 (7195):615–619. doi:10.1038/nature06945

Ramey AM, Pearce JM, Flint PL, Ip HS, Derkson DV, Franson JC, Petrula MJ, Scotton BD, Sowl KM, Wege ML, Trust KA (2010) Intercontinental reassortment and genomic variation of low pathogenic avian influenza viruses isolated form Nothern Pintails (*Anas acuta*) in Alaska: examining the evidence through space and time. Virology 401:179–189

Ramey AM, Reeves AB, Ogawa H, Ip HS, Imai K, Bui VN, Yamaguchi E, Silko NY, Afonso CL (2013) Genetic diversity and mutation of avian paramyxovirus serotype 1 (Newcastle disease virus) in wild birds and evidence for intercontinental spread. Arch Virol 158(12):2495–2503. doi:10.1007/s00705-013-1761-0

Ramey AM, Poulson RL, Gonzalez-Reiche AS, Wilcox BR, Walther P, Link P, Carter DL, Newsome GM, Muller ML, Berghaus RD, Perez DR, Hall JS, Stallknecht DE (2014) Evidence for seasonal patterns in the relative abundance of avian influenza virus subtypes in blue-winged teal (*Anas discors*). J Wildl Dis 50(4):916–922. doi:10.7589/2013-09-232

Ramey AM, Pearce JM, Reeves AB, Poulson RL, Dobson J, Lefferts B, Spragens K, Stallknecht DE (2016) Surveillance for Eurasian-origin and intercontinental reassortant highly pathogenic influenza A viruses in Alaska, spring and summer 2015. Virol J 13(1):55. doi:10.1186/s12985-016-0511-9

Ramos R, Garnier R, Gonzalez-Solis J, Boulinier T (2014) Long antibody persistence and transgenerational transfer of immunity in a long-lived vertebrate. Am Nat 184(6):764–776. doi:10.1086/678400

Ren H, Jin Y, Hu M, Zhou J, Song T, Huang Z, Li B, Li K, Zhou W, Dai H, Shi W, Yue J, Liang L (2016) Ecological dynamics of influenza A viruses: cross-species transmission and global migration. Sci Rep 6:36839. doi:10.1038/srep36839

Roche B, Lebarbenchon C, Gauthier-Clerc M, Chang CM, Thomas F, Renaud F, van der Werf S, Guegan JF (2009) Water-borne transmission drives avian influenza dynamics in wild birds: the case of the 2005–2006 epidemics in the Camargue area. Infect Genet Evol 9(5):800–805. doi:10.1016/j.meegid.2009.04.009

Roche B, Drake JM, Brown J, Stallknecht DE, Bedford T, Rohani P (2014) Adaptive evolution and environmental durability jointly structure phylodynamic patterns in avian influenza viruses. PLoS Biol 12(8):e1001931. doi:10.1371/journal.pbio.1001931

Roger F, Caron A, Morand S, Pedrono M, de Garine-Wichatitsky M, Chevalier V, Tran A, Gaidet N, Figuie M, de Visscher MN, Binot A (2016) One Health and EcoHealth: the same wine in different bottles? Infect Ecol Epidemiol 6:30978. doi:10.3402/iee.v6.30978

Rosseel T, Lambrecht B, Vandenbussche F, van den Berg T, Van Borm S (2011) Identification and complete genome sequencing of paramyxoviruses in mallard ducks (*Anas platyrhynchos*) using random access amplification and next generation sequencing technologies. Virol J 8:463. doi:10.1186/1743-422X-8-463

Runstadler J, Happ G, Slemons RD, Sheng Z-M, Gundlach N, Petrula M, Senne D, Nolting J, Evers DL, Modrell A, Huson H, Hills S, Rothe T, Marr T, Taubenberger JK (2007) Using RRT-PCR analysis and virus isolation to determine the prevalence of avian influenza virus infections in ducks at Minto Flats Refuge, Alaska, during August 2005. Arch Virol 152 (10):1901–1910

Samji T (2009) Influenza A: understanding the viral life cycle. Yale J Biol Med 82:153–159

Samuel MD, Shadduck DJ, Goldberg DR (2005a) Avian cholera exposure and carriers in greater white-fronted geese breeding in Alaska, USA. J Wildl Dis 41(3):498–502

Samuel MD, Shadduck DJ, Goldberg DR, Johnson WP (2005b) Avian cholera in waterfowl: the role of lesser snow and ross's geese as disease carriers in the Playa Lakes Region. J Wildl Dis 41(1):48–57

Samuel M, Botzler R, Wobeser G (2007) Avian cholera. In: Thomas N, Hunter B, Atkinson C (eds) Infectious diseases of wild birds. Blackwell, Iowa, USA, pp 239–269

Schmid Hempel P (2011) Evolutionary parasitology. The integrated study of infections, immunology, ecology, and genetics. Oxford University Press, Oxford, UK

Scholtissek C, Rohde W, Von Hoyningen C, Rott R (1978) On the origin of human influenza virus subtypes H2N2 and H3N2. Virology 87:13–20

Segovia KM, Stallknecht DE, Kapczynski DR, Stabler L, Berghaus RD, Fotjik A, Latorre-Margalef N, Franca MS (2017) Adaptive heterosubtypic immunity to low pathogenic avian influenza viruses in experimentally infected mallards. Plos One 12(1):e0170335. doi:10.1371/journal.pone.0170335

Sharp GB, Kawaoka Y, Jones DJ, Bean WJ, Pryor SP, Hinshaw V, Webster RG (1997) Coinfection of wild ducks by influenza A viruses: distribution patterns and biological significance. J Virol 71(8):6128–6135

Shin H-J, Njenga MK, McComb B, Halvorson DA, Nagaraja KV (2000) Avian pneumovirus (APV) RNA from wild and sentinel birds in the United States has genetic homology with RNA from APV isolates from domestic turkeys. J Clin Microbiol 38(11):4282–4284

Shin H-J, Nagaraja KV, McComb B, Halvorson DA, Jirjis FF, Shaw DP, Seal BS, Njenga MK (2002) Isolation of avian pneumovirus from mallard ducks that is genetically similar to viruses isolated from neighboring commercial turkeys. Virus Res 83:207–212

Slemons RD, Easterday BC (1978) Virus replication in the digestive tract of ducks exposed by aerosol to type-A influenza. Avian Dis 22:367–377

Slusher MJ, Wilcox BR, Lutrell MP, Poulson RL, Brown JD, Yabsley MJ, Stallknecht DE (2014) Are passerine birds reservoirs for influenza a viruses? J Wildl Dis. doi:10.7589/2014-02-043

Smith DJ, Lapedes AS, de Jong JC, Bestebroer TM, Rimmelzwaan GF, Osterhaus AD, Fouchier RA (2004) Mapping the antigenic and genetic evolution of influenza virus. Science 305 (5682):371–376. doi:10.1126/science.1097211

Songserm T, Amonsin A, Jam-on R, Sae-Heng N, Meemak N, Pariyothorn N, Payungporn S, Theamboonlers A, Poovorawan Y (2006a) Avian influenza H5N1 in naturally infected domestic cat. Emerg Infect Dis 12(4):681–683. doi:10.3201/eid1204.051396

Songserm T, Amonsin A, Jam-on R, Sae-Heng N, Pariyothorn N, Payungporn S, Theamboonlers A, Chutinimitkul S, Thanawongnuwech R, Poovorawan Y (2006b) Fatal avian influenza A H5N1 in a dog. Emerg Infect Dis 12(11):1744–1747. doi:10.3201/eid1211.060542

Stallknecht DE, Brown JD (2007) Wild birds and the epidemiology of avian influenza. J Wildl Dis 43:S15–S20

Stedt J, Bonnedahl J, Hernandez J, McMahon BJ, Hasan B, Olsen B, Drobni M, Waldenstrom J (2014) Antibiotic resistance patterns in Escherichia coli from gulls in nine European countries. Infect Ecol Epidemiol 4. doi:10.3402/iee.v4.21565

Steel J, Lowen AC (2014) Influenza a virus reassortment. Curr Top Microbiol Immunol 385:377–401. doi:10.1007/82_2014_395

Strong T, Dowd S, Gutierrez AF, Molnar D, Coffman J (2013) Amplicon pyrosequencing and ion torrent sequencing of wild duck eubacterial microbiome from fecal samples reveals numerous species linked to human and animal diseases. F1000Research 2(224). doi:10.12688/f1000research.2-224.v2

Suarez DL (2000) Evolution of avian influenza viruses. Vet Microbiol 74(1-2):15–27

Swayne DE, Suarez DL (2000) Highly pathogenic avian influenza. Rev Sci Tech 19:463–482

Taubenberger JK, Reid AH, Lourens RM, Wang R, Jin G, Fanning TG (2005) Characterization of the 1918 influenza virus polymerase genes. Nature 437:889–893

Thangavel RR, Reed A, Norcross EW, Dixon SN, Marquart ME, Stray SJ (2011) "Boom" and "Bust" cycles in virus growth suggest multiple selective forces in influenza a evolution. Virology J 8:180. doi:10.1186/1743-422X-8-180

Todd D, Smyth VJ, Ball NW, Donnelly BM, Wylie M, Knowles NJ, Adair BM (2009) Identification of chicken enterovirus-like viruses, duck hepatitis virus type 2 and duck hepatitis virus type 3 as astroviruses. Avian Pathol 38(1):21–30. doi:10.1080/03079450802632056

Tolf C, Latorre-Margalef N, Wille M, Bengtsson D, Gunnarsson G, Grosbois V, Hasselquist D, Olsen B, Elmberg J, Waldenström J (2013a) Individual variation in influenza A virus infection histories and long-term immune responses in mallards. PLoS One 8(4):e61201. doi:10.1371/journal.pone.0061201

Tolf C, Wille M, Haidar A-K, Avril A, Zohari S, Waldenström J (2013b) Prevalence of avian paramyxovirus type 1 in mallards during autumn migration in the western Baltic Sea region. Virol J 10:285. doi:10.1186/1743-422X-10-285

USFWS (2003) Review of captive-bred mallard regulatons on shooting preserves. Division of Migratory Bird Management, U. S. Fish and Wildlife Service, Washington, USA.

USFWS (2011) Waterfowl population status, 2011. U. S. Department of the Interior, USA

USGS (1999) Aspergillosis. In: Franson MFaJC (ed) Field manual of wildlife diseases: birds. USGS, pp 129–134

van Dijk JGB, Hoye BJ, Verhagen JH, Nolet BA, Fouchier RAM, Klaassen M (2014) Juveniles and migrants as drivers for seasonal epizootics of avian influenza virus. J Anim Ecol 83 (1):266–275. doi:10.1111/1365-2656.12131

van Dijk J, Kleyheeg E, Soons M, Nolet B, Fouchier R, Klaassen M (2015a) Weak negative associations between avian influenza virus infection and movement behaviour in a key host species, the mallard Anas platyrhynchos. Oikos 10:1293–1303. doi:10.1111/oik.01836

van Dijk JGB, Fouchier RAM, Klaassen M, Matson KD (2015b) Minor differences in body condition and immune status between avian influenza virus-infected and noninfected mallards: a sign of coevolution? Ecol Evol 5(2):436–449. doi:10.1002/Ece3.1359

van Gils JA, Munster VJ, Radersma R, Liefhebber D, Fouchier RAM, Klaassen M (2007) Hampered foraging and migratory performance in swans infected with low-pathogenic avian influenza A virus. PLoS One 2(1):e184. doi:10.1371/journal.pone.0000184

Vanderven HA, Petkau K, Ryan-Jean KE, Aldridge JR Jr, Webster RG, Magor KE (2012) Avian influenza rapidly induces antiviral genes in duck lung and intestine. Mol Immunol 51 (3-4):316–324. doi:10.1016/j.molimm.2012.03.034

Verhagen JH, Herfst S, Fouchier RAM (2015) How a virus travels the world. Science 347:616–617

Vincent A, Awada L, Brown I, Chen H, Claes F, Dauphin G, Donis R, Culhane M, Hamilton K, Lewis N, Mumford E, Nguyen T, Parchariyanon S, Pasick J, Pavade G, Pereda A, Peiris M, Saito T, Swenson S, Van Reeth K, Webby R, Wong F, Ciacci-Zanella J (2014) Review of influenza A virus in swine worldwide: a call for increased surveillance and research. Zoonoses Public Health 61(1):4–17. doi:10.1111/zph.12049

Wang G, Zhan D, Li L, Lei F, Liu B, Liu D, Xiao H, Feng Y, Li J, Yang B, Yin Z, Song X, Zhu X, Cong Y, Pu J, Wang J, Liu J, Gao GF, Zhu Q (2008) H5N1 avian influenza re-emergence of Lake Qinghai: phylogenetic and antigenic analyses of the newly isolated viruses and roles of migratory birds in virus circulation. J Gen Virol 89:697–702

Warr GW, Magor KE, Higgins DA (1995) IgY: clues to the origins of modern antibodies. Immunol Today 16:392–398

Webster RG, Kawaoka Y (1988) Avian influenza. CRC Crit Rev Poult Biol 1:211–246

Webster RG, Rott R (1987) Influenza virus A pathogenicity: the pivotal role of hemagglutinin. Cell 50:665–666

Webster RG, Morita M, Pridgen C, Tumova B (1976) Ortho- and paramyxoviruses from migrating feral ducks: characterization of a new group of influenza A viruses. J Gen Virol 32:317–225

Webster RG, Yakhno M, Hinshaw VS, Bean WJ, Murt KC (1978) Intestinal influenza: replication and characterization of influena viruses in ducks. Virology 84(2):268–278

Webster RG, Bean WJ, Gorman OT, Chambers TM, Kawaoka Y (1992) Evolution and ecology of influenza A viruses. Microbiol Rev 56(1):152–179

Wetlands International (2012) Waterbird population estimates, 5th edn. Summary report. Wetlands International, Wageningen. ISBN: 978-90-5882-000-6

White CL, Ip HS, Meteyer CU, Walsh DP, Hall JS, Carstensen M, Wolf PC (2014) Spatial and temporal patterns of avian paramyxovirus-1 outbreaks in double-crested cormorants (*Phalacrocorax auritus*) in the USA. J Wildl Dis. doi:10.7589/2014-05-132

Wilcox BR, Knutsen GA, Berdeen J, Goekjian VH, Poulson R, Goyal S, Sreevatsan S, Cardona C, Berghaus V, Swayne DE, Yabsley MJ, Stallknecht DE (2011) Influenza A viruses in ducks in Northwestern Minnesota: fine scale spatial and temporal variation in prevalence and subtype diversity. PLoS One 6:e24010. doi:10.1371/journal.pone.0024010

Wille M, Robertson GJ, Whitney H, Bishop MA, Runstadler J, Lang AS (2011a) Extensive geographic mosaicism in avian influenza viruses from gulls in the northern hemisphere. PLoS One 6:e20664. doi:10.1371/journal.pone.0020664

Wille M, Robertson GJ, Whitney H, Ojkic D, Lang AS (2011b) Reassortment of American and Eurasian genes in an influenza A virus isolated from a great black-backed gull (*Larus marinus*), a species demonstrated to move between these regions. Arch Virol 156(1):107–115. doi:10.1007/s00705-010-0839-1

Wille M, Tolf C, Avril A, Latorre-Margalef N, Wallerstrom S, Olsen B, Waldenstrom J (2013) Frequency and patterns of reassortment in natural influenza A virus infection in a reservoir host. Virology 443(1):150–160. doi:10.1016/j.virol.2013.05.004

Wille M (2015) Viruses on the wing: evolution and dynamics of influenza A virus in the Mallard reservoir. Doctoral thesis, Linnaeus University Press. ISBN: 978-91-87925-56-6

Wille M, Latorre-Margalef N, Tolf C, Stallknecht DE, Waldenström J (2016) No evidence for homosubtypic immunity to influenza H3 in mallards following vaccination in a natural experimental system. Mol Ecol. doi:10.1111/mec.13967

Winker K, Gibson DD (2010) The Asia-to-America influx of avian influenza wild bird hosts is large. Avian Dis 54:477–482

Winker K, Spackman E, Swayne DE (2008) Rarity of influenza A virus in spring shorebirds, Southern Alaska. Emerg Infect Dis 14:1314–1316

Wobeser G (1992) Avian cholera and waterfowl biology. J Wildl Dis 28(4):674–682

Wobeser GA (1997) Diseases of wild waterfowl, 2nd edn. Plenum Press, New York
Wolfe ND, Dunavan CP, Diamond J (2007) Origins of major human infectious diseases. Nature 447(7142):279–283. doi:10.1038/nature05775
Woo PC, Lau SK, Lam CS, Lai KK, Huang Y, Lee P, Luk GS, Dyrting KC, Chan KH, Yuen KY (2009) Comparative analysis of complete genome sequences of three avian coronaviruses reveals a novel group 3c coronavirus. J Virol 83(2):908–917. doi:10.1128/JVI.01977-08
Woo PC, Lau SK, Huang Y, Lam CS, Poon RW, Tsoi HW, Lee P, Tse H, Chan AS, Luk G, Chan KH, Yuen KY (2010) Comparative analysis of six genome sequences of three novel picornaviruses, turdiviruses 1, 2 and 3, in dead wild birds, and proposal of two novel genera, Orthoturdivirus and Paraturdivirus, in the family Picornaviridae. J Gen Virol 91(Pt 10):2433–2448. doi:10.1099/vir.0.021717-0
Woolhouse ME, Gowtage-Sequeria S (2005) Host range and emerging and reemerging pathogens. Emerg Infect Dis 11(12):1842–1847. doi:10.3201/eid1112.050997
Worobey M, Han GZ, Rambaut A (2014) A synchronized global sweep of the internal genes of modern avian influenza virus. Nature 508(7495):254–257. doi:10.1038/nature13016
Wu H, Peng X, Xu L, Jin C, Cheng L, Lu X, Xie T, Yao H, Wu N (2014a) Novel reassortant influenza A(H5N8) viruses in domestic ducks, eastern China. Emerg Infect Dis 20 (8):1315–1318. doi:10.3201/eid2008.140339
Wu Y, Wu Y, Tefsen B, Shi Y, Gao GF (2014b) Bat-derived influenza-like viruses H17N10 and H18N11. Trends Microbiol 22(4):183–191. doi:10.1016/j.tim.2014.01.010
Yu K, Li Y, Han H, Song M, Ma X, Liu C, Huang B, Li F (2014) Complete genome sequence of an avian reovirus isolated from wild mallard ducks in china. Genome Announc 2(5). doi:10.1128/genomeA.00813-14
Yun T, Yu B, Ni Z, Ye W, Chen L, Hua J, Zhang C (2013) Isolation and genomic characterization of a classical Muscovy duck reovirus isolated in Zhejiang, China. Infect Genet Evol 20:444–453. doi:10.1016/j.meegid.2013.10.004
Zhang Y, Guo D, Geng H, Liu M, Hu Q, Wang J, Tong G, Kong X, Liu N, Liu C (2007) Characterization of M-class genome segments of Muscovy duck reovirus S14. Virus Res 125 (1):42–53. doi:10.1016/j.virusres.2006.12.004
Zohari S, Neimanis A, Harkonen T, Moraeus C, Valarcher JF (2014) Avian influenza A(H10N7) virus involvement in mass mortality of harbour seals (Phoca vitulina) in Sweden, March through October 2014. Euro Surveill 19(46)

FSC
www.fsc.org
MIX
Papier | Fördert gute Waldnutzung
FSC® C083411

Kolibri 360 GmbH
Ferdinand-Jühlke-Straße 7
99095 Erfurt, Deutschland
produktsicherheit@kolibri360.de